ALLE ZEIT WACH
1842

42. Colloquium der Gesellschaft für Biologische Chemie
11.–13. April 1991 in Mosbach/Baden

Molecular Aspects of Inflammation

Edited by
H. Sies, L. Flohé and G. Zimmer

With 124 Figures

Springer-Verlag
Berlin Heidelberg New York
London Paris Tokyo
Hong Kong Barcelona
Budapest

Prof. Dr. Helmut Sies
Institut für Physiologische Chemie I
Universität Düsseldorf
Moorenstr. 5
4000 Düsseldorf, FRG

Prof. Dr. Leopold Flohé
Gesellschaft für Biotechnologische Forschung
Mascheroder Weg 1
3300 Braunschweig, FRG

Prof. Dr. Guido Zimmer
Gustav-Embden-Zentrum der Biologischen Chemie
Universität Frankfurt
Theodor-Stern-Kai 7
6000 Frankfurt am Main, FRG

ISBN-13:978-3-642-76414-1 e-ISBN-13:978-3-642-76412-7
DOI: 10.1007/978-3-642-76412-7

Library of Congress Cataloging-in-Publication Data. Molecular aspects of inflammation: 11.–13. April 1991 in Mosbach/Baden/[42. Colloquium Mosbach 1991]. Ed. by H. Sies ... – Berlin; Heidelberg; New York; London; Paris; Tokyo; Hong Kong; Barcelona; Budapest: Springer, 1991 (... Colloquium der Gesellschaft für Biologische Chemie; 42)
ISBN-13:978-3-642-76414-1

NE: Sies, Helmut [Hrsg.]; Colloquium Mosbach ⟨42, 1991⟩; Gesellschaft für Biologische Chemie: ... Colloquium der ...

Softcover reprint of the hardcover 1st edition 1991

31/3145-543210 – Printed on acid-free paper

Preface

Understanding the molecular basis of complex biological processes has been a major goal of biological chemistry from early on. Inflammation is one such entity, and recent years have seen exciting progress in the understanding of molecular interactions; there has been a long way from *dolor, rubor, calor,* and *tumor* as a fundamental description of the phenomenon to current knowledge, e.g., on the control of the respiratory burst of the granulocyte, the atomic details of protease regulation, or the interaction of cytokines.

We were glad to have been given the opportunity by the Gesellschaft für Biologische Chemie to call upon experts in this bustling field of research for this Colloquium. The sessions were organized in the order of increasing complexity, starting with the key phenomena of the inflammatory response and its modulation by cytokines to intravascular events and shock and sepsis; thus, the current attempts to apply basic knowledge on mediators of inflammation to the clinical situation were also considered.

In-depth chapters presenting the state of the art in these areas are collected in this book, and we thank the authors for their efforts. We also thank the Chairmen of the sessions, Profs. C. Sorg (Münster), D. Roos (Amsterdam), S. Bhakdi (Mainz), H. J. Müller-Eberhard (Hamburg), H. G. Schwick (Marburg), K. Resch (Hannover), W. Schaper (Bad Nauheim), D. Keppler (Heidelberg), and O. Trentz (Zürich) for their input, and it is a great pleasure to acknowledge the support by Drs. K. Beaucamp and E. Truscheit and their teams in the organization of the Colloquium. The crisp and sunny spring weather and the beautiful town of Mosbach did their part to make the exciting meeting as pleasant as it was.

September 1991 HELMUT SIES, LEOPOLD FLOHÉ, and GUIDO ZIMMER

Contents

Contributors

You will find the addresses at the beginning of the respective contribution

Contributors

Basic Mechanisms of the Inflammatory Response*

K. DECKER[1]

1 Introduction

Inflammation is experienced by people as a pandora box of nasty things like fever and pain. In essence, however, it is one of the organism's most powerful emergency measures designed to mobilize whatever forces can be mustered to overcome a life-threatening situation. As with many emergency actions, it sometimes happens that it exceeds its purpose and poses by itself a grave danger to the organism. That is the time when inflammation is just seen as a "bad", pathological process. Nevertheless, it is basically a network of reactions indispensable for the survival of higher organisms.

In this context, *inflammation* shall be understood as a response elicited by injured tissue or toxic agents (Table 1). Typical consequences are dilation and increased permeability of blood vessels, leukocyte invasion into the afflicted areas, and loss or impairment of cellular functions.

Table 1. Inflammation

A *response* elicited by toxic agents or injured cells.	
The *Agents*	may be
	microorganisms
	viruses
	toxins
	substances released by injured cells
The *Defense*	consists of
	the reticulo-endothelial system
	immunocompetent cells
	the complement system
	the contact activation system
The *Response*	may involve
	specific reactants
	(e.g., immunoglobulins)
	unspecific mechanisms
	(lytic enzymes, inhibitors,
	immunomodulators, mediators,
	complement components, oxygen species)

Table 2. Inflammatory agents

* Dedicated to Prof. Wolfgang Gerok at the occasion of his 65th birthday

[1] Biochemisches Institut der Albert-Ludwigs-Universität, Hermann-Herderstr. 7, D-7800 Freiburg i. Br., FRG

42. Colloquium Mosbach 1991
Molecular Aspects of Inflammation

Table 2. Inflammatory agents

Microorganisms	
	Viruses
	Bacteria
	Eukaryotic parasites
Cytotoxins	
	Exotoxins
	Endotoxins
	Carrageenan
	Phorbol esters
	D-Galactosamine
	Carbon tetrachloride
Signals from injured cells	
	Histamine
	Nucleotides
	Oligopeptides
	Proteolytic enzymes

The most frequent elicitors of inflammation are of three different types (Table 2): Microorganisms, viruses and eukaryotic parasites; toxins; cytotoxic chemicals and substances released from cells injured by biological, chemical, or physical (mechanical or radiation) insults. Chemically elicited inflammations are often useful models because they are easily and reproducibly provoked, aseptical, and more likely to be elucidated as to their mechanisms.

Table 3. Examples of inflammatory diseases

Disease	Immune component	Major mediators
Immunological overreaction		
Anaphylaxis, atopy	IgE	Histamine, eicosanoids
Lupus erythematosus	IgM, IgG	Multiple
Rheumatoid arthritis	IgM, IgG	Multiple
Granulomatosis	T lymphocytes	IL-4
RES-mediated disorders		
Septic shock	(IgE)	TNF, IL-1
Asthma		Histamine, leukotrienes
Adult respiratory distress syndrome		TNF-α, eicosanoids, oxygen radicals (?)
Diabetes		IL-1
Proteinase-related disorders		
Emphysema		Elastase, lack of α_1-antitrypsin
Chronicity of inflammation		
Fibrosis cirrhosis		TGF-β

The inflammatory response of an organism involves a network of effector and target cells. Their interactions may and most often do result in an inconspicuous handling of the damaging agent. But it may also lead to manifestation of the clinical symptoms of inflammation described already by Galen: dolor (pain), calor (fever), rubor (redness), and tumor (swelling). Inability to trigger a sufficiently strong inflammatory response renders the organism compromised and open for the continued action of the intruder. On the other hand, an excessive reaction may lead to forms of inflammatory disease (Table 3). Thus, it is obvious that better knowledge of these complex systems will provide new routes for pharmacological intervention.

2 Results and Discussion

2.1 Cells and Elicitors of the Inflammatory Process

Cells participating in the inflammatory response are primarily those belonging to the reticulo-endothelial system (RES), a term introduced by the Pathologist Ludwig Aschoff in 1922. It includes what is now called the mononuclear phagocyte system (MPS) (Metcalf and Burgess 1982) comprising circulating and sessile tissue macrophages, monocytes, and natural killer (NK) cells; but also polymorphonuclear leukocytes (PMN, neutrophils, granulocytes), lymphocytes, mast cells (basophils), endothelial cells, fibroblasts, and blood platelets (thrombocytes). All of them fulfill important functions in the inflammatory response.

The subcellular and soluble components participating in the inflammatory response of the organism are of different type and origin. Some offender-specific responses are mediated by the immune system, i.e., by circulating antibodies produced by B-lymphocytes and also by T-cell-involving mechanisms. The Contact Activation System (Cochrane and Griffin 1982) comprises parts of the coagulation systems (Factors XI and XII, prekallikrein) and involves activating (often negatively charged) surfaces of neutrophils, endothelial cells, and thrombocytes. Release of mediators, enzymes, and reactive inorganic compounds can contribute to the spread and severity of an inflammatory insult, e.g., in arthritis or in bacteremic shock.

2.2 Activation of Phagocytes

Phagocytes learn of the presence of noxious material by contact through receptors and peri- or intracellular binding sites that convey the cell specificity and bring the trans- and intracellular signal transduction pathways into action. Aside from organismic offenders a number of activators or immunomodulators stimulate resting macrophages. Table 4 lists some well-defined elicitors of phagocytes. Of these agents, interferon-γ, a product of activated lymphocytes, is the most important macrophage-activating factor (MAF). It acts not only and not so much by itself but also as an enhancer or co-stimulator of immunomodulators like lipopolysaccharides (LPS). The concept of the two-stage activation of phagocytes is still popular among immunologists (for review,

Table 4. Typical elicitors of macrophages

Viruses (e.g., Sendai, Newcastle disease)
Interferon-γ
Interleukin-2
Tumor necrosis factor-α (cachectin)
Complement factor 5a
N-Formyl-met-leu-phe
Muramyldipeptide
Lipopolysaccharides
Platelet-activating factor
Phorbol 12-myristate 13-acetate
Nucleotide di- and triphosphates
β-1,3-D-Glucan (soluble derivative)
Ionophores for Ca^{2+} or H^+

see Adams and Hamilton 1984). Interferon-γ activates specific DNA-binding proteins that exist in a latent form in the cytoplasm of untreated cells. The activated factor translocates to the nucleus and promotes the expression of some characteristic genes (Decker et al. 1991). The mechanism of activation by interferon-γ may be exemplary for the action of other cytokines.

Interleukin-2 (IL-2) was hitherto mainly known for its role in the B/T cell interaction and for its ability to activate one type of natural killer cells [lymphokine-activated killer (LAK) cells] (for review, see Herberman 1988). Now it turned out to also be an elicitor of macrophages (Benveniste and Merrill 1986) including Kupffer cells (Bouwens et al. 1991; Latham et al. 1991). IL-2 receptors have been identified on some of these phagocytes (Herrmann et al. 1985; Adachi et al. 1991).

Activation of macrophages occurs also as a consequence of contact with phagocytosable particles. Several types of receptors are involved in recognition and binding; signal transduction pathways ares activated that may be different from those used by immunomodulators and cytokines (Dieter et al. 1988). Also, the substances produced in this process are not those typical for the inflammatory reaction such as PGE_2 and cytokines. This is true at least for the Kupffer cells that release mainly superoxide and PGD_2 during phagocytosis (Decker 1985). It appears that in the inflammatory process phagocytosis by macrophages serves primarily the removal and destruction of organisms, cell debris, and immune complexes.

It should be emphasized at this point that the inflammatory response is pleiotropic: a given cell may produce several mediators, respond to more than one stimulus, and respond in different ways to different stimuli; furthermore, a given mediator may be produced by several cell types and interact with more than one cell type. Thus, we face a multidimensional network of elicitors, responsive cells, mediators, and target cells in the overall inflammatory process.

2.3 Mediators of the Inflammatory Response

A good number of mediator molecules and other soluble components are involved in the inflammatory response (Table 5). They signal the presence of an offender that has been recognized by the phagocyte and stimulate the cells that are able to destroy it; they also help in targeting and inactivating the culprit.

The intricate system of interactions may be exemplified by the network of cells, compounds, and reactions that is set into motion when an organism gets exposed to endotoxin. Specific aspects of the shock syndrome following massive endotoxin poisoning will not be elaborated here as they are dealt with elsewhere in this volume.

2.4 The Role of Endotoxins

The endotoxins are among the strongest elicitors of the inflammatory reaction. They are a family of lipopolysaccharides (LPS) and a major component of the outer membrane of Gram-negative enteric bacteria (Westphal et al. 1986). Endotoxin reaches the circulation either with intact germs as in bacteremia or as material from decaying intestinal microorganisms after resorption from the gut. LPS seems to leak into the portal blood *all the time*, an important fact to remember! An increased input of LPS occurs as a consequence of intestinal malfunction. Stimulation of the mast cells associated with the intestinal mucosa (histamine release) has been discussed to facilitate LPS uptake from the gut (Liehr et al. 1978).

Table 5. Signals from elicited phagocytes

Interleukin-1
Interleukin-2
Interleukin-6
Interferon-α/β
Tumor necrosis factor-α (cachectin)
Granulocyte/macrophage colony-stimulating factor
Macrophage colony stimulating factor
Erythropoietin
Transforming growth factor-α
Transforming growth factor-β
Fibroblast growth factor
Plasminogen activator
Plasminogen activator inhibitor
Prostaglandins
Thromboxane A_2
Leukotrienes
Platelet-activating factor
Superoxide, H_2O_2, $^{\bullet}OH$, $^1O_2$
Nitric oxide

Table 6. The endotoxin-refractory state of liver macrophages

Pretreatment	Stimulus	TNF (U/10^6 cells)
none	None	< 1
none	LPS	720 ± 210
none	NDV	2480 ± 740
LPS	None	< 1
LPS	LPS	< 1
LPS	NDV	2160 ± 680
PGE2	None	< 1
PGE2	LPS	24 ± 16
PGE2	NDV	98 ± 38
Indomethacin + LPS	LPS	< 1
Indomethacin + LPS	NDV	2180 ± 610

Which cells will be the first to deal with the resorbed endotoxin? Some blood cells, of course, but mainly the liver-associated tissue macrophages, the Kupffer cells. Almost all of the incoming LPS is taken up by this cell type (Freudenberg et al. 1982) that constitutes the largest pool of macrophages in the whole body.

Endotoxin induces a refractory state in responsive cells after prolonged exposure. This phenomenon has been known for many years and accounts for the LPS desensitization observed in vivo (Zuckerman et al. 1989). Animals pretreated with sublethal doses of endotoxin become tolerant to the injection of otherwise lethal amounts of LPS. Concomitantly, they do not produce the mediators that the unprotected animals release. It has now become evident that the inhibition of the response is restricted to the elicitor (e.g., LPS) itself. LPS pretreatment does not prevent the cells from synthesizing the mediators if the stimulant is, for example, a virus (Table 6). It means for the in vivo situation that the presence of an infection with an endotoxin-producing bacterium may not be able to suppress the inflammatory reaction against a following superinfection by a virus.

Contact of liver macrophages with endotoxin starts the inflammatory response. What happens in the Kupffer cell after uptake of LPS can also happen in other macrophages though one finds some characteristic differences among this family of cells both qualitatively and quantitatively. The sessile macrophages of the rat liver, for example, respond to LPS with the synthesis and release of cytokines, enzymes, several biologically active metabolites of arachidonic acid (eicosanoids), and of the arginine-derived nitric oxide (for review, see Decker 1990; Table 5). Interestingly, a very similar response is encountered after infection with certain viruses (Busam et al. 1990a).

Superoxide abundantly produced by Kupffer cells after phorbol ester or zymosan treatment is not synthesized in cells exposed to LPS alone. However, endotoxin strongly enhances the zymosan-triggered release of that anion radical (Grewe et al. 1990).

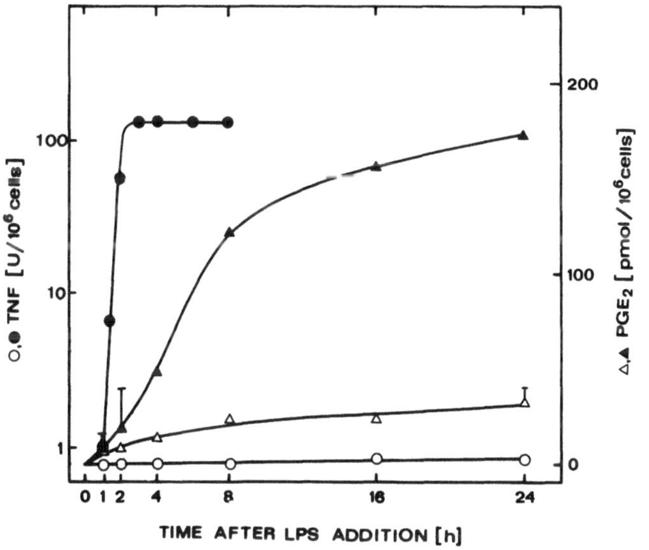

Fig. 1. Time course of endotoxin-induced cytokine and eicosanoid synthesis. Primary cultures of rat liver macrophages were exposed to LPS (400 ng/ml) for 24 h. *Circles*, TNF; *triangles*, PGE_2. *Open symbols* denote untreated controls. (Peters et al. 1990)

The time courses of these processes vary widely. The LPS-elicited eicosanoid production is slow but long-lasting compared to the release of a cytokine (Fig. 1). Also, LPS- and virus-provoked syntheses of tumor necrosis factor-α (TNF) show different kinetics in primary cultures of rat Kupffer cells (Fig. 2).

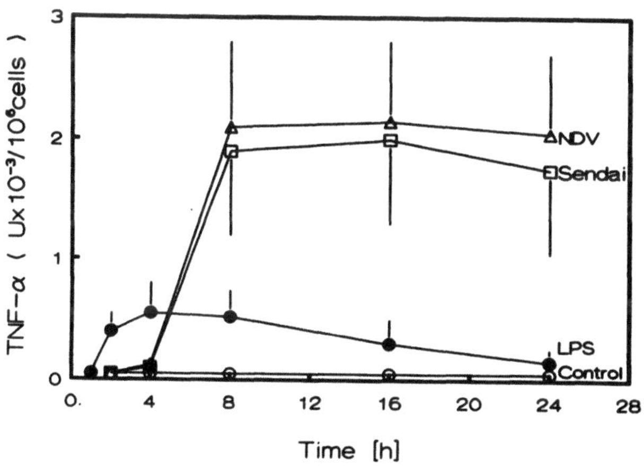

Fig. 2. Different time courses of tumor necrosis factor-α formation after endotoxin and virus. Primary cultures of rat liver macrophages were treated with LPS (100 ng/ml), Newcastle disease virus (NDV, 26 HAU/ml), and Sendai virus (1250 HAU/ml), respectively. (Busam et al. 1990a)

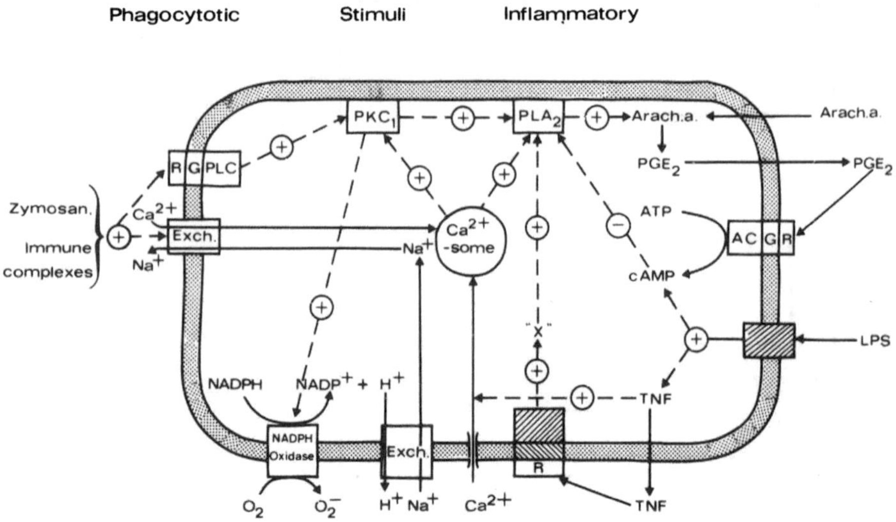

Fig. 3. Scheme of the signal transducing pathways in rat liver macrophages. AC Adenylate cyclase; *Arach. a.* arachidonic acid; *Exch.* exchanger; *G* G-proteins; *PKC* proteinkinase C; *PLA_2* phospholipase A_2; *R* receptors

These variations suggest the use of different signal transduction paths and/or mechanisms of action. Some of them have been elucidated already. Figure 3 tries to summarize schematically the present state of information with regard to rat Kupffer cells. A transient influx of Ca^{2+} appears to play a crucial role in the PLA_2-mediated eicosanoid synthesis, but not in superoxide or cytokine formation (Dieter et al. 1988).

2.5 The Eicosanoids

Among the eicosanoids released by phagocytes (Fig. 4) PGE_2 is the most typical for the inflammatory response; it is, together with interleukin-1, involved in the propagation of pain and the induction of fever (for review, see Needleman et al. 1986). PGE_2 is the preferred arachidonate metabolite, both relatively and absolutely, of Kupffer cells that were exposed to LPS or viruses (Peters et al. 1990). Thromboxane is known as a powerful vasoconstrictor and platelet activator; it is able to influence the hemodynamics close to its production site as shown for rat liver (Fig. 5). Some macrophages release leukotrienes following LPS exposure (Lüderitz et al. 1986). The biological roles of the chemotactic LTB_4, the vasoactive peptidoleukotrienes, their receptors, and various other aspects of the 5-lipoxygenase family of eicosanoids will be discussed in several contributions of this volume.

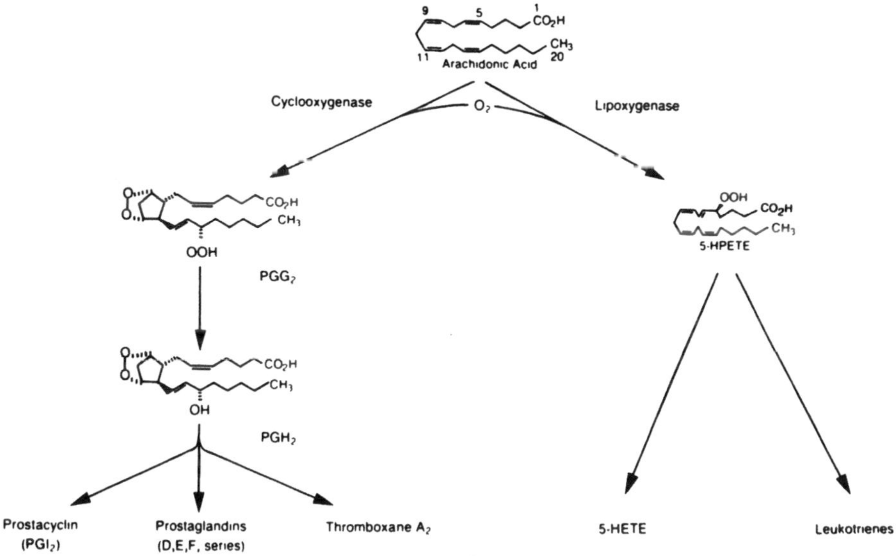

Fig. 4. The eicosanoid family.

The 1-O-Alkyl-2-acyl-sn-glycero-3-phosphocholine family, also known as platelet-activating factor (PAF), is in many instances involved in the inflammatory process, although effects in noninflammation states are also reported. PAF can be produced by a great variety of cells, e.g., activated leukocytes, and is able to activate inflammatory cells including platelets, but also vascular smooth muscle and endothelial cells (for review, see Hanahan 1986).

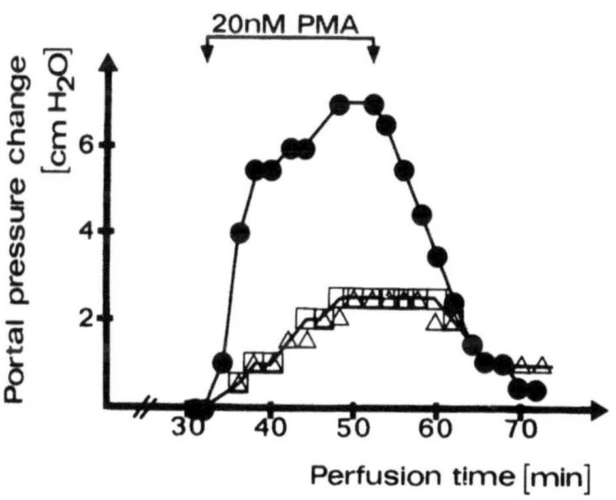

Fig. 5. Thromboxane increases the perfusion pressure in the isolated liver. Mediator release was induced with 20 nM phorbol 12-myristate 13-acetate (PMA, *full circles*). The suppressive effects of the simultaneous addition of a thromboxane synthase inhibitor (CGS 13080, 20 nM; *triangles*) or a thromboxane receptor antagonist (BM 13177, 20 μM; *squares*) indicate the predominant role of thromboxane in the observed pressure increase. (Tran-Thi et al. 1988)

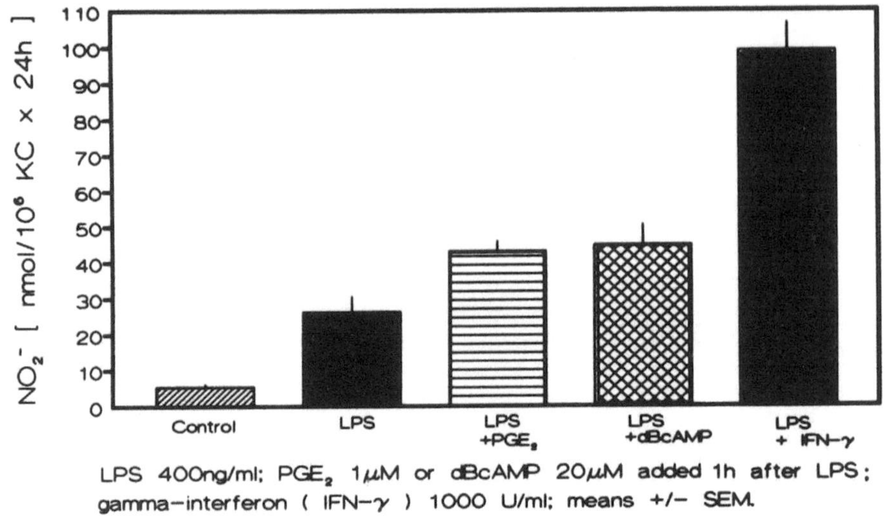

Fig. 6. Nitric oxide formation by liver macrophages after endotoxin treatment. Primary cultures of rat liver macrophages were exposed to LPS (400 ng/ml), PGE_2 (1 μM), dibutyryl cAMP (20 μM), and interferon-γ (1000 U/ml), respectively. The nitrite measured is proportional to the amount of NO formed.(Gaillard et al. 1990)

The major inorganic compounds generated by activated macrophages are superoxide (O_2^-) and nitric oxide (NO), both highly reactive compounds of radical nature. Nitric oxide is produced from L-arginine (Marletta et al. 1988) by a wide variety of properly stimulated cells. The radical (or a derivative thereof) possesses a cytotoxic potential (Hibbs et al. 1988) but it is better known as the endothelium-derived relaxing factor (EDRF) (Palmer et al. 1987). Unlike superoxide, nitric oxide is produced by Kupffer cells after exposure to LPS (Gaillard et al. 1990; Fig. 6). The rate of its synthesis is enhanced by PGE_2 (and by cAMP); this is remarkable insofar as the inflammatory signal, PGE_2, tends to inhibit the production of other macrophage products including the vasoconstrictive thromboxane (Peters et al. 1990). It remains to be seen whether this regulatory behavior is of physiological significance for maintaining unobstructed blood flow in inflamed areas.

2.6 The Role of the Tumor Necrosis Factor-α

A major mediator produced by monocytes and macrophages that were exposed to LPS or to some viruses is the tumor necrosis factor-α (TNF) also known as cachectin (Beutler and Cerami 1986). The soluble, biologically active form is a homotrimeric polypeptide of a subunit M_r of 17 kDa (Beutler and Cerami 1988). It is synthesized as a precursor of 26 kDa and – in rat but not in human macrophages – partially glycosylated. TNF, together with IL-1, mediates most of the biological effects seen in LPS

Table 7. Major effects of tumor necrosis factor-α (cachectin)

Lysis of certain tumor cells, necrosis of some solid tumors
Cytotoxicity for some transformed cells (L929)
Fever
Loss of weight
Inhibition of proliferation of some cells
Activation of polymorphonuclear neutrophils
Binding to endothelial cells
Mediation of most endotoxin effects including bacteremic shock
Induction of enzymes of triglyceride catabolism
Suppression of enzymes of triglyceride formation
Enhancement of prostaglandin E_2 synthesis
Enhancement of interleukin-6 synthesis

poisoning: injection of TNF leads to very similar symptoms, anti-TNF antibodies suppress the effects of LPS exposure (Beutler et al. 1985). This close relationship between the endotoxin-related symptoms and TNF puts the TNF-producing macrophages in a central position of the inflammatory process.

Numerous effects of TNF on a variety of cells have been reported (Table 7). The induction of catabolic enzymes explains, at least in part, the consuming effects experienced in severe infectious diseases. It appears that the name "cachectin" would be better suited to describe the biologically relevant properties of this cytokine, while the ability to induce tumor cell killing seems to be restricted to a limited number of neoplastic cell types.

Evidence accumulated in recent years indicates that signals from Kupffer cells are instrumental in the transformation of the fat-storing (Ito) cells into myofibroblast-like collagen-producing cells. Ito cells are located in the perisinusoidal space (space of Disse) of the liver lobule; in the resting state they are characterized as the main storage site for retinol and retinyl esters. Following inflammatory processes in the liver they undergo a metamorphosis into cells that synthesize very actively collagen and other extracellular matrix proteins (Shiratori et al. 1986). This remarkable ability is elicited by transforming growth factor-α (TGF-α) in conjunction with TGF-β (Davis and Vucic 1989; Friedman and Arthur 1989; Meyer et al. 1990). Fibrosis and cirrhosis are well-established consequences of chronic inflammation of the liver. The interplay of Kupffer cells that produce TGF-α and -β and Ito cells appears to be the decisive factor in the initiation of the fibrotic restructuring of the liver lobule following chronic inflammation.

2.7 The Interleukins

Several other cytokines are involved in the body's response to inflammatory signals (Table 8). Some of the interleukins and the colony-stimulating and growth factors participate in the differentiation and propagation of the bone marrow (stem cell)-derived blood and tissue cells; they include the various types of macrophages and the polymorphonuclear neutrophils. In the context of the inflammatory response the pos-

Table 8. Cytokines involved in the inflammatory process

	Major functions
Interleukins	
IL-1α	B-cell activation
IL-1β	Mediation of pain and fever
IL-2	T-cell proliferation, activation of macrophages, LAK cells
IL-3	Differentiation of hematopoietic stem cells
IL-4	Induction of IgE, inhibition of macrophage migration
IL-5	Differentiation of eosinophils
IL-6	Induction of acute-phase proteins, terminal maturation of B-cells
IL-8	Neutrophil chemotaxis
IL-9	Activation of B-cells
IL-10	Repression of interferon synthesis in T-cells
Colony-stimulating and growth factors	
M-CSF	Expansion of macrophage populations
G-CSF	Expansion of granulocyte populations
GM-CSF	Expansion of granulocyte, macrophage and eosinophil populations
TGF-α	Regulator of cell growth, induction of Ito cell transformation,
TGF-β	Regulator of cell growth (counteracting TGF-α)
Interferons and tumor necrosis factors	
Interferon-α/β	Antiviral activity
Interferon-γ	Activation of macrophages, co-stimulator of cytokine synthesis, stimulation of antigen presentation
TNF-α (cachectin)	see Table 7
TNF-β (lymphotoxin)	Similar to TNF-α

itive feedback cycle involving macrophage or granulocyte/macrophage colony stimulating factors (M-CSF, GM-CSF) must be mentioned. These mediators are formed by endothelial cells and fibroblasts exposed to inflammatory stimuli, e.g., LPS, IL-1. They lead to the propagation and maturation of macrophages that release more IL-1 or transforming growth factors to activate the endothelium and the collagen-producing fibroblasts, respectively.

The elucidation of the DNA coding for the cytokines revealed common motifs in the 5'-flanking region of their structural genes and thus common features in the mechanisms of their regulation. The pleiotropic transcription factor, NF-κB, plays a prominent role in the regulation of cytokine expression (Lenardo and Baltimore 1989). Even more conspicuous is the striking homology of the amino acid sequences of the respective receptors. It appears that all or most of the cytokine receptors belong to a genetic superfamily. The accumulating data (for review, see Arai et al. 1990) begin to reveal a coherent picture of the genetic relationships of the structural elements and functions of these important mediators of the inflammatory response.

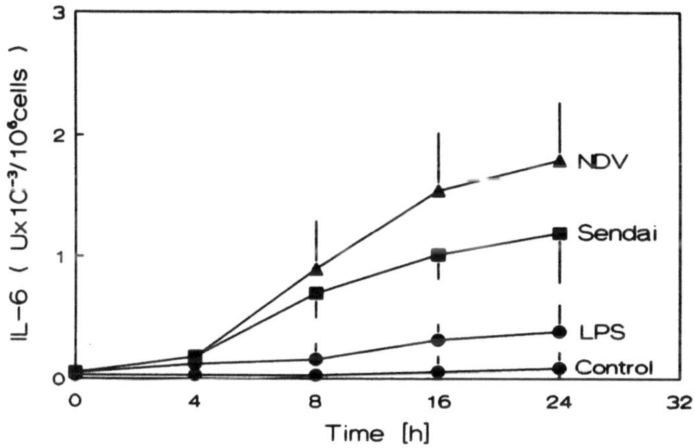

Fig. 7. Interleukin-6 production after virus or endotoxin. Primary cultures of rat liver macrophages were exposed to LPS (100 ng/ml; *circles*), Newcastle disease virus (NDV, 26 HAU/ml; *triangles*), or Sendai virus (1250 HAU/ml; *squares*). (Busam et al. 1990a)

Of particular interest for the inflammatory process are the interleukins-1 and -6 (IL-1, IL-6). They are the topics of special chapters of this volume; it may suffice, therefore, to mention that they are also (but not exclusive) products of macrophages. One of the major effects of IL-6 is the induction in hepatocytes of the acute-phase reactants (Gauldie et al. 1987; Bauer 1989) that are regularly released following inflammatory insults. The close topological association in the liver gives great importance to the ability of Kupffer cells to produce substantial amounts of IL-6 after exposure to LPS or viruses (Fig. 7).

A description of the responses of the body to inflammatory stimuli can be presented in several ways: for example, according to the cell types primarily involved, to the types of the major mediators produced, or to the target cells. Of great biological importance is yet another distinction: the *time course of action*. Some reactions are elicited rapidly, i.e., within minutes, sometimes even seconds. Contact with phagocytosable material, phorbol esters, or Ca^{2+} ionophores triggers responses rapidly. One can safely assume – as has been shown in the case of proteinkinase C, phospholipase C, or adenylate cyclase – that activation of pacemaker enzymes through interconversion or translocation determines the speed of the response. Usually, these responses are also of rather short duration and may be considered as warning signals to other cells or organs, e.g., the release of PGE_2.

Another group of processes can be seen as emergency responses primarily against toxins such as endotoxin and against some viral infections; the release of cytokines is a typical example. These responses are somewhat slower in occurrence and appear to involve changes in the translatability of the respective mRNAs, though not necessarily de novo transcription. The response may last for extended periods of time as seen in the cytokine production following contact with Newcastle disease virus (Busam et al. 1990a).

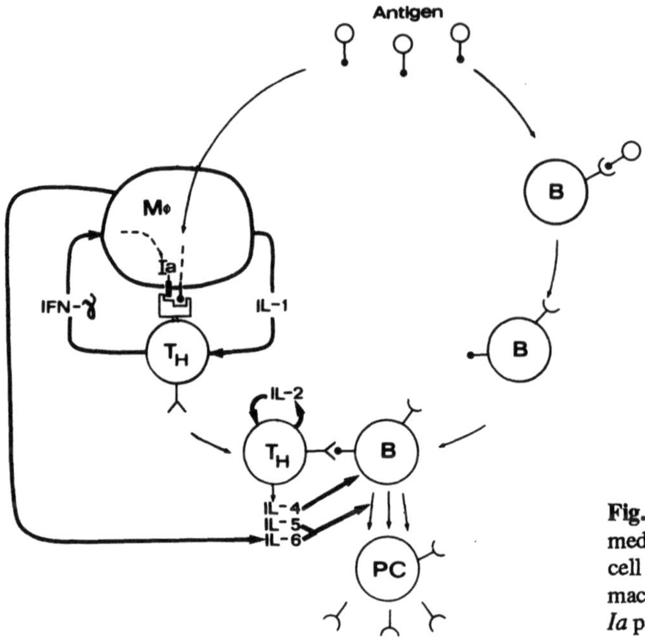

Fig. 8. Scheme of the cells and mediators participating in B-cell activation. *B* B-cell; *MΦ* macrophage; T_H T-helper cell; *Ia* part of the MHC complex

2.8 The Specific Immune Response

Finally, the body has a late (in case of first encounter with the antigenic material) response available through the activation of the immune system. Among the myriads of potential antigens, it is mainly – with regard to inflammatory processes – the contact with viruses and bacteria that provokes the production of highly specific antibodies against the offender. Preceding the transformation of B-cells into immunoglobulin-releasing plasma cells is an intricate interplay with T-cells, macrophages, and the noxious agent (Fig. 8).

Interleukins signal the presence of the enemy and activate the responsive cells. Macrophages, particularly when activated by interferon-γ to express the surface antigen Ia (Steeg et al. 1982), are able to present antigenic material to the immunocompetent cells. The process of antigen presentation involves the endocytosis of the noxious agent, its partial degradation, the association of the fragments (epitopes) with proteins of the major histocompatibility complex (MHC) within the endosomes, and the reintegration into the plasma membrane (Unanue 1984; Teyton et al. 1990). The presence of the Ia region of the MHC complex on the surface of macrophages is necessary for the interaction with T-cells; determinant selection occurs by recognition of the antigen-Ia complex by the T-cells. Mediators formed by IL-1- and macrophage-activated T-lymphocytes include IL-4, the cytokine first established as an effector of the transformation of B-cells to plasma cells (Ralph et al. 1984; Arai et al. 1990).

Antibodies elaborated by B-cells will form immune complexes with their respective antigens. Removal of these highmolecular weight conglomerates from the circulation is accomplished by the macrophages with the aid of their F_c receptors. The

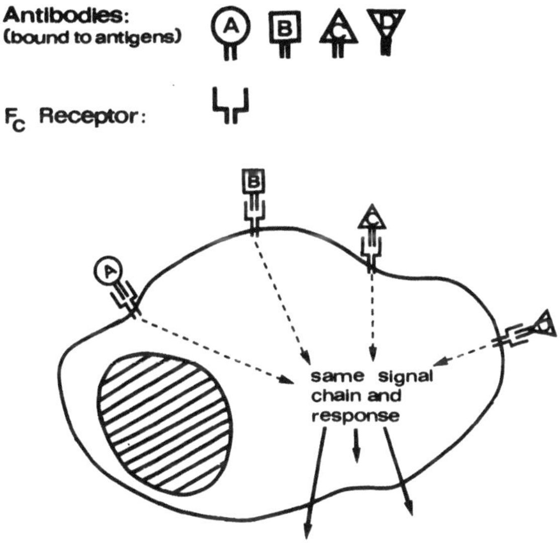

Fig. 9. The F_c receptors on macrophages channel the variety of specific antibodies and their immune complexes into a uniform pattern of signal transduction and response

structure of the immunoglobulins allows the combination of high specificity and affinity of the F_{ab} part with the universality of the F_c component (Fig. 9).

In this way a great number of noxious agents can be efficiently channeled to a general process that uses F_c receptors and the mechanisms of phagocytosis or lysis by the complement system. Obviously, macrophages and granulocytes are the cells predisposed for these terminal stages of the inflammatory reaction cascade.

2.9 The Complement System in the Inflammatory Response

In the course of an inflammatory insult the complement system becomes activated; this may occur either by the classical (antibody-dependent) or the alternative pathway. Bacteria, parasites, or damaged cells may undergo final destruction by the use of complement components. Cytotoxic lymphocytes use the "membrane attack complex" (Müller-Eberhard 1986) consisting of several proteins of the complement system to start the cytolysis of the "foreign" cell and to provide a C9-related proteinase (pore-forming protein, perforin) to perforate its cell membrane. Without damaging the target cell the membrane attack complex can initiate signal transduction; in macrophages the arachidonate cascade is stimulated following contact with the circulating complex (Hansch et al. 1984).

Single components of the complement system are instrumental in inflammation-associated phagocytosis. Although phagocytes are able to recognize epitopes of the F_c part of IgG bound to foreign cells or particles and to internalize them, this process is enhanced by binding of the C3b (C3bi) component of the complement system to the phagocytosable material (opsonization). C3b also induces neutrophils and monocytes to release reactive oxygen intermediates and leukotrienes. The component C5a is known as a very strong chemotactic agent for leukocytes and C5-derived peptides augment the adhesiveness of neutrophils (for review, see Goldstein 1988).

2.10 Regulatory Mechanisms of Mediator Synthesis and Action

In view of the intricate interactions of the mediators, the regulation of their synthesis, degradation, and action is of particular importance. Some of these mechanisms have been mentioned already (Table 6, Fig. 8). A few more examples of typical intra- and transcellular regulations will be added.

A phenomenon mostly seen with unicellular components of the reticulo-endothelial system is autostimulation. The term "autacoids" is frequently used for such autokrine processes. The positive feedback loop consists of an inducible machinery for the generation of a mediator, its release from the cell, and a receptor for this mediator on the same cell. In most instances it can be shown that the mediator has to pass through the extracellular space.

Different cell types may employ different pathways in response to the same stimulus. For multiple myeloma cells, IL-6 is a growth factor and IL-6-treated cells produce more IL-6 receptors, thereby amplifying their responsiveness. In contrast, IL-6 released from LPS-stimulated monocytes suppresses the synthesis of IL-6 receptors on these cells, leading to a downregulation of this potent mediator in the blood (Bauer et al. 1989). Still a different type of interaction is observed in the liver. IL-6 is released from LPS- or virus-stimulated Kupffer cells (Busam et al. 1990a, b). While

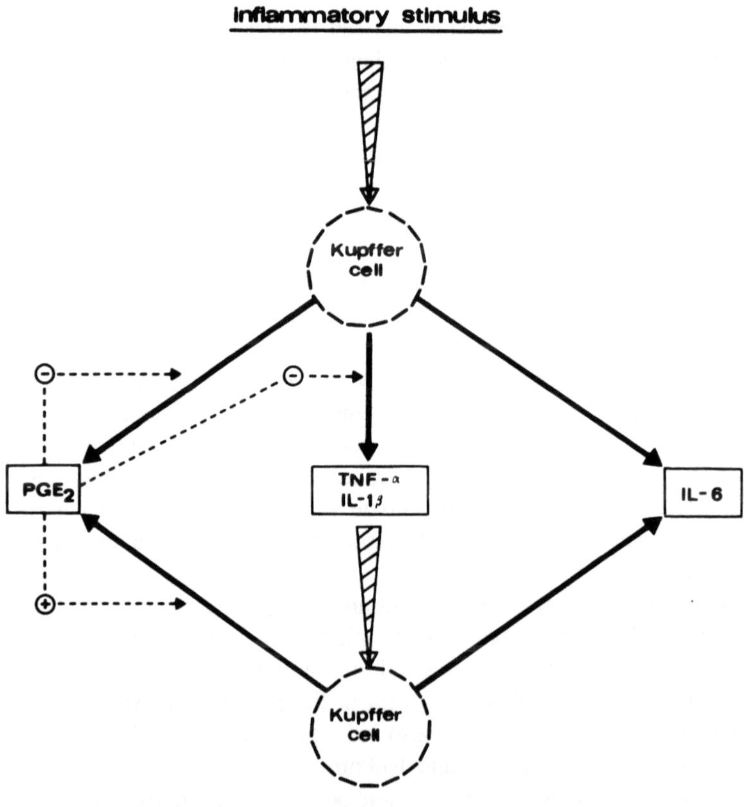

Fig. 10. Tumor necrosis factor-α and PGE_2 form a regulatory circuit

normally the number of IL-6 receptors on hepatocytes is low, it is greatly increased after exposure to IL-6, especially in the presence of glucocorticoids that are released from the adrenals under conditions of stress (Castell et al. 1989). These mechanisms synergize to provide the acute phase reactants necessary to cope with inflammatory insults (Bauer et al. 1989).

A regulatory circuit employing different mediators operates in macrophages (Fig. 10). LPS-exposed Kupffer cells synthesize TNF which, in turn, elicits PGE_2 release by the same macrophages (Peters et al. 1990). This prostaglandin, however, is an efficient inhibitor of TNF synthesis (Karck et al. 1988) and an additional enhancer of its own TNF-provoked formation. Thus, one observes in these cells after LPS stimulation a rather rapid synthesis of TNF that stops after a short time. On the other hand, PGE_2 synthesis begins slowly but continues steadily for a long period of time.

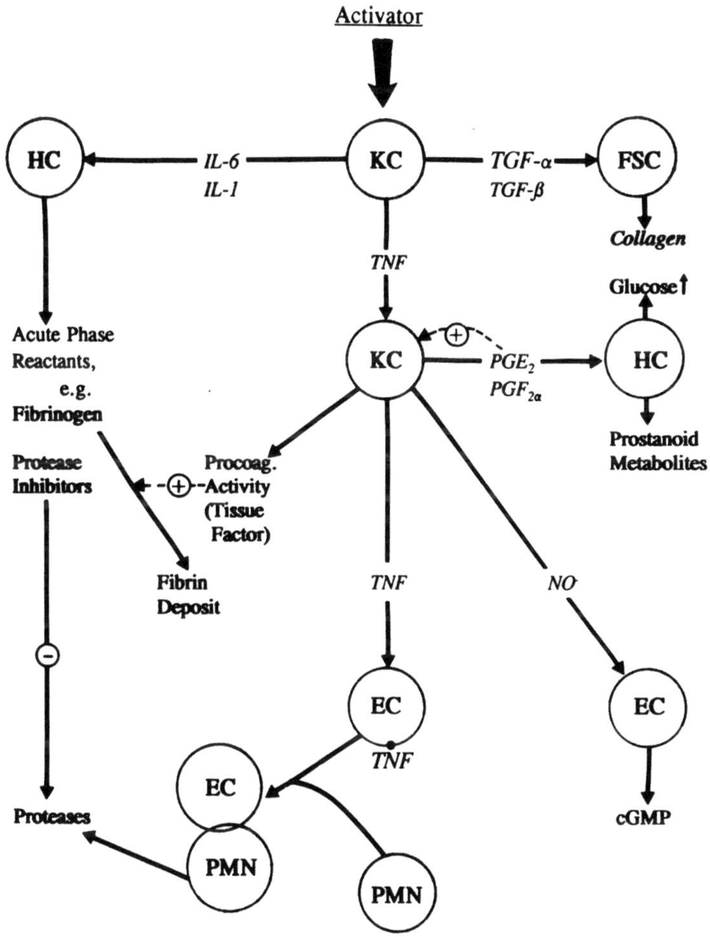

Fig. 11. Consequences of cytokine release in the liver. *EC* (Sinusoidal) endothelial cells: *FSC* fat-storing (Ito) cells; *HC* hepatocytes; *KC* Kupffer cells; *PMN* polymorphonuclear neutrophils (granulocytes)

A network of different mediators and cell types participates in the pathological sequelae following an inflammatory insult of the liver (Fig. 11). In mice, TNF was shown to bind strongly to the surface – probably to receptors – of endothelial cells (Gamble et al. 1985; Schlayer et al. 1987). In this form it is able to induce granulocytes to stick to the endothelium, leading to their activation and to the release of lytic enzymes (elastase, cathepsins, lysozyme) and reactive oxygen intermediates (Shalaby et al. 1985). This activation together with the effects of thromboxane A_2 on blood platelets and the concomitant release of procoagulant activity (Bevilacqua et al. 1986) initiate the formation of localized inflammatory foci (Schlayer et al. 1988). The hepatic consequences of endotoxinemia, thus, appear to be traceable to the primary contact of LPS with the Kupffer cells (Schlayer et al. 1988).

3. Lines of Pharmacological Intervention in the Inflammatory Response

Intervention in the network of reactants and cells involved in the inflammatory process is possible at many points. Obviously, those located at strategic regulatory steps and those easily accessible from the extracellular space bear the greatest promise. In detail, several target areas are feasible:

1. Intervention at the process of *synthesis*; this strategy is being adopted by using (synthetic) glucocorticoids. The hormone is believed to induce proteins such as lipocortin (Flower 1986) that inhibit the action of phospholipases, e.g., phospholipase A_2. This enzyme liberates the arachidonic acid required for eicosanoid synthesis. However, the mechanisms of glucocorticoid suppression of the inflammatory response are far from being clear. The very important inhibition by these steroids of the production of TNF or IL-6 appears to work differently from that of phospholipase A_2,

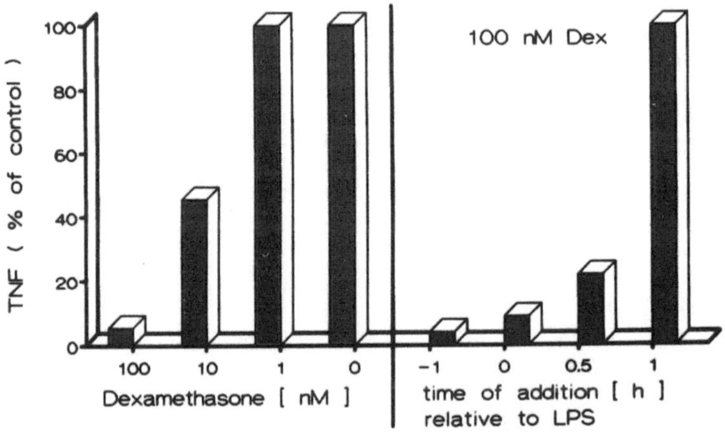

Fig. 12. Inhibition of tumor necrosis factor-α production by dexamethasone. Primary cultures of rat liver macrophages were exposed to LPS (400 ng/ml) and rINF-γ (1000 U/ml) at time zero. TNF-α in the medium was determined 4 h later. *Left* dexamethasone was added 1 h before LPS in the final concentrations indicated. *Right* dexamethasone (final conc. 100 nM) was given at the times indicated relative to the time of LPS addition. (Karck et al. 1988)

as it does not require the time consistently observed in steroïd-induced protein synthesis. The suppression of TNF formation by rat Kupffer cells is fully operative if the hormone is given concomitant with the elicitor (Karck et al. 1988; Fig. 12).

2. Many of the nonsteroidal antiinflammatory drugs have as target an *enzyme reaction* specifically involved in eicosanoid synthesis. Acetyl salicylic acid irreversibly inactivates the key enzyme of prostanoid synthesis (Fig. 5), prostaglandin H synthase (cyclooxygenase). The enzyme is also inhibited by indomethacin, although less specifically and reversibly. The pathway leading to the chemotactic 5-HETE and LTB_4 as well as to the powerful vasoconstrictive peptidoleukotrienes (Fig. 5) starts with arachidonic acid oxygenation by 5-lipoxygenase. The activity of this enzyme is quite specifically suppressed by benoxaprofen or the newly developed compound MK 88675 (Gillard et al. 1989).

3. In the latter case, neither the synthesis nor the activity of the enzyme as such but its *attachment to the membrane anchor* (Kargman and Rouzer 1989) is blocked. A similar mechanism is discussed also for the lipocortin-like inhibitors of phospholipase A_2 (Davidson et al. 1989) and, with some modification, for the inhibition of proteinkinase C by phorbol 12-myristate 13-acetate (Parker et al. 1986). These processes, which have the requirement of Ca^{2+} in common, offer new possibilities for specific pharmacological intervention at specific points in the signal transduction pathways.

4. In some instances, it appears feasible to modulate the *processing and secretion or insertion of proteins*. Examples of this approach are the use of colchicine and vinca alkaloids in chemotherapy; tunicamycin and the nojirimycins would also fall into this class, but proved too cytotoxic for therapeutic use. Cyclosporin A, a standard drug in present-day immunosuppression, is bound to responsive cells by cyclophilin (Handschumacher et al. 1984). This receptor is identical with the enzyme prolyl cis-trans isomerase (Fischer et al. 1984). Thus, an important aspect of cylosporin A action is the inhibition of the proper folding of protein chains (Fischer et al. 1989). The possibility to interfere quite specifically with the higher-order structure of proteins certainly deserves further attention.

5. A different approach to control the inflammatory process involves the use of *immunosuppressive drugs*. Their application in the aftermath of tissue transplantation notwithstanding, this treatment aims at blocking the interplay of the immunocompetent cells (Fig. 10). The multitude of molecules produced upon stimulation and highly effective in the inflammatory response can be prevented from overshooting. Autoimmune diseases, such as rheumatoid arthritis, are prone to immunopharmacological intervention. On the other hand, lack of endogenous immunocompetence as in various immunodeficiencies may be alleviated by *stimulants of the immune system* like levamisole. However, it is necessary to find more specific ways to deal with individual cell types and subpopulations of immunocompetent cells and with the relevant phases in their life cycle to achieve a satisfactory intervention in the inflammatory process without undue side effects.

6. Substances or cells that circulate freely in the extracellular space of the body may be reached by *enzymes, inhibitors, or antibodies*. The use of specific enzymes, e.g., superoxide dismutase and catalase for the removal of O_2^- and H_2O_2, respectively, or of other factors of protein nature has to deal with immune tolerance and the

restricted half-life in the circulation. The introduction of monoclonal antibodies and the feasibility of their deliberate modification opens a new dimension to the immunological intervention in inflammatory diseases.

7. The advent of *molecular biology* provided the tools to tailor effector or inhibitor molecules designed to interfere at specific points in the inflammatory processes. The construction of analogs to biological effectors with a high efficiency but narrow range of action is everybody's wish, but progress is slow. The success with the inhibitor of the angiotensin-converting enzyme II and some newly developed inhibitors of the renin system offer great hopes but are still rarities.

8. Other lines of approach include the manipulation of the *receptor-effector interaction*. The construction of soluble fragments of receptors or adhesion molecules with retained binding capability – for various reasons combined with carrier molecules – may open the way for the diversion of effector molecules or even cells from their targets in vivo. Recently, a soluble chimera was constructed (Watson et al. 1991) consisting of the extracellular, lectin-binding domain of the leukocyte homing receptor (adhesion molecule $gp^{90\ MEL}$) and part of the IgG heavy chain. Administration of this biologically stabilized form of the homing receptor decreased the number of infiltrating neutrophils in the thioglycollate-irritated peritoneum of mice.

9. *Antireceptor antibodies* to which destructive enzymes or toxins are attached are part of a strategy that has been tried extensively though yet with mixed success in the fight against tumors.

4 Conclusion

In recent years, we have learned much about the cells participating in the inflammatory response, their interaction with invaders, and their capability to produce mediators and to initiate defensive measures. Cytokines and eicosanoids have been identified as major signal substances involving macrophages, monocytes, and granulocytes as the primary cells responsive to inflammatory insults. The importance of the immune and the complement systems in dealing with the noxious material will be better understood in the future and more likely to be exploited for external intervention.

The availability of isolated cells in culture and of perfused single organs has greatly advanced these studies. Identification and cloning of receptors involved in signal transmission is in rapid progress. Molecular biology has once more proved to be an indispensable instrument for the elucidation of regulatory mechanisms involved in the expression of signals and metabolic responses; it opened new vistas for more precisely targeted pharmaceuticals.

Much is still to be learned, particularly with respect to the mechanisms that damage the target cell after contact with the inflammatory agent or mediator. In view of the delicate network of cellular interactions, any attempt to interfere with one reaction is likely to upset the responsive system as a whole. To minimize unwanted side effects the highest possible specificity of the substances used is mandatory.

Acknowledgement. The work performed in the author's laboratory was supported by grants from the Deutsche Forschungsgemeinschaft, Bonn through SFB 154 and from Fonds der Chemischen Industrie, Frankfurt, Germany.

References

Arai K, Lee F, Miyajima A, Miyatake S, Arai N & Yokota T (1990) Cytokines: coordinators of immune and inflammatory responses. Annu Rev Biochem 59:783–836

Adachi V, Arii S, Monden K, Sasaoki T, Itai S, Funaki N, Higashitsuji H, Tamura J & Tobe T (1991) Interleukin-2 receptor expression on Kupffer cells: the mechanism of activation of Kupffer cells by BRMS. In: Wisse E, Knook DL & McCuskey RS (eds) Cells of the hepatic sinusoid Vol 3, Kupffer Cell Foundation Rijswijk pp 16–18

Bauer J (1989) Interleukin-6 and its receptor during homeostasis, inflammation, and tumor growth. Klin Wochenschr 67:697–706

Bauer J, Bauer TM, Kalb T, Taga T, Lengyel G, Hirano T, Kishimoto T, Acs G, Mayer L & Gerok W (1989) Regulation of interleukin-6 receptor expression in human monocytes and monocyte-derived macrophages. J Exp Med 170:1537–1549

Benveniste EN & Merrill JE (1986) Stimulation of oligodendroglial proliferation and maturation by interleukin-2. Nature (London) 321:610–613

Beutler C & Cerami A (1986) Cachectin and tumor necrosis factor are two sides of the same biological coin. Nature (London) 320:584

Beutler B & Cerami A (1988) Cachectin (tumor necrosis factor): a macrophage hormone governing cellular metabolism and inflammatory response. Endocrine Rev 9:57–66

Beutler B, Milsark IW & Cerami A (1985) Passive immunization against cachectin/tumor necrosis factor protects mice from lethal effect of endotoxin. Science 29:869–871

Bouwens L, Marinelli A, Kuppen PJK, Eggermont AMM, van de Velde CJH & Wisse E (1991) Pit cell and Kupffer cell reactions in rats treated with systemic or hepatic continuous infusion of interleukin-2. In: Wisse E, Knook DL & McCuskey RS (eds) Cells of the hepatic sinusoid, Vol 3. Kupffer Cell Foundation Rijswijk pp 306–311

Busam KJ, Homfeld A, Zawatzky R, Kästner S, Bauer J, Gerok W & Decker K (1990a) Virus- versus endotoxin-induced activation of liver macrophages. Eur J Biochem 191:577–582

Busam KJ, Bauer TM, Bauer J, Gerok W & Decker K (1990b) Interleukin-6 release by rat liver macrophages. J Hepatol 11:367–373

Busam KJ, Schulze-Specking A & Decker K (1991) Endotoxin-refractory macrophages secrete tumor necrosis factor-α upon viral infection. Biol Chem Hoppe-Seyler 372:157–162

Castell JV, Gomez-Lechon MJ, David M, Andus T, Geiger T, Trullenque R, Fabra R & Heinrich PC (1989) Interleukin-6 is the major regulator of acute phase protein synthesis in adult human hepatocytes. FEBS Lett 242:237–239

Cochrane CG & Griffin JH (1982) The biochemistry and pathophysiology of the contact system of plasma. Adv Immunol 33:241–304

Davidson FF, Dennis EA, Powell M & Glenney JR Jr (1987) Inhibition of phospholipase A_2 by "lipocortins" and calpactins. An effect of binding to substrate phospholipids. J Biol Chem 262:1698–1705

Davis BH & Vucic A (1989) Transforming growth factor-beta modulates hepatic Ito cell proliferation, collagen synthesis and vitamin A metabolism in vitro. In: Wisse E, Knook DL & Decker K (eds) Cells of the hepatic sinusoid, Vol 2. Kupffer Cell Foundation Rijswijk pp 39–42

Decker K (1985) Eicosanoids, signal molecules of liver cells. In: Berk PD (ed) Seminars in liver disease. Vol 5. Thieme-Stratton New York pp 175–190

Decker K (1990) Biologically active products of stimulated liver macrophages (Kupffer cells). Eur J Biochem 192:245–261

Decker T, Lew DJ, Mirkovitch J & Darnell JE Jr (1991) Cytoplasmic activation of GAF, an interferon-γ regulated DNA-binding factor. EMBO J 10:927–932

Dieter P, Schulze-Specking A & Decker K (1988) Ca^{2+} requirement of prostanoid but not of superoxide production by rat Kupffer cells. Eur J Biochem 177:61–67

Fischer G, Bang H & Mech C (1984) Nachweis einer Enzymkatalyse für die cis-trans-Isomerisierung der Peptidbindung in prolinhaltigen Peptiden. Biomed Biochim Acta 43:1101–1111

Fischer G, Wittmann-Liebold B, Lang K, Kiefhaber T & Schmid F (1989) Cyclophilin and peptidyl-prolyl cis-trans isomerase are probably identical proteins. Nature (London) 337:476–478

Flower RJ (1986) Background and discovery of lipocortins. Agents Actions 17:255–262

Freudenberg MA, Freudenberg N, Galanos C (1982) Time course of cellular distribution of endotoxin in liver, lungs and kidneys of rats. Br J Exp Pathol 63:56–65

Friedman SL & Arthur MJP (1989) Activation of cultured rat hepatic lipocytes by Kupffer cell conditioned medium. Direct enhancement of matrix synthesis and stimulation of cell proliferation via induction of platelet-derived growth factor receptors. J Clin Invest 84:1780–1785

Gaillard T, Mülsch A, Busse R, Klein H & Decker K (1990) PGE_2-abhängige Reaktion der NO-Synthese in stimulierten Kupfferzellen. Z Gastroenterol 28:692

Gauldie J, Richard C, Harnish D, Lansdorp P & Baumann H (1987) Interferon beta 2/B-cell stimulatory factor type 2 shares identity with monokine-derived hepatocyte-stimulating factor and regulates the major acute phase protein response in liver cells. Proc Natl Acad Sci USA 84:7251–7255

Goldstein IM (1988) Complement: biologically active products. In: Gallin JI, Goldstein IM & Snyderman R (eds) Inflammation: basic principles and clinical correlates. Raven Press New York pp 55–74

Grewe M, Estler HC, Ballhorn A & Decker K (1990) Wirkungen von Interleukin-2 auf Kupfferzellen der Ratte. Z Gastroenterol 28:695–696

Hanahan DJ (1986) Platelet activating factor: a biologically active phosphoglyceride. Annu Rev Biochem 55:483–509

Handschumacher RE, Harding MW, Rice J, Drugge RJ & Speicher DW (1984) Cyclophilin: a specific cytosolic binding protein for Cyclosporin A. Science 226:544–547

Hansch GM, Seitz M, Martinotti G, Betz M, Rauterberg EW & Gemsa D (1984) Macrophages release arachidonic acid, prostaglandin E_2, and thromboxane in response to late complement components. J Immunol 133:2145–2150

Herberman RB (1988) Lymphocytes: cytotoxic activities. In: Gallin JI, Goldstein IM & Snyderman R (eds) Inflammation: basic principles and clinical correlates. Raven Press New York pp 613–630

Herrmann F, Cannistra SA, Levine H & Griffin JD (1985) Expression of interleukin-2 receptors and binding of interleukin-2 by gamma interferon-induced human leukemic and normal monocytic cells. J Exp Med 162:1111–1116

Hibbs JB, Taintor RR, Vavrin Z & Rachlin EM (1988) Nitric oxide: a cytotoxic activated macrophage effector molecule. Biochem Biophys Res Commun 157:87–94

Karck U, Peters T & Decker K (1988) The release of tumor necrosis factor from endotoxin-stimulated rat Kupffer cells is regulated by prostaglandin E_2 and dexamethasone. J Hepatol 7:352–361

Kargman S & Rouzer CA (1989) Studies on the regulation, biosynthesis and activation of 5-lipoxygenase in differentiated HL 60 cells. J Biol Chem 264:13313–13320

Latham PS, Pilaro AM, Echeagaray PL & Varesio L (1991) Interleukin-2 induced cytotoxicity of rat Kupffer cells in vitro. In: Wisse E, Knook DL & McCuskey RS (eds) Cells of the hepatic sinusoid, Vol III. Kupffer Cell Foundation Rijswijk pp 312–314

Lenardo MJ & Baltimore D (1989) NF-kB: a pleiotropic mediator of inducible and tissue-specific gene control. Cell 58:227–229

Liehr H, Grün M, Seelig H-P, Seelig R, Reutter W & Heine W-D (1978) On the pathogenesis of galactosamine hepatitis. Indications of extrahepatocellular mechanisms responsible for liver cell death. Virchows Arch B Cell Path 26:331–344

Lüderitz T, Schade U & Rietschel ETh (1986) Formation and metabolism of leukotriene C_4 in macrophages exposed to bacterial lipopolysaccharide. Eur J Biochem 155:377–382

Metcalf D & Burgess AW (1982) Clonal analysis of the progenitor cell committment of granulocyte and macrophage production. J Cell Physiol 111:275–283

Meyer DH, Bachem MG & Gressner AM (1990) Modulation of hepatic lipocyte proteoglycan synthesis and proliferation by Kupffer cell-derived transforming growth factors type β1 and type α. Biochem Biophys Res Comm 171:1122–1129

Müller-Eberhard HJ, (1986) The membrane attack complex of complement. Annu Rev Immunol 4:503–528

Needleman P, Turk J, Jakschik BA, Morrison AR & Lefkowitz JB (1986) Arachidonic acid metabolism. Annu Rev Biochem 55:69–102

Palmer RM, Ferrige AG & Moncada S (1987) Nitric oxide release accounts for the biological activity of endothelium-derived relaxing factor. Nature (London) 327:524–526

Parker PJ, Coussens L, Totty N, Rhee L, Young S, Chen E, Stabel S, Waterfield MD & Ullrich A (1986) The complete primary structure of protein kinase C – the major phorbol ester receptor. Science 233:853–859

Peters T, Gaillard T & Decker K (1990) Tumor necrosis factor-alpha stimulates prostaglandin but not superoxide synthesis in rat Kupffer cells. Eicosanoids 3:115–120

Peters T, Karck U & Decker K (1990) Interdependence of tumor necrosis factor, prostaglandin E_2, and protein synthesis in LPS-exposed rat Kupffer cells. Eur J Biochem 191:583–589

Ralph P, Jeong G, Welte K, Mertelsmann R, Rabin H, Henderson LE, Souza LM, Boone TC & Robb RJ (1984) Stimulation of immunoglobulin secretion in human B lymphocytes as a direct effect of high concentrations of IL-2. J Immunol 133:2442–2445

Schlayer H-J, Karck U, Ganter U, Hermann R & Decker K (1987) Enhancement of neutrophil adherence to isolated rat liver sinusoidal endothelial cells by supernatants of lipopolysaccharide-activated monocytes – role of tumor necrosis factor. J Hepatol 5:322–331

Schlayer H-J, Laaf H, Peters T, Woort-Menker M, Estler C, Karck U, Schaefer HE & Decker K (1988) Involvement of tumor necrosis factor in endotoxin-triggered neutrophil adherence to sinusoidal endothelial cells of mouse liver and its modulation in acute phase. J Hepatol 7:239–249

Shalaby MR, Aggarwal BB, Rinderknecht E, Svedersky LP, Finkle BS & Palladino MA Jr (1985) Activation of human polymorphonuclear neutrophil functions by interferon-γ and tumor necrosis factor. J Immunol 135:2069–2073

Shiratori Y, Geerts A, Ichida T & Wisse E (1986) Collagen production and Kupffer cell-modulated proliferation of fat-storing cells in culture. In: Kirn A, Knook DL & Wisse E (eds) Cells of the hepatic sinusoid. Kupffer Cell Foundation Rijswijk pp 239–245

Teyton L, O'Sullivan D, Dickson PW, Lotteau V, Sette A, Fink P & Peterson PA (1990) Invariant chain distinguishes between the exogenous and endogenous antigen presentation pathways. Nature (London) 348:39–44

Tran-Thi TA, Gyufko K, Reinke M & Decker K (1988) Output and effects of thromboxane produced by the liver perfused with phorbol myristate acetate. Biol Chem Hoppe-Seyler 369:1179–1184

Watson SR, Fennie C & Lasky LA (1991) Neutrophil influx into an inflammatory site inhibited by a soluble homing receptor-IgG chimaera. Nature (London) 349:164–167

Westphal O, Lüderitz O, Galanos C, Mayer H & Rietschel ETh (1986) The story of bacterial endotoxin. Adv Immunopharmacol, Vol 3. Pergamon Press Oxford pp 13–34

Zuckerman SH, Evans GF, Snyder YM & Roeder WD (1989) Endotoxin-macrophage interaction: post-translational regulation of tumor necrosis factor expression. J Immunol 143:1223–1227

Neutrophils, Interleukin 8, and Related Chemotactic Cytokines

M. BAGGIOLINI[1]

1 Introduction

Neutrophils are the most frequent immigrant cells in inflammatory lesions. They are usually present in very high numbers and are often incriminated as effectors of tissue damage because of their high contents in neutral and acid proteases and their ability to generate superoxide and other reactive oxygen derivatives.

In order to release enzymes and other products, and to generate oxygen-derived radicals, neutrophils must be activated. The circulating cells are quiescent and most neutrophils are eliminated from the circulation and substituted by new ones without becoming involved in a defense function, i.e., without apparent activation. Upon inflammation (and often independently of its course) neutrophils are recruited into the affected site, and this process involves multiple activation events.

Initially, circulating neutrophils sense chemotactic signals that are generated in the affected tissue site and diffuse into the local microvessels. The type of chemotactic molecule that initiates recruitment depend on the type of inflammatory injury. N-formylmethionyl peptides are likely to be preponderant upon infection, while C5a is the initial attractant when immune complexes are formed. In other situations, like trauma or ischemia, the initial stimulus for neutrophil mobilization is unknown.

While one must assume, as we do, that recruitment starts with the formation of a single chemotactic agonist, the process is invariably going to expand through the involvement of additional agonists that can act in concert on the same cell since they are directed to distinct receptors.

Once the neutrophils begin accumulating in the inflamed tissue, chemotactic stimulation as well as phagocytosis concur in the triggering of product release.

2 The Neutrophils

The neutrophils mature in the bone marrow for about 2 weeks before being ready for release into the blood (Bainton et al. 1971). The most striking event in maturation is the formation of the *azurophil* or *primary granules* at the promyelocyte stage and the *specific* or *secondary granules* later on, at the myelocyte stage. The process can be followed in promyelocytes by cytochemistry of myeloperoxidase which is detected in the rough endoplasmic reticulum where it is produced, in the Golgi apparatus, where

[1] Theodor-Kocher Institute, University of Bern, P.O. Box, CH-3000 Bern 9, Switzerland

42. Colloquium Mosbach 1991
Molecular Aspects of Inflammation

it is packaged, and in the granules, which are its storage site (Bainton et al. 1971). It is assumed by analogy that all azurophil granule enzymes, including the neutral proteases, which are particularly important in inflammation, are formed and packaged together with the myeloperoxidase, and that specific granules arise in a similar way in myelocytes. When granule formation ceases, the rough endoplasmic reticulum is largely shed and the Golgi apparatus decreases markedly in size. From this stage on the neutrophils are unable to form new granules and gradually acquire the properties that are essential for their function: the ability to migrate in response to chemotactic stimuli, to phagocytose, and to activate the respiratory burst.

3 The Granules

The two granules are easily distinguished in electron micrographs reacted for peroxidase. The electron-dense staining of oxidized diaminobenzidine in the azurophil granules contrasts with the pale, uniform matrix of the specific granules. The biochemical properties of both types of granules were established following subcellular fractionation which also revealed the presence of subcellular storage organelles that are considerably smaller than the granules (Baggiolini 1982; Baggiolini and Dewald 1984). The azurophil granules contain a remarkable series of lytic enzymes including lysosomal hydrolases, three neutral serine proteases (elastase, cathepsin G, and proteinase 3), lysozyme, and myeloperoxidase, which is the only nonhydrolytic enzyme. Proteinase 3 was identified more than 10 years ago (Baggiolini et al. 1978, 1980), and was almost forgotten, when Kao et al. (1988) described its purification. More recently, this serine protease was found to be a major autoantigen in Wegener's granulomatosis (Goldschmeding et al. 1989). The specific granules contain type-I collagenase, lysozyme, lactoferrin, and vitamin B_{12}-binding protein(s). Nothing more is known, and it is difficult to speculate about the role of this granule population in neutrophil function. The existence of storage organelles that are distinct from the granules was originally suggested by the subcellular distribution profiles of acid glycosidases (Bretz and Baggiolini 1974). They were later shown to contain cathepsin B, cathepsin D, and proteinase 3 (Baggiolini 1982; Baggiolini and Dewald 1984). In a subsequent study, we found that gelatinase is localized in such small storage organelles (Dewald et al. 1982).

4 Neutrophil Activation

Chemotactic agonists are the stimuli that recruit neutrophils from the blood into infected or inflamed tissues. In the tissues, the neutrophils usually experience a second process of activation induced by particles (e.g., bacteria or immune complexes) which they phagocytose. This second activation event is directed against the offending agent to be eliminated and is considerably more extensive.

4.1 Activation by Phagocytosis

I shall first describe this final event in activation since it shows the full extent of the response of which the neutrophils are capable. The process is triggered by opsonized particles that are recognized through Fc and C3b receptors. At the site of contact the cell extrudes ectoplasmic lamellae which bind to the opsonin-covered surface of the particle and gradually envelop it. Upon phagocytosis there is an immediate discharge of granules. It begins long before the phagocytic vacuoles are sealed off from the outside and is restricted to the area of the neutrophil membrane which is in contact with the particle and forms a phagocytic vacuole (Baggiolini and Dewald 1984). The release of enzymes into the pericellular environment through forming (and thus still open) vacuoles is thought to contribute to inflammation-dependent tissue damage.

As in other leukocytes and other excitable cells, neutrophils release preformed products (which are stored in the granules and other secretory organelles) and products that are generated de novo upon stimulation. The preformed species, mainly enzymes, are liberated upon fusion of the membrane of the storage organelles with the plasma membrane. The newly synthesized species are small molecules, e.g., superoxide, arachidonic acid oxygenation products, and platelet-activating factor. In these instances, product formation depends on the activation of a key enzyme, like NADPH-oxidase, phospholipase A_2, or the acetyl-transferase that generates platelet-activating factor.

4.2 Activation by Chemotaxins

A large number of molecules have been proposed in the past as chemoattractants to explain the often massive accumulation of neutrophils in infected tissues. The mechanism of chemotaxis and neutrophil activation, however, began to unravel with the work on N-formylmethionyl peptides (Showell et al. 1976) and on the complement fragment C5a (Fernandez et al. 1978), which were the first chemotactic agonists to be defined structurally. It was subsequently found that leukotriene B_4 (LTB_4) and platelet-activating factor (PAF) also act as chemoattractants (Ford Hutchinson et al. 1980; Baggiolini et al. 1988), and a new class of chemotactic cytokines, including interleukin 8 (neutrophil-activating peptide 1) (Baggiolini et al. 1989; Leonard and Yoshimura 1990), neutrophil-activating peptide 2 (NAP-2) (Walz et al. 1989), and the so-called melanocyte growth stimulatory activity (*gro*/MGSA) (Moser et al. 1990), was identified most recently.

Except for IL-8, NAP-2, and *gro*/MGSA, which are homologues, the chemotactic agonists differ in structure, origin, and mode of formation and act via distinct receptors. Despite this diversity, however, they all induce a similar pattern of biological responses, and activate neutrophils through a similar signal transduction process. IL-8 was originally identified in the culture supernatants of stimulated human monocytes and macrophages, and was subsequently found to be produced by many different tissue cells (Baggiolini et al. 1989). A similar situation and tissue distribution was found for *gro*/MGSA which appears to be produced under the same conditions as IL-8. NAP-2, by contrast, is formed in the extracellular space through the protelytic cleavage of inactive platelet-derived precursor proteins, platelet basic proteins, and connective tissue-activating peptide III (Walz et al. 1989; Walz and Baggiolini 1990). The precursors are released from stimulated platelets, and the enzymes that process them are released from monocytes (Walz and Baggiolini 1990). Of several enzymes studied, cathepsin G, a chymotrypsin-like neutral protease of neutrophil and monocyte granules, was found to cleave selectively connective tissue-activating peptide III into NAP-2 (Car et al. 1991).

IL-8, NAP-2, and *gro*/MGSA are generally considered as chemotactic receptor agonists for neutrophils on the basis of the results of extensive comparisons with established chemotaxins, fMet-Leu-Phe and C5a in particular. They were found to elicit all the responses that are characteristic of chemotactic stimulation, including activation of the motile apparatus, directional migration, expression of surface adhesion molecules, exocytosis from different storage organelles, and the production of superoxide and hydrogen peroxide through the respiratory burst (Baggiolini et al. 1989). By contrast, the precursors of NAP-2, platelet basic protein, and connective tissue-activating peptide III, as well as the structure homologue, platelet factor 4 are inactive at 100 to 10 000 times higher concentrations (Walz et al. 1989).

5 Mechanism of Activation

Two elements are believed to be essential for neutrophil activation and, in particular, for the process of transduction of chemotactic agonist signals, a rise of the cytosolic free calcium concentration ($[Ca^{2+}]_i$), and the activation of protein kinase C. A role for protein kinase C. A role for protein kinase C is suggested by the fact that phorbol esters and permeant diacylglycerols elicit the respiratory burst (Dewald et al. 1984). The requirement for enhanced $[Ca^{2+}]_i$ is indicated by the fact that neutrophils depleted of intracellular storage pool Ca^{2+} do not respond with exocytosis or the respiratory burst to agonist stimulation (Grzeskowiak et al. 1986).

Taking the respiratory burst as the response, the simplest signal transduction sequence for activation includes the agonist receptor, a GTP-binding protein, phospholipase C, and protein kinase C (see Baggiolini and Wymann 1990 for references). It has been shown that the activity of the NADPH-oxidase depends on the ongoing interaction of the agonist with its receptor, the interaction between the receptor and a GTP-binding protein, and the activation of phosphatidylinositol-specific phospholipase C. This enzyme delivers the products which bring about the translocation of protein kinase C to the plasma membrane and its subsequent activation.

Signal transduction and activation can also be studied on the basis of the kinetics of the response. Using a chemiluminescence assay for H_2O_2, which directly assesses NADPH-oxidase activity, we found that the onset time of H_2O_2 formation as well as the apparent rate of activation of the NADPH-oxidase were similar upon stimulation with different chemotactic agonists, suggesting that signals coming from distinct receptors are transduced by the same machinery (Wymann et al. 1987). Much longer onset times and lower rates of NADPH-oxidase activation were observed upon stimulation with ionomycin or PMA, indicating that receptor-dependent activation is mediated through a different, more efficient transduction process.

6 Interleukin 8 and Homologues as Mediators of Inflammation

The initial studies had already shown that the biological profile of IL-8 is very similar to that of established chemotactic peptides like C5a and fMet-Leu-Phe (Peveri et al. 1988). IL-8 activates the motile apparatus of neutrophils and induces migration, expression of surface adhesion molecules, release of storage enzymes, and the production of reactive oxygen metabolites. The responses are all transient, which is characteristic for chemotactic agonists (Thelen et al. 1988; Baggiolini et al. 1989). IL-8 appears to be unusually selective for neutrophils. It has only borderline activities on eosinophils (Kernen et al. J. Clin. Invest. 1991, in press), basophils (Dahinden et al. 1989) and monocytes (Walz et al., J. Leukocyte Biol. 1991, in press) and does not activate blood platelets (Thelen et al. 1988). IL-8 was reported to have chemotactic activity for lymphocytes (Larsen et al. 1989). No lymphocyte infiltration was observed, however, upon intracutaneous injection in human subjects (Leonard and Yoshimura 1990).

Since the action of IL-8 is not species restricted, it was possible to study its effects in animals. Intradermal injection in rabbits induces plasma exudation and massive neutrophil accumulation. The infiltration is long-lasting and apparently limited to neutrophils, as no other leukocytes are found (Colditz et al. 1989). At the same molar doses, fMet-Leu-Phe induces similar, but less severe and more transient effects. Neutrophil infiltration is also induced by *E. coli* lipopolysaccharide (LPS) or interleukin-1. In this case, however, the effect is indirect, and depends on the induction of a secondary mediator, presumably IL-8 itself (Colditz et al. 1989, 1990). Massive neutrophil infiltration is also observed upon intradermal injection of IL-8 in rats, mice, guinea pigs, and dogs (Zwahlen et al., unpubl.). The long duration of action could be a consequence of the remarkable resistance of IL-8 to enzymatic inactivation and denaturation (Peveri et al. 1988).

7 Pathophysiological Considerations

Host defense must be commensurate to the initiating injury, and a balance must exist between protective action and inflammatory damage. The mechanisms of neutrophil-dependent issue damage have been reviewed repeatedly (Baggiolini et al. 1980; Baggiolini and Dewald 1985), and I shall limit my comments to the process of neutrophil invasion which must be viewed in a different light now after the discovery of tissue-derived chemotactic cytokines.

All chemotaxins described may be involved in attracting neutrophils into inflamed tissues, and it is likely that several agonists contribute to the overall chemotactic response since they act through different receptors. The recent observations that IL-8 is produced by a wide variety of tissue cells following exposure to interleukin-1 or tumor necrosis factor suggest that chemotactic cytokines may play a primary role in the attraction of neutrophils into homeostatically disturbed tissues (Baggiolini et al. 1989). The response may be amplified by resident and immigrant phagocytes since monocytes and macrophages (Baggiolini et al. 1989) and even the neutrophils themselves, as recently reported (Bazzoni et al. 1991), produce and release significant amounts of IL-8.

The actual involvement of IL-8 in disease is suggested by the demonstration of is occurrence in the skin lesions of patients with psoriasis. Such lesions also contain high levels of interleukin-1 which could function as inducer. In fact, the characteristic neutrophilic microabscesses in this and other skin diseases are likely to depend on IL-8 which is more selective toward neutrophils than any other known chemotaxin (Baggiolini et al. 1989). IL-8 is also believed to be a major case of neutrophil invasion into the synovial fluid in inflammatory joint diseases, as it is produced by interleukin-1-stimulated synovial cells (Watson et al. 1988) and is released in high amounts from the mononuclear cells of arthritic joints upon stimulation with inflammatory cytokines and immune complexes (Seitz et al. 1991). Neutrophil accumulation is also prominent in lung diseases like asbestosis and idiopathic pulmonary fibrosis, and in the adult respiratory distress syndrome, where elastase and other neutral proteases released from the neutrophils are considered the main effectors of lung damage. The presence of neutrophil attractants in the bronchoalveolar lavage fluid of patients with various types of inflammatory lung diseases has been reported before the discovery of IL-8 (see Peveri et al. 1988 for references).

Further information on the pathophysiological effects of IL-8 must now be gained by studying the production of this mediator in normal and diseased tissues. Its widespread expression in response to inflammatory cytokines like interleukin-1 and tumor necrosis factor suggests that IL-8 may represent a common tissue-specific neutrophil attractant. From these considerations it is clear that the modulation of IL-8 production and the inhibition of IL-8 action on neutrophils must be considered a worthy goal of future pharmacotherapeutic research.

References

Baggiolini M (1982) Phagocytes and phagocytosis 100 years after Metchnikoff. A current picture of the neutrophil leukocyte. Schweiz Med Wochenschr 112:1304–1411

Baggiolini M, Bretz U, Dewald B & Feigenson ME (1978) The polymorphonuclear leukocyte. Agents Actions 8:3–10

Baggiolini M & Dewald B (1984) Exocytosis by neutrophils. Contemp Top Immunobiol 14:221–246

Baggiolini M & Dewald B (1985) The neutrophil. Int Arch Allergy Appl Immunol 76:13–20

Baggiolini M, Dewald B & Thelen M (1988) Effects of PAF on neutrophils and mononuclear phagocytes. Prog Biochem Pharmacol 22:90–105

Baggiolini M, Schnyder J, Bretz U, Dewald B & Ruch W (1980) Cellular mechanisms of proteinase release from inflammatory cells and the degradation of extracellular proteins. Ciba Found Symp 75:105–121

Baggiolini M, Walz A & Kunkel SL (1989) Neutrophil-activating peptide-1/interleukin 8, a novel cytokine that activates neutrophils. J Clin Invest 84:1045–1049

Baggiolini M & Wymann MP (1990) Turning on the respiratory burst. Trends Biochem Sci 15:69–72

Bainton DF, Ullyot JL & Farquhar MG (1971) The development of neutrophilic polymorphonuclear leukocytes in human bone marrow. Origin and content of azurophil and specific granules. J Exp Med 134:907–934

Bazzoni F, Cassatella MA, Rossi F, Ceska M, Dewald B & Baggiolini M (1991) Phagocytosing neutrophils produce and release high amounts of the neutrophil-activating peptide 1/interleukin 8. J Exp Med 173:771–774

Bretz U & Baggiolini M (1974) Biochemical and morphological characterization of azurophil and specific granules of human neutrophilic polymorphonuclear leukocytes. J Cell Biol 63:251–269

Car BD, Baggiolini M & Walz A (1991) Formation of neutrophil-activating peptide 2 from platelet-derived connective-tissue-activating peptide III by different tissue proteinases. Biochem J (in press)

Colditz I, Zwahlen R, Dewald B & Baggiolini M (1989) In vivo inflammatory activity of neutrophil-activating factor, a novel chemotactic peptide derived from human monocytes. Am J Pathol 134:755–760

Colditz IG, Zwahlen RD & Baggiolini M (1990) Neutrophil accumulation and plasma leakage induced in vivo by neutrophil-activating peptide-1. J Leukocyte Biol 48:129–137

Dahinden CA, Kurimoto Y, De Weck AL, Lindley I, Dewald B & Baggiolini M (1989) The neutrophil-activating peptide NAF/NAP-1 induces histamine and leukotriene release by interleukin 3-primed basophils. J Exp Med 170:1787–1792

Dewald B, Bretz U & Baggiolini M (1982) Release of gelatinase from a novel secretory compartment of human neutrophils. J Clin Invest 70:518–525

Dewald B, Payne TG & Baggiolini M (1984) Activation of NADPH oxidase of human neutrophils. Potentiation of chemotactic peptide by a diacylglycerol. Biochem Biophys Res Commun 125:367–373

Fernandez HN, Henson PM, Otani A & Hugli TE (1978) Chemotactic response to human C3a and C5a anaphylatoxins. I. Evaluation of C3a and C5a leukotaxis in vitro and under simulated in vivo conditions. J Immunol 120:109–115

Ford Hutchinson AW, Bray MA, Doig MV, Shipley ME & Smith MJ (1980) Leukotriene B, a potent chemokinetic and aggregating substance released from polymorphonuclear leukocytes. Nature (London) 286:264–265

Goldschmeding R, Van der Schoot CE, ten Bokkel Huinink D & von dem Borne, AEGKr (1989) Wegener's granulomatosis autoantibodies identify a novel diisopropylfluorophoshate-binding protein in the lysosomes of normal human neutrophils. J Clin Invest 84:1577–1587

Grzeskowiak M, Della Bianca V, Cassatella MA & Rossi F (1986) Complete dissociation between the activation of phosphoinositide turnover and of NADPH oxidase by formyl-methionyl-leucyl-phenylalanine in human neutrophils depleted of Ca^{2+} and primed by subthreshold doses of phorbol 12, myristate 13, acetate. Biochem Biophys Res Commun 135:785–794

Kao RC, Wehner NG, Skubitz KM, Gray BH & Hoidal JR (1988) Proteinase 3. A distinct human polymophonuclear leukocyte proteinase that produces emphysema in hamsters. J Clin Invest 82:1963–1973

Larsen CG, Anderson AO, Appella E, Oppenheim JJ & Matsushima K (1989) The neutrophil-activating protein (NAP-1) is also chemotactic for T lymphocytes. Science 243:1464–1466

Leonard EJ & Yoshimura T (1990) Neutrophil attractant/activation protein-1 (NAP-1 (interleukin-8)). Am J Respir Cell Mol Biol 2:479–486

Moser B, Clark-Lewis I, Zwahlen R & Baggiolini M (1990) Neutrophil-activating properties of the melanoma growth-stimulatory activity. J Exp Med 171:1797–1802

Peveri P, Walz A, Dewald B & Baggiolini M (1988) A novel neutrophil-activating factor produced by human mononuclear phagocytes. J Exp Med 167:1547–1559

Seitz M, Dewald B, Gerber N & Baggiolini M (1991) Enhanced production of neutrophil-activating peptide-1/interleukin-8 in rheumatoid arthritis. J Clin Invest 87:463–469

Showell HJ, Freer RJ, Zigmond SH, Schiffmann E, Aswanikumar S, Corcoran B & Becker EL (1976) The structure-activity relations of synthetic peptides as chemotactic factors and inducers of lysosomal enzyme secretion for neutrophils. J Exp Med 143:1154–1169

Thelen M, Peveri P, Kernen P, von Tscharner V, Walz A & Baggiolini M (1988) Mechanism of neutrophil activation by NAF, a novel monocyte-derived peptide agonist. FASEB J 2:2702–2706

Walz A & Baggiolini M (1990) Generation of the neutrophil-activating peptide NAP-2 from platelet basic protein or connective tissue-activating peptide III through monocyte proteases. J Exp Med 171:449–454

Walz A, Dewald B, von Tscharner V & Baggiolini M (1989) Effects of the neutrophil-activating peptide NAP-2, platelet basic protein, connective tissue-activating peptide III and platelet factor 4 on human neutrophils. J Exp Med 170:1745–1750

Watson ML, Westwick J, Fincham NJ & Camp RD (1988) Elevation of PMN cytosolic free calcium and locomotion stimulated by novel peptides from IL-1-treated human synovial cell cultures. Biochem Biophys Res Commun 155:1154–1160

Wymann MP, von Tscharner V, Deranleau DA & Baggiolini M (1987) The onset of the respiratory burst in human neutrophils. Real-time studies of H_2O_2 formation reveal a rapid agonist-induced transduction process. J Biol Chem 262:12048–12053

Leukotrienes and Chemotaxis – 5-Lipoxygenase Activation and Control

A. W. FORD-HUTCHINSON[1]

1 Introduction

The leukotrienes are a group of mediators derived from arachidonic acid through the action of the 5-lipoxygenase enzyme system (Ford-Hutchinson 1990a). This enzyme is present in cells of the myeloid lineage, such as polymorphonuclear leukocytes, eosinophils, mast cells, basophils and macrophages. Following cellular activation, which induces a rise in intracellular calcium, free arachidonic acid is liberated from membrane phospholipids through the action of phospholipases. The rise in intracellular calcium also causes activation of 5-lipoxygenase which can oxygenate arachidonic acid to produce the intermediate, 5-hydroperoxyeicosatetraenoic acid. This intermediate then undergoes a dehydrase step, also catalyzed by 5-lipoxygenase, to produce the unstable epoxide intermediate, leukotriene A_4. There are two enzymic routes of metabolism of leukotriene A_4. The first involves the addition of water, through the action of leukotriene A_4 hydrolase, to produce the dihydroxy fatty acid, leukotriene B_4. The second involves the action of another unique enzyme, leukotriene C_4 synthase, which catalyzes the conjugation of glutathione with leukotriene A_4 to produce the peptidolipid leukotriene C_4. Leukotriene C_4 is further metabolized through the action of γ-glutamyl transferase through loss of L-glutamic acid to probably the most important peptidolipid leukotriene, leukotriene D_4. Leukotriene D_4 in turn can be metabolized by a dipeptidase through loss of glycine to produce leukotriene E_4.

Leukotriene B_4 is a potent stimulator of leukocyte activation and can modulate lymphocyte functions (Ford-Hutchinson et al. 1980; Ford-Hutchinson 1990b). Thus, leukotriene B_4 binds to high affinity, structurally specific G protein-coupled receptors on the leukocyte cell surface. Following activation of these receptors a number of leukocyte functions are stimulated including chemotaxis, chemokinesis, adhesion and expression of surface receptors. This is reflected in vivo in the induction of leukocyte accumulation at the sites of application of leukotriene B_4 and through secondary leukocyte-mediated events, such as modulation of pain responses and changes in vascular permeability. Leukotriene B_4 has also been shown to have effects on lymphocyte function and, for example, will augment the action of various cytokines. Because of this, leukotriene B_4 has been proposed as an important mediator of inflammation. In man considerable attention has focused on the potential role of this mediator in inflammatory bowel disease (Fretland et al. 1990).

[1] Merck Frosst Centre for Therapeutic Research, P.O. Box 1005, Pointe Claire – Dorval, Quebec H9R 4P8, Canada

42. Colloquium Mosbach 1991
Molecular Aspects of Inflammation

Leukotriene D_4 also interacts with high affinity, structurally specific G protein-coupled receptors (Lewis et al. 1985). These receptors are present in a number of preparations including human bronchial smooth muscle. There is little evidence for the presence of separate leukotriene C_4 and leukotriene E_4 receptors in this preparation and these autocoids on human bronchial smooth muscle interact with the leukotriene D_4 receptor. Activation of the leukotriene D_4 receptor has been proposed to be an important event in the initiation of allergic reactions (Piper 1984) and leukotriene D_4 appears to be a key mediator of human bronchial asthma. Thus, peptidolipid leukotrienes have been demonstrated in vitro and in vivo following antigen challenge of allergic patients and elevated excretion of leukotriene E_4 can be observed in human urine following antigen and aspirin provocation of sensitive asthmatic subjects (Taylor et al. 1989; Christie et al., in press). Leukotriene D_4 induces multiple airway responses including smooth muscle contraction, increased bronchial mucus secretion and the formation of airway mucosal edema (Piper 1984). When compared to other vasoactive mast cell-derived mediators, such as histamine, leukotriene D_4 is approximately 1000 times more potent and causes prolonged bronchoconstriction in man. There is also evidence that leukotrienes may be involved in the induction of bronchial hyperreactivity. Thus, administration of a single aerosol dose of leukotriene D_4 to normal individuals caused a greater degree of airway narrowing than observed with methacholine and, in addition, leukotriene D_4 heightened the level of maximal response to methacholine for a prolonged period of time without affecting the position of the dose response curve (Bel et al. 1987). In another study administration of aerosols of leukotriene E_4 to asthmatic individuals, but not to normal subjects, induced not only bronchoconstriction but also a subsequent increase in responsiveness to histamine and methacholine which persisted for up to 1 week after challenge (Arm et al. 1988). These observations indicate that leukotrienes can mimic many of the pathological aspects of human bronchial asthma.

2 Leukotriene D_4 Receptor Antagonist

One approach to controlling the actions of leukotriene D_4 in diseases, such as human bronchial asthma, has been to use selective, potent leukotriene D_4 receptor antagonists. An example of such a compound is MK-571 (3-(((3(2-(7-chloro-2-quinolinyl)ethenyl)phenyl)(((3-(dimethyl-amino)-3-oxopropyl)thio)methyl)thio)-propanoic acid) (Jones et al. 1989). MK-571 is a competitive and specific antagonist of LTD_4 (IC_{50} values on guinea pig and human lung preparations $\approx$ 1 and 8 nM respectively). This potent in vitro activity has been extended to in vivo studies which have shown that MK-571 is a potent inhibitor of leukotriene D_4-induced bronchoconstriction in the guinea pig and the squirrel monkey and ascaris-induced bronchoconstriction in the squirrel monkey and allergic sheep.

Early clinical studies with leukotriene antagonists of modest potency yielded unimpressive clinical results. This is not the case, however, with MK-571. Thus, this agent has been shown to be a potent antagonist of leukotriene D_4-induced bronchoconstriction in asthmatic men (Kips et al. in press). Also, it has been shown to inhibit exercise-induced asthma (Manning et al. 1990) as well as the early- and late-phase bronchoconstrictor responses to inhaled antigen (Hendeles et al. 1990). In patients with moderate asthma (baseline FEV_1 50–80% predicted) MK-571 caused a clinically significantly bronchodilatation (≈ 20% increase in FEV_1) 20 min after the start of MK-571 infusion, this bronchodilatation being maintained throughout the study (Gaddy et al. 1990). The bronchodilatation induced by inhaled albuterol appeared to be additive with MK-571. In addition, baseline airway obstruction could be correlated with the degree of bronchodilatation achieved with MK-571, suggesting that LTD_4 receptor activation may be proportional to the severity of airway obstruction in asthmatic patients. Finally, a chronic asthma study has been carried out with MK-571 and over a 6-week period significant improvements in FEV_1 as well as reductions in symptom scores and the use of β-agonist inhalers were observed (Margolskee et al. 1991). All these results are consistent with leukotriene D_4 being a central mediator of asthma and suggest that both leukotriene D_4 receptor antagonists, as well as leukotriene biosynthesis inhibitors, may find a place in the therapy and management of human bronchial asthma, as well as other allergic diseases.

3 Leukotriene Biosynthesis Inhibitors

An alternative therapeutic approach to the use of leukotriene D_4 receptor antagonists would be to develop inhibitors of the biosynthesis of leukotrienes. Such compounds should have advantages over leukotriene D_4 receptor antagonists, the most important being their ability to inhibit the production of leukotriene B_4. This suggests that such inhibitors may have an antiinflammatory activity and may be useful in the treatment of various human diseases, such as inflammatory bowel disease (Fretland et al. 1990). A large number of 5-lipoxygenase inhibitors of diverse structures have been described in the literature. The majority of these can be classified as "redox" compounds and presumably act on 5-lipoxygenase by reducing the iron at the active site of the enzyme which results in the generation of radical intermediates by a pseudoperoxidase cycle. The more potent the inhibitor the more likely it is that the radical species generated by the pseudoperoxidase cycle may be highly reactive and may form covalent adducts with macromolecules. Such inhibitors may also interact with other redox-based biological systems. A common toxicological problem with such compounds has been the formation of methemoglobin in dogs. Because of this, and other problems associated with such agents, alternative approaches have been examined. MK-0886 (L-663,536); (3-[-(4-chlorobenzyl)-3-t-butyl-thio-5-isopropylindol-2-yl]-2,2-dimethylpropanoic acid) is a potent inhibitor of leukotriene biosynthesis in a variety of intact cell preparations (IC_{50} values in the low nM range) (Gillard et al. 1989). This compound is also a potent inhibitor of leukotriene production in vivo which is reflected in functional inhibition of antigen-induced bronchoconstriction in a number of animal models. Despite the fact that MK-0886 inhibits leukotriene production in all intact cells examined to date, the compound has no effect on either purified 5-lipoxygenase nor on leukotriene biosynthesis in broken cell preparations. It also does not appear to affect the availability of the substrate, arachidonic acid. A number of studies have addressed the mechanism of action of this compound. One possibility was that in the intact cell 5-lipoxygenase required an activation step before it could catalyze leukotriene biosynthesis. Evidence in support of this was obtained by Rouzer and Kargman (1988) who demonstrated that following neutrophil activation, 5-lipoxygenase underwent a translocation from the cytosol to the cell membrane where it could be detected as dead enzyme, presumably having undergone turnover-associated suicide inactivation. When MK-0886 was added to these incubation mixtures, inhibition of this translocation process occurred (Rouzer et al. 1990). This inhibition of translocation was shown to be correlated with inhibition of leukotriene synthesis over a range of drug concentrations using either MK-0886 or structurally related analogs. These results strongly suggested that translocation to the membrane was a key activation step for 5-lipoxygenase.

Because of the high potency of MK-0886, it was suggested that this compound might bind with high affinity to a specific molecular target. In order to search for such a putative binding protein two series of chemical tools were synthesized. The first were radioactive photoaffinity probes based on the structure of MK-0886 and the second a series of affinity columns to which MK-0886, or structurally related analogs, were coupled through a variety of linkages. Through the use of a photoaffinity probe it was possible to show that specific binding occurred to a single 18-kDa protein, present in the membranes of leukocytes, but which was not present in a variety of other cells types not known to contain 5-lipoxygenase (Miller et al. 1990). This protein could be purified from CHAPS-solubilized extracts of rat membranes through the use of affinity columns. The 18-kDa protein was eluted from the columns by MK-0886 and subsequent purification through size exclusion columns led to the isolation of sufficient protein to obtain sequence information (Miller et al. 1990). Sequence data obtained both from the native rat protein as well as tryptic and cynogen bromide cleavage products were used to construct oligonucleotide probes. These in turn were used to screen rat neutrophil and rat basophil leukemia cell line libraries to obtain the corresponding rat cDNA (Dixon et al. 1990). The rat cDNA was then used to screen a human HL-60 cell library to obtain the human cDNA. Both cDNA clones encoded proteins of 101 amino acid residues which were 92% identical and included the peptide sequences derived from the previously purified rat protein. It has been proposed that the structure of the protein consists of three transmembrane spanning regions connected by two hydrophilic groups with the C- and N-terminus on opposite sides of the membrane. There appear to be no significant sequence homologies with other known proteins.

In order to correlate the presence of this novel 18-kDa protein with cellular leukotriene biosynthesis a series of transfection experiments were carried out in osteosarcoma cells (Dixon et al. 1990). In cells transfected with either 5-lipoxygenase or the 18-kDa protein alone and stimulated with ionophore A23187, no leukotriene synthesis could be observed. However, in similar cells transfected with both proteins and then exposed to ionophore, cellular leukotriene synthesis was observed and this synthesis could be inhibited by MK-0886. These results suggest that both the 18-kDa protein and 5-lipoxygenase are required for cellular leukotriene synthesis and because of this the new 18-kDa protein has been termed 5-lipoxygenase activating protein (FLAP).

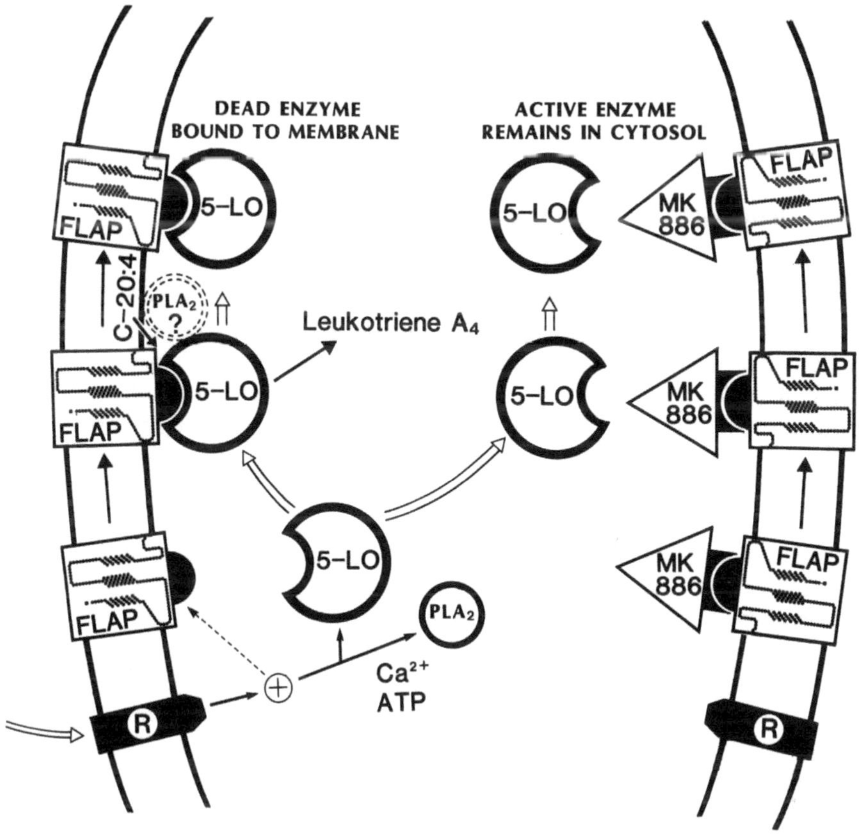

Fig. 1. Proposed scheme for the involvement of FLAP, 5-lipoxygenase and MK-0886 in the regulation of cellular leukotriene biosynthesis. *5-LO* 5-Lipoxygenase; *FLAP* 5-lipoxygenase-activating protein; PLA_2 phospholipase A_2; *c20:4* arachidonic acid; *R* receptor. (A. W. Ford-Hutchinson, R. Young and J. Gillard Drug News and Perspectives in press)

The current hypothesis for the mechanism of action of MK-0886 is summarized in Fig. 1. Following leukocyte activation (e.g. through a receptor-mediated event) the levels of intracellular Ca^{2+} rise significantly. This results in the activation of 5-lipoxygenase (as well as phospholipases such as phospholipase A_2), a process associated with translocation of the enzyme from the cytosol to the "docking" protein (FLAP) in the membrane. This model requires that a stable complex be formed in the membrane between activated 5-lipoxygenase and FLAP; this complex could also include other components of the leukotriene biosynthetic pathway, such as phospholipase A_2. The complex regulates the interaction of the enzyme with its substrate, arachidonic acid, resulting in efficient leukotriene biosynthesis. The 5-lipoxygenase enzyme then undergoes turnover-associated suicide inactivation, resulting in "dead" enzyme remaining bound to the membrane. Leukotriene biosynthesis inhibitors, such as MK-0886, prevent 5-lipoxygenase translocation by binding to FLAP; thus, the enzyme remains in the cytosol as active, intact protein and no leukotriene biosynthesis occurs.

4 Conclusions and Further Studies

The availability of potent leukotriene D_4 receptor antagonists, such as MK-571, and their clinical testing have established that leukotriene D_4 must now be considered as an important mediator of human bronchial asthma. Such compounds can be also used to define the role of leukotriene D_4 in other human diseases such as allergic rhinitis and allergic conjunctivitis. Development of compounds, such as MK-0886, with novel mechanisms of action may overcome many of the problems associated with direct redox-based 5-lipoxygenase inhibitors. If such compounds can be shown to be biochemically effective in man, they can then be used in the treatment of human bronchial asthma as well as to define the role of leukotriene B_4 in various human diseases, such as inflammatory human bowel disease and psoriasis. A number of basic studies need to be carried out to further characterize the mechanism of action of leukotriene biosynthesis inhibitors and many questions remain to be resolved. For example, there is no direct proof that FLAP and 5-lipoxygenase directly associate together and it is unclear why 5-lipoxygenase should undergo translocation following activation by rises in intracellular Ca^{2+}. In addition, basic studies need to be carried out on the purification and characterization of leukotriene receptors, on receptor subtypes and secondary transduction mechanisms as well as on the purification of other enzymes in the biosynthetic pathways, such as leukotriene C_4 synthase.

References

Arm JP, Spur BW & Lee TH (1988) The effects of inhaled leukotriene E_4 on the airway responsiveness to histamine in subjects with asthma and normal subjects. J Allergy Clin Immunol 82:654–660

Bel EH, Van Der Veen H, Kramps JA, Dijkman JH & Sterk PJ (1987) Maximal airway narrowing to inhaled leukotriene D_4 in normal subjects. Comparison and interaction with methacholine. Am Rev Respir Dis 136:979–984

Christie PE, Tagari P, Ford-Hutchinson AW, Charleson S, Chee P, Arm JP & Lee TH (1991) Urinary leukotriene E_4 concentrations increase after aspirin challenge in aspirin-sensitive asthmatic subjects. Am Rev Respir Dis (in press)

Dixon RAF, Diehl RE, Opas E, Rands E, Vickers PJ, Evans JF, Gillard JW & Miller DK (1990) Requirement of a 5-lipoxygenase activating protein for leukotriene synthesis. Nature (London) 343:282–284

Ford-Hutchinson AW (1990a) Arachidonic acid metabolism – enzymatic pathways. In: Ruzicka T (ed) Eicosanoids and the skin. CRC Press, Boca Raton, Florida, pp 3–22

Ford-Hutchinson AW (1990b) Leukotriene B_4 in inflammation. Crit Rev Immunol 10:1–12

Ford-Hutchinson AW, Bray MA, Doig MV, Shipley ME & Smith MJH (1980) Leukotriene B, a potent chemokinetic and aggregating substance released from polymorphonuclear leukocytes. Nature (London) 286:264–265

Fretland DJ, Djuric SW & Gaginella TS (1990) Eicosanoids and inflammatory bowel disease: regulation and prospects for therapy. Prostaglandins Leukot Essent Fatty Acids 41:215–233

Gaddy J, Bush RK, Margolskee D, Williams VC & Busse W (1990) The effects of a leukotriene D_4 (LTD_4) antagonist (MK-571) in mild to moderate asthma. J Allergy Clin Immunol 85:197 (Abstr)

Gillard J, Ford-Hutchinson AW, Chan C, Charleson S, Denis D, Foster A, Fortin R, Leger S, McFarlane CS, Morton H, Piechuta H, Riendeau D, Rouzer CA, Rokach J, Young R, MacIntyre DE, Peterson L, Bach T, Eiermann G, Hopple S, Humes J, Hupe L, Luell S, Metzger J, Meurer R, Miller DK, Opas E & Pacholok S (1989) L-663,536 (MK-886) (3-[1-[4-chlorobenzyl)-3-t-butyl-thio-5-isopropylindol-2-yl]-2,2-dimethylpropanoic acid), an novel, orally active leukotriene biosynthesis inhibitor. Can J Physiol Pharmacol 67:456–464

Hendeles L, Davison D, Blake K, Harman E, Cooper R & Margolskee D (1990) Leukotriene D_4 is an important mediator of antigen-induced bronchoconstriction: attenuation of dual response with MK-571, a specific LTD_4 receptor antagonist. J Allergy Clin Immunol 85:197(Abstr)

Jones TR, Zamboni R, Belley M, Champion E, Charette L, Ford-Hutchinson AW, Frenette R, Gauthier JY, Leger S, Masson P, McFarlane CS, Piechuta H, Rokach J, Williams H, Young RN, DeHaven RN & Pong SS (1989) Pharmacology of L-660,711 (MK-571): a novel, potent and selective leukotriene D_4 receptor antagonist. Can J Physiol Pharmacol 67:17–28

Kips J, Joos G, DeLepeleire I, Margolskee D, Buntinx A, Pauwels R & Van der Straten M MK-571: a potent antagonist of LTD_4 induced bronchoconstriction in man. Am Rev Respir Dis (in press)

Lewis MA, Mong S, Vessella RL & Crooke ST (1985) Identification and characterization of leukotriene D_4 receptors in adult and fetal human lung. Biochem Pharmacol 34:4311–4317

Manning PJ, Watson RM, Margolskee DJ, Williams VC, Schwartz JI & O'Byrne PM (1990) Inhibition of exercise-induced bronchoconstriction by MK-571, a potent leukotriene D_4 receptor antagonist. N Engl J Med 323:1736–1739

Margolskee D, Bodman S, Dockhorn R, Israel E, Kemp J, Mansmann H, Minotti DA, Spector S, Stricker W, Tinkelman D, Townley R, Winder J & Williams V (1991) The therapeutic effects of MK-571, a potent selective leukotriene (LT) D_4 receptor antagonist, in patients with chronic asthma. J Allergy Clin Immunol 87:309 (Abstr)

Miller DK, Gillard JW, Vickers PJ, Sadowski S, Léveillé C, Mancini JA, Charleson P, Dixon RAF, Ford-Hutchinson AW, Fortin R, Gauthier JY, Rodkey JC, Rosen R, Rouzer C, Sigal IS, Strader CD & Evans JF (1990) Identification and isolation of a membrane protein necessary for leukotriene production. Nature (London) 343:278–281

Rouzer CA & Kargman S (1988) Translocation of 5-lipoxygenase to the membrane in human leukocytes challenged with ionophore A23187. J Biol Chem 263:10980–10988

Taylor GW, Maltby NH, Black P, Taylor I, Turner N, Fuller RW & Dollery CT (1989) Urinary leukotriene E_4 after antigen challenge and in acute asthma and allergic rhinitis. Lancet I:584–588

The Respiratory Burst Oxidase

B. M. BABIOR[1]

Introduction

The respiratory burst oxidase is an enzyme found principally in phagocytes (Babior 1978; Weiss et al. 1978; Tauber et al. 1979) that catalyzes the 1-electron reduction of oxygen to O_2^- at the expense of NADPH:

$$2\,O_2 + NADPH \longrightarrow 2\,O_2^- + NADPH^+ + H^+.$$

It is dormant in resting cells, but comes to life when the cells are exposed to any of a large number of activating agents, including opsonized bacteria (Curnutte and Babior 1974), soluble inflammatory mediators (Sklar et al. 1987; Huizinga et al. 1989) and miscellaneous activators such as phorbol myristate acetate (Markert et al. 1984). The O_2^- produced by these cells is the precursor of a large and complex group of reactive oxidants that are used by the phagocytes as microbicidal agents.

Recent work has revealed that the respiratory burst oxidase is a highly complex enzyme consisting of many components, some yet to be discovered. Its complexity is probably related to the very tight spatial and temporal control to which its activity is subject. In the resting cell, the components of the oxidase are distributed between the cytosol and the plasma membrane, but when the cell is exposed to a stimulus the cytosolic components (at least those identified to date) move to the plasma membrane to assemble the active oxidase.

This chapter provides a brief review of what is known about the individual components of the respiratory burst oxidase as determined through studies with human neutrophils. The cytosolic components will be described first, then the membrane-associated components. In each category, both known and postulated components will be discussed.

[1] From the Department of Molecular and Experimental Medicine, Scripps Research Institute, 10666 North Torrey Pines Road, La Jolla, CA 92037, USA

42. Colloquium Mosbach 1991
Molecular Aspects of Inflammation

Cytosolic Components

The NADPH-Binding Component: p32^phox2^. The only electron-transporting component of the oxidase that has been found to reside in the cytosol is the NADPH-binding component, a 32K protein that is transferred from the cytosol to the plasma membrane when the cell is activated (Umei et al. 1991). In the resting cell it exists in the form of a complex with at least two other cytosolic components ($p47^{phox}$ and $p67^{phox}$; these are discussed below) (Heyworth et al. 1990). In the absence of $p47^{phox}$, transfer of $p32^{phox}$ to the plasma membrane does not take place. $P32^{phox}$ is inactivated by NADPH dialdehyde, a property upon which the studies of this component have been based (Umei et al. 1986; Smith et al. 1989a, b). Purification and cloning of $p32^{phox}$ is presently under way.

An Oxidase-Related Phosphoprotein: $p47^{phox}$. $P47^{phox}$ is a cytosolic oxidase component that undergoes phosphorylation when neutrophils are activated. Unphosphorylated $p47^{phox}$ is very basic (pI ≈ 10), but during neutrophil activation the protein can acquire up to eight phosphates, all attached to serine residues, with an accompanying drop in its pI to a value as low as ≈ 6.8 (Hayakawa et al. 1986; Okamura et al. 1988; Rotrosen and Leto 1990). $P47^{phox}$ acquires most of its phosphates in the cytosol, but before the final phosphate can be attached the protein must form a complex with one of the membrane-associated components of the respiratory burst oxidase ($gp91^{phox}$, see below) (Okamura et al. 1988; Rotrosen and Leto 1990). The respiratory burst oxidase is deactivated when neutrophils are removed from the stimulating agent; deactivation is accompanied by the dephosphorylation of $p47^{phox}$ (Hayakawa et al. 1986).

The relation of $p47^{phox}$ to the respiratory burst oxidase was demonstrated by studies of patients with chronic granulomatous disease (CGD). In this inherited disorder, respiratory burst oxidase activity is low or absent (Curnutte and Babior 1987). Furthermore, whereas in normal neutrophil homogenates the respiratory burst oxidase can be activated by the addition of certain anionic detergents, oxidase activation does not take place in homogenates from CGD neutrophils (Curnutte 1985; McPhail et al. 1985). Studies with antibody have shown that neutrophils from about 30% of patients with CGD lack $p47^{phox}$ (Clark et al. 1989). In homogenates from these neutrophils, O_2^- production in response to anionic detergents can be restored by the addition of recombinant $p47^{phox}$ or a truncated form of recombinant $p47^{phox}$ in which 15K of amino-terminal polypeptide has been deleted (Wilde et al. 1990).

A Possible Lipid-Binding Component: $p67^{phox}$. A third cytosolic component of the respiratory burst oxidase is $p67^{phox}$, a protein discovered through its absence from a small minority of patients with CGD. This protein, essential for the activity of the oxidase, is associated with the cytoskeleton in both resting and activated neutrophils

---

[2] According to an informal but widely observed convention, the components of the respiratory burst oxidase are named by attaching the superscript suffix *phox* (*ph*agocyte *ox*idase) to the component's M_r (expressed in thousands). Thus, $p32^{phox}$ represents an oxidase component that migrates on SDS-PAGE with an M_r of 32K.

(Woodman et al. 1991), and may therefore be partly responsible for the restriction of oxidase activation to regions of membrane in direct contact with the activating agent (Ohno et al. 1982). In the fully soluble cell-free oxidase activating system, $p67^{phox}$ is destroyed by anionic detergent at the same rate that the oxidase is activated, suggesting that oxidase activation in this system may be accomplished by a detergent-induced conformational change in $p67^{phox}$ that renders the subunit active and labile at the same time (Fujimoto et al. 1989):

This finding explains the complex kinetics observed in this cell-free system (Babior et al. 1988). Like $p47^{phox}$, $p67^{phox}$ has been cloned and expressed, and the recombinant material can restore O_2^- production to cell-free oxidase activating systems from $p67^{phox}$ deficient neutrophils (Leto et al. 1990).

A G-Protein. GTP (or its nonhydrolyzable analog, GTPγS) is an absolute requirement for O_2^- production by the cell-free oxidase activating system (Gabig et al. 1987; Peveri and Curnutte 1989). This finding strongly suggests the participation of a G-protein in the activation of the respiratory burst oxidase. Many G-proteins have been reported to occur in neutrophils (Gierschik et al. 1986; Kikuchi et al. 1986; Goldsmith et al. 1987, 1988; Oinuma et al. 1987; Uhing et al. 1987; Bokoch et al. 1988; Polakis et al. 1988; Stasia et al. 1989; Khachatrian et al. 1990); which of these is the oxidase-associated G-protein remains to be determined.

Membrane-Associated Components

Cytochrome b_{558}: $gp91^{phox}$ and $p22^{phox}$. Cytochrome b_{558}, the first respiratory burst oxidase component to be identified, is an unusual heme protein with a b-type optical spectrum and a very low redox potential (– 245 mV) (Segal 1988). It is composed of two subunits: gp 91, a glycoprotein consisting of a 53K polypeptide heavily glycosylated with N-linked oligosaccharides, and p22, a sugar-free protein that appears to carry the heme (Dinauer et al. 1987; Parkos et al. 1987; Yamaguchi et al. 1989; Kleinberg et al. 1991). Both subunits must be synthesized for either to appear in the membrane of cytochrome b_{558}-containing cells (Parkos et al. 1989; Verhoeven et al. 1989). The intact cytochrome has an M_r of ≈ 110–130K, and each molecule appears to contain two heme groups, suggesting that the subunit stoichiometry of the cytochrome is $\alpha\beta_2$ (α = gp91, β = p22) (Pember et al. 1984; Parkos et al. 1987). Like $p67^{phox}$, the plasma membrane-bound cytochrome is associated with the neutrophil cytoskeleton (Woodman et al. 1991).

A recent examination of the rates of electron transfer to and from cytochrome b_{558} (Cross et al. 1985) leaves little doubt that the cytochrome participates as a redox component of the respiratory burst oxidase, probably serving as the terminal electron carrier, though earlier studies were not compatible with this idea (Cross et al. 1982; Gabig et al. 1982), and O_2^- can be produced at a rapid rate by oxidase preparations devoid of the cytochrome (Glass et al. 1986). Both subunits of the cytochrome have been cloned and sequenced (Dinauer et al. 1987; Parkos et al. 1988), but as yet neither has been expressed in cells from CGD patients whose phagocytes lack the cytochrome.

The cytochrome must be present for the phosphorylation of $p47^{phox}$ to reach completion during neutrophil activation. The most acidic forms of ^{32}P-labeled $p47^{phox}$ are not found in activated neutrophils that lack the cytochrome, although they do appear in cells that contain normal amounts of a functionally defective cytochrome (Hayakawa et al. 1986; Okamura et al. 1988a, b). In the cell-free oxidase activating system, the complete phosphorylation of $p47^{phox}$ is inhibited by a peptide corresponding to a portion of the intracytoplasmic C-terminal tail of gp91, suggesting that phosphorylation only takes place when $p47^{phox}$ is associated with this region of gp91 (Rostrosen et al. 1990)

A Membrane-Associated Flavoprotein: $p45^{phox}$ (?). The participation of FAD in electron transport by the respiratory burst oxidase has been recognized for over 10 years (Babior and Kipnes 1977). Recently it was found that the respiratory burst oxidase, like certain other flavoenzymes, is powerfully and irreversibly inhibited by the arylating agent diphenyleneiodonium (Cross and Jones 1986). When radioactive diphenyleneiodonium was used, the major labeled band corresponded to a 45K protein that appeared predominantly in the particulate fraction of neutrophil homogenates (Yea et al. 1990). Earlier work strongly suggested the presence of a membrane-associated flavoprotein as a component of the respiratory burst oxidase (Curnutte and Babior 1987). It may be that this flavoprotein component is the 45K protein identified by diphenyleneiodonium labeling.

A ras Protein. A 22K member of the *ras* family of GTP-binding proteins was found to copurify with cytochrome b_{558} (Quinn et al. 1989). Its provenance suggests that this *ras* protein has something to do with the regulation of the respiratory burst oxidase, but further evidence regarding the nature of its involvement, if any, remains to be obtained.

Acknowledgment. Supported in part by Grants AI-24227, AI-24879, and AI-30742 from the National Institutes of Health.

References

Babior BM (1978) Oxygen-dependent microbial killing by phagocytes. N Engl J Med 298:659–668

Babior BM & Kipnes RS (1977) The superoxide-formimg enzyme from human neutrophils. Evidence for a flavin requirement. Blood 59:517–524

Babior BM, Kuver R & Curnutte JT (1988) Kinetics of activation of the respiratory burst oxidase in a fully soluble system from human neutrophils. J Biol Chem 263:1713–1718

Bokoch GM, Parkos CA & Mumby SM (1988) Purification and characterization of the 22,000-dalton GTP-binding protein substrate for ADP-ribosylation by botulinum toxin, G22k. J Biol Chem 263:16744–16749

Clark RA, Malech HL, Gallin JI, Nunoi H, Volpp B, Pearson DW, Nauseef WM & Curnutte JT (1989) Genetic variants of chronic granulomatous disease: prevalence of deficiencies of two discrete cytosolic components of the NADPH oxidase system. N Engl J Med 321:647–652

Cross AR, Higson FR, Jones OTG, Harper AM & Segal AW (1982) The enzymic reduction and kinetics of oxidation of cytochrome b-245 of neutrophils. Biochem J 204:479–485

Cross AR & Jones OTG (1986) The effect of the inhibitor diphenylene iodonium on the superoxide-generating system of neutrophils. Biochem J 237:111–116

Cross AR, Parkinson JF & Jones OTG (1985) Mechanism of the superoxide-producing oxidase of neutrophils. O_2 is necessary for the fast reduction of cytochrome b-245 by NADPH. Biochem J 226:881–884

Curnutte JT (1985) Activation of human neutrophil nicotinamide adenine dinucleotide phosphate, reduced (triphosphopyridine nucleotide, reduced) oxidase by arachidonic acid in a cell-free system. J Clin Invest 75:1740–1743

Curnutte JT & Babior BM (1974) Biological defense mechanisms: the effect of bacteria and serum on superoxide production by granulocytes. J Clin Invest 53:1662–1672

Curnutte JT & Babior BM (1987) Chronic granulomatous disease. In: Harris H & Hirschhorn K (eds) Advances in human genetics. Plenum, New York, pp 229–297

Dinauer MC, Orkin SH, Brown R, Jesaitis AJ & Parkos CA (1987) The glycoprotein encoded by the X-linked chronic granulomatous disease locus is a component of the neutrophil cytochrome b complex. Nature (London) 327:717–720

Fujimoto S, Smith RM, Curnutte JT & Babior BM (1989) Evidence that activation of the respiratory burst oxidase in a cell-free system from human neutrophils is accomplished in part through an alteration of the oxidase-related 67-kDa cytosolic protein. J Biol Chem 264:21629–21632

Gabig TG, English D, Akard LP & Schell MJ (1987) Regulation of neutrophil NADPH oxidase activation in a cell-free system by guanine nucleotides and fluoride. Evidence for participation of a pertussis and cholera toxin-insensitive G protein. J Biol Chem 262:1685–1690

Gabig TG, Schervish EW & Santinga JT (1982) Functional relationship of the cytochrome b to the superoxide-generating oxidase of human neutrophils. J Biol Chem 257:4114–4119

Gierschik P, Falloon J, Milligan G, Pines M, Gallin JI & Spiegel A (1986) Immunochemical evidence for a novel pertussis toxin substrate in human neutrophils. J Biol Chem 261:8058–8062

Glass GA, DeLisle DM, DeTogni P, Gabig TG, Magee BH, Markert M & Babior BM (1986) The respiratory burst oxidase of human neutrophils. Further studies of the purified enzyme. J Biol Chem 261:13247–13251

Goldsmith P, Gierschik P, Milligan G, Unson CG, Vinitsky R, Malech HL & Spiegel AM (1987) Antibodies directed against synthetic peptides distinguish between GTP-binding proteins in neutrophil and brain. J Biol Chem 262:14683–14688

Goldsmith P, Rossiter K, Carter A, Simonds W, Unson CG, Vinitsky R & Spiegel AM (1988) Identification of the GTP-binding protein encoded by Gi3 complementary DNA. J Biol Chem 263:6476–6479

Hayakawa T, Suzuki K, Suzuki S, Andrews PC & Babior BM (1986) A possible role for protein phosphorylation in the activation of the respiratory burst in human neutrophils. J Biol Chem 261:9109–9115

Heyworth PG, Tolley JO, Smith RM & Curnutte JT (1990) Blood 76:183a

Huizinga TWJ, Van Kemenade F, Koenderman L, Dolman KM, Von Dem Borne AEGKr, Tetteroo PA & Roos D (1989) The 40-kDa Fc-gamma receptor (FcRII) on human neutrophils is essential for the IgG-induced respiratory burst and IgG-induced phagocytosis. J Immunol 142:2365–2369

Khachatrian L, Rubins JB, Manning, EC, Dexter D, Tauber AI & Dickey BF (1990) Subcellular distribution and characterization of GTP-binding proteins in human neutrophils. Biochim Biophys Acta Mol Cell Res 1054:237–245

Kikuchi A, Kozawa O, Kaibuchi K, Katada T, Ui M & Takai Y (1986) Direct evidence for involvement of a guanine nucleotide-binding protein in chemotactic peptide-stimulated formation of inositol bisphosphate and trisphosphate in differentiated human leukemic (HL-60) cells: reconstitution with G_i or G_o of the plasma membranes ADP-ribosylated by pertussis toxin. J Biol Chem 261:11558–11562

Kleinberg ME, Rotrosen D & Malech HL (1990) Asparagine-linked glycosylation of cytochrome b558 large subunit varies in different human phagocytic cells. J Immunol 143;4152–4157

Leto TL, Lomax KJ, Volpp BD, Nunoi H, Sechler JMG, Nauseef WM, Clark RA, Gallin JI & Malech HL (1990) Cloning of a 67-kD neutrophil oxidase factor with similarity to a noncatalytic region of $p60^{c-src}$. Science 248:727–730

Markert M, Andrews PC & Babior BM (1984) Measurement of O_2^--production by human neutrophils; the preparation and assay of NADPH oxidase-containing particles from human neutrophils. Methods Enzymol 105:358–365

McPhail LC, Shirley PS, Clayton CC & Snyderman R (1985) Activation of the respiratory burst enzyme from human neutrophils in a cell-free system: evidence for a soluble cofactor. J Clin Invest 75:1735–1739

Ohno YI, Hirai KI, Kanoh T, Uchino H & Ogawa K (1982) Subcellular localization of H_2O_2 production in human neutrophils stimulated with particles and an effect of cytochalasin-B on the cells. Blood 60:253–260

Oinuma M, Katada T & Ui M, (1987) A new GTP-binding protein in differentiated human leukemic (HL-60) cells serving as the specific substrate of islet-activating protein, pertussis toxin. J Biol Chem 262:8347–8353

Okamura N, Curnutte JT, Roberts RL & Babior BM (1988a) Relationship of protein phosphorylation to the activation of the respiratory burst in human neutrophils. Defects in the phosphorylation of a group of closely related 48K proteins in two forms of chronic granulomatous disease. J Biol Chem 263:6777–6782

Okamura N, Malawista SE, Roberts RL, Rosen H, Ochs HD, Babior BM & Curnutte JT (1988b) Phosphorylation of the oxidase-related 48K phosphoprotein family in the unusual autosomal cytochrome-negative and X-linked cytochrome-positive types of chronic granulomatous disease. Blood 72:811–816

Parkos CA, Allen RA, Cochrane CG & Jesaitis AJ (1987) Purified cytochrome b from human granulocyte plasma membrane is comprised of two polypeptides with relative molecular weights of 91,000 and 22,000. J Clin Invest 80:732–742

Parkos CA, Dinauer MC, Walker LE, Allen RA, Jesaitis AJ & Orkin SH (1988) Primary structure and unique expression of the 22-kilodalton light chain of human neutrophil cytochrome b. Proc Natl Acad Sci USA 85:3319–3323

Parkos CA, Dinauer MC, Jesaitis AJ, Orkin SH & Curnutte JT (1989) Absence of both the 91kD and 22kD subunits of human neutrophil cytochrome b in two genetic forms of chronic granulomatous disease. Blood 73:1416–1420

Pember SO, Heyl BL, Kinkade Jr JM & Lambeth JD (1984) Cytochrome b_{558} from (bovine) granulocytes. Partial purification from Triton X-114 extracts and properties of the isolated cytochrome. J Biol Chem 259:10590–10595

Peveri P & Curnutte JT (1989) Absolute requirement for GTP or ATP in NADPH oxidase activation in human neutrophils: evidence for two pathways of activation. Blood 74:107a

Polakis PG, Uhing RJ & Snyderman R (1988) The formylpeptide chemoattractant receptor copurifies with a GTP-binding protein containing a distinct 40-kDa pertussis toxin substrate. J Biol Chem 263:4969–4976

Quinn MT, Parkos CA, Walker L, Orkin SH, Dinauer MC & Jesaitis AJ (1989) Association of a Ras-related protein with cytochrome b of human neutrophils. Nature (London) 342:198–200

Rotrosen D, Kleinberg ME, Nunoi H, Leto T, Gallin JI & Malech HL (1990) Evidence for a functional cytoplasmic domain of phagocyte oxidase cytochrome b_{558}. J Biol Chem 265:745–8750

Rotrosen D & Leto TL (1990) Phosphorylation of neutrophil 47-kDa cytosolic oxidase factor. Translocation to membrane is associated with distinct phosphorylation events. J Biol Chem 265:19910–19915

Segal AW (1988) Cytochrome b-245 and its involvement in the molecular pathology of chronic granulomatous disease. In: Curnutte JT (ed) Hematology/oncology clinics of North America. Phagocytic defects II, vol 2. WB Saunders, Philadelphia, pp 213–223

Sklar LA, Bokoch GM, Button D & Smolen JE (1987) Regulation of ligand-receptor dynamics by guanine nucleotides. Real-time analysis of interconverting states for the neutrophil formyl peptide receptor. J Biol Chem 262:135–139

Smith RM, Curnutte JT & Babior BM (1989a) Affinity labeling of the cytosolic and membrane components of respiratory burst oxidase by the 2, 3-dialdehyde derivative of NADPH. Evidence for a cytosolic location of the nucleotide-binding site in the resting cell. J Biol Chem 264:1958–1962

Smith RM, Curnutte JT, Mayo LA & Babior BM (1989b) Use of an affinity label to probe the function of the NADPH binding component of the respiratory burst oxidase of human neutrophils. J Biol Chem 264:12243–12248

Stasia M-J, Dianoux A-C & Vignais PV (1989) A 23-kDa protein as a substrate for protein kinase C in bovine neutrophils. Purification and partial characterization. Biochemistry 28:9659–9667

Tauber AI, Goetzl EJ & Babior BM (1979) Unique characteristics of superoxide production by human eosinophils in eosinophilic states. Inflammation 3:261–272

Uhing RJ, Polakis PG & Snyderman R (1987) Isolation of GTP-binding proteins from myeloid HL-60 cells. Identification of two pertussis toxin substrates. J Biol Chem 262:15575–15579

Umei T, Babior BM, Curnutte JT & Smith RM (1991) Identification of the NADPH-binding subunit of the respiratory burst oxidase. J Biol Chem (in press)

Umei T, Takeshige K & Minakami S (1986) NADPH binding component of neutrophil superoxide-generating oxidase. J Biol Chem 261:5229–5232

Verhoeven AJ, Bolscher BG, Meerhof LJ, van Zwieten R, Keijer J, Weening RS & Roos D (1989) Characterization of two monoclonal antibodies against cytochrome b_{558} of human neutrophils. Blood 73:1686–1694

Weiss SJ, King GW & LoBuglio AF (1978) Superoxide generation by human monocytes and macrophages. Am J Hematol 4:1–8

Wilde CG, Snable JL, Griffith JE & Scott RW (1990) Characterization of two azurophil granule proteases with active-site homology to neutrophil elastase. J Biol Chem 265:2038–2041

Woodman RC, Ruedi JM, Okamura N, Jesaitis AJ, Quinn MT, Curnutte JT & Babior BM (1991) The respirtory burst oxidase and 3 of 4 oxidase-related polypeptides are associated with the cytoskeleton of human neutrophils. J Clin Invest (in press)

Yamaguchi T, Hayakawa T, Kaneda M, Kakinuma K & Yoshikawa A (1989) Purification and some properties of the small subunit of cytochrome b_{558} from human neutrophils. J Biol Chem 264:112–118

Yea CM, Cross AR & Jones OTG (1990) Purification and some properties of the 45 kDa diphenylene iodonium-binding flavoprotein of neutrophil NADPH oxidase. Biochem J 265:95–100

Cellular Activation Mechanisms: The Blood Platelet as a Model

W. SIESS[1]

1 Introduction

Human platelets are a useful model to study cellular activation mechanisms: their isolation from peripheral venous blood is simple and rapid, and their physiological responses such as shape change, aggregation and secretion can be easily measured. It is therefore not surprising that some of the pioneering work on cellular signal transduction mechanisms such as phospholipase C-induced diglyceride production from phosphatidylinositol or phorbol ester/diglyceride-induced protein kinase C-activation was first described in platelets (Rittenhouse-Simmons 1979; Nishizuka 1984). Some of the main signal transduction mechanisms such as phospholipase C-activation, Ca^{2+} mobilization, and protein kinase C-activation which are related to platelet activation, as well as adenylate cyclase stimulation and protein kinase A-activation which are related to platelet inhibition, have been demonstrated in many other cell types, for example, in neutrophils and smooth muscle cells. The detailed regulation of these pathways might however be different from one cell type to another. For example, the expression of various phospholipase C-enzymes, protein kinase C-isozymes, and protein kinase substrates will be regulated in a manner specific to each cell.

In this chapter, I would like to outline some of the actual concepts of the mechanisms of platelet activation and inhibition. The reader interested in more comprehensive reviews on platelets or in new important topics of platelet research such as platelet tyrosine phosphorylation and activation of the Na^+/H^+-antiporter is referred to the literature (Ferrell and Martin 1989; Siess 1989, 1990, 1991; Siffert and Akkermann 1989; Haslam et al. 1990; Rink and Sage 1990).

2 Platelet Activation

2.1 Receptors, G-Proteins, and Phospholipase C

A wide variety of physiological substances activates platelets (Table 1). These are: von Willebrand factor and collagen, which stimulate platelet adhesion; thrombin, generated by the coagulation system; prostaglandin endoperoxides, thromboxane A_2, and ADP, which are released from activated platelets; epinephrine and vasopressin,

[1] Institut für Prophylaxe und Epidemiologie der Kreislaufkrankheiten b. d. Universität München, Pettenkoferstr. 9, 8000 München 2, FRG

42. Colloquium Mosbach 1991
Molecular Aspects of Inflammation

Table 1. Platelet stimuli and their receptors

Stimulus		Receptor	Molecular mass of binding site (kDa)	Sites per platelet	Inhibition by aspirin[a]
Weak:	ADP	Unknown			+
	Epinephrine	α_2-adrenergic	64	150–450	+
Intermediary:	Platelet-activating factor		160	200–400	(+)
	PG-endoperoxides Thromboxane A2		200	1700–2200	–
	Vasopressin	V_1-receptor		100	(+)
Strong:	Collagen, low	Glycoprotein Ia	160		+
	high				–
	Thrombin, low	Glycoprotein Ib (α-chain)	143	50	+
	high				–

[a] Secondary aggregation and secretion are completely +, partially (+) or not at all (–), inhibited by aspirin.

which are present in the circulation; and platelet-activating-factor, which is liberated from stimulated neutrophils and macrophages. Although the platelet stimuli and platelet responses are quite diverse, it is emerging that platelet activation is regulated by a distinct number of mechanisms. All physiological platelet stimuli bind to their specific receptors on the platelet surface. The signal emitted from the activated receptor is transduced through the plasma membrane by GTP-binding proteins (G-proteins). These in turn stimulate and regulate specific effector systems, such as phospholipase C-induced inositol phospholipid hydrolysis and ion channels. The effector systems modulate the levels of mobile intracellular messengers such as inositol 1,4,5-trisphosphate, Ca^{2+}, and diacylglycerol which then trigger the physiological response by changing the phosphorylation, enzymatic activities, and structural properties of proteins.

The release of prostaglandin-endoperoxides and thromboxane A_2, and the secretion of ADP are important amplification mechanisms of platelet activation: they bind to specific receptors and mediate secretion and irreversible aggregation of some of the platelet stimuli (Table 1). However, they also hinder the elucidation of intracellular signal transduction pathways coupled to one receptor. Therefore these two major positive feedback mechanisms are usually eliminated by irreversibly inhibiting platelet cyclooxygenase with aspirin and by using ADP-scavenging enzymes or ADP-receptor blockers.

Receptors of the various platelet agonists are coupled differently to guanine-nucleotide binding (G)-proteins (Table 2). Receptors of prostacyclin, prostaglandin E_1 and D_2 that stimulate adenylate cyclase in human platelets, are coupled to G_s (Edwards et al. 1987; Tsai et al. 1989). α_2-Adrenergic receptors which inhibit adenylate cyclase in human platelets are coupled to G_i (Jakobs et al. 1978; Cerione et al. 1986). At least two $G_{i\alpha}$-subunits ($G_{i\alpha-2}$ and $G_{i\alpha-3}$) are expressed in platelets (Nagata et al. 1988; Crouch et al. 1989; Williams et al. 1990): $G_{i\alpha-2}$ mediates the α_2-adrenergic inhibition of adenylate cyclase (Simonds et al. 1989). In addition to membrane $G_{i\alpha}$, a soluble $G_{i\alpha}$ with smaller molecular weight was recently detected in platelets (Gennity and Siess 1991). Its function is unknown at the moment. G_i-proteins are substrates for pertussis toxin, and $G_{i\alpha}$-subunits are only ADP-ribosylated by pertussis

Table 2. Heterotrimeric GTP-binding proteins in platelets

Name of α-subunit	Mw (kDa)	ADP-ribosylating toxin	Effector	Comment
$G_{s\alpha-1}$	45	Cholera toxin	Adenylate cyclase +	
$G_{i\alpha-2}$	41	Pertussis toxin	Adenylate cyclase –	Major
$G_{i\alpha-3}$	41	Pertussis toxin		Minor
$G_{i\alpha}$	40	Pertussis toxin	?	Minor, soluble
"$G_{z\alpha}$-like"	40	---	?	Phosphorylation by protein kinase C

toxin, if associated with β- and γ-subunits (Ui 1984). Therefore, a decreased pertussis toxin-dependent ADP-ribosylation of membrane $G_{i\alpha}$ upon prior exposure of platelets to thrombin has been thought to reflect the thrombin receptor-mediated activation of G_i-proteins (Lapetina et al. 1986; Brass et al. 1988; Crouch and Lapetina 1988). However, recent studies show that the Ca^{2+}-ionophore A23187 reduces also the pertussis toxin-dependent ADP-ribosylation of G_i-proteins in platelets; furthermore, the decrease of pertussis toxin-dependent ADP-ribosylation after A23187 and thrombin stimulation of platelets is not confined to $G_{i\alpha}$, but involves other proteins as well (Siess et al. 1991). Therefore, the decrease of subsequent pertussis toxin-dependent ADP-ribosylation of $G_{i\alpha}$ in stimulated platelets cannot be taken as only evidence for a specific receptor/G_i-protein coupling. Furthermore, stimulation of platelet α_2-adrenoceptors known to be coupled to G_i-proteins does not lead to a subsequent decrease of pertussis toxin-dependent ADP-ribosylation (Siess and Lapetina 1989).

Receptors that stimulate phospholipase C in platelets are coupled most likely to a G-protein, since GTP and its stable analogs stimulate phospholipase C in permeabilized platelets (Haslam et al. 1990). The G-protein involved in the activation of phospholipase C has not yet been identified in platelets but seems to be most likely a G-protein insensitive to pertussis toxin. Unfortunately, the effect of pertussis toxin on phospholipase C-activation cannot be studied in intact platelets, since platelets do not possess the pertussis toxin receptor required for its cellular uptake. The effect of pertussis toxin on phospholipase C-activation in saponized platelets is controversial (Siess 1989). One candidate which might be involved in the activation of phospholipase C is a G-protein similar or identical to G_z (Matsuoka et al. 1988). Its α-subunit is phosphorylated by protein kinase C in platelets (Carlson et al. 1989) that could be responsible for the protein kinase C-dependent inhibition of coupling between receptor activation and phospholipase C (see below). Another candidate is the recently identified G_q that is also a pertussis toxin-insensitive G-protein. Purified G_q has recently been shown to activate phospholipase C (Smrcka et al. 1991).

Membrane-bound and cytosolic phospholipases C of various molecular weight have been purified from human platelets (Manne and Kung 1987; Banno et al. 1988; Moriyama et al. 1990). Phospholipase C in intact activated platelets hydrolyzes polyphosphoinositides first and later degrades phosphatidylinositol (Siess 1989). Prior mobilization of Ca^{2+} might be required for the phospholipase C action on phosphatidylinositol which accounts for the bulk of diglyceride formation. The two messengers derived from inositol phospholipid hydrolysis by phospholipase C account for the two main pathways mediating platelet activation: 1,2-diacylglycerol activates protein kinase C; inositol 1,4,5-trisphosphate binds to a specific receptor on the endoplasmatic reticulum and evokes mobilization of Ca^{2+}.

2.2 Synergistic Interaction of Signal Transduction/Effector Systems

At least three signal transduction pathways are present in platelets which synergize with each other: calcium mobilization, protein kinase C-activation, and an unknown pathway triggered by G_i protein activation (Table 3). Recent studies indicate that the *Ca^{2+}-pathway*, which leads to myosin light chain phosphorylation and shape change,

Table 3. Synergistic interaction of signal transduction pathways

– Ca^{2+} mobilization and protein kinase C-activation
– G_i-protein and protein kinase C-activation
– Ca^{2+} mobilization and G_i-protein activation

is also directly involved in the activation of the *protein kinase C pathway* which regulates platelet aggregation and secretion. It was shown that low concentrations of phorbol ester, which by themselves only cause a slight and slow protein kinase C-activation, elicit a rapid and large protein kinase C-activation when added in combination with low concentration of Ca^{2+}-mobilizing platelet stimuli such as A 23187, vasopressin, ADP, platelet-activating factor, and prostaglandin endoperoxide analogs. In parallel, these agonists increased the binding of phorbol ester to intact platelets (Siess and Lapetina 1988). The potentiation of protein kinase C-activation was associated with increased platelet aggregation and secretion (Siess and Lapetina 1987, 1988). These results are best explained by a model in which Ca^{2+} mobilization induces a translocation of protein kinase C to the platelet membrane or an affinity change of the kinase, thereby priming the enzyme for activation by phorbol ester. Epinephrine, which does not activate phospholipase C or increase Ca^{2+} in the platelet cytosol to a significant degree did not enhance phorbol ester binding to the platelets or prime protein kinase C to activation by phorbol ester.

The observations so far fit into a scheme in which two steps of platelet activation can be distinguished (Siess 1989). The first step of platelet activation is inositol 1,4,5-trisphosphate formation, Ca^{2+} mobilization, and myosin light chain phosphorylation. These events cause platelet shape change. The mobilization of Ca^{2+} primes protein kinase C either by an affinity change or by a translocation of protein kinase C. The second step involves the activation of protein kinase C by diacylglycerol, the bulk of which is formed later by hydrolysis of phosphatidylinositol. Protein kinase C activation is a positive signal for platelet aggregation and secretion and suppresses the signal transduction between receptor and phospholipase C.

Epinephrine as a physiological platelet stimulus is unique. It does not induce activation of phospholipase C or protein kinase C (Siess et al. 1984). It binds to α_2-adrenergic receptors that are coupled to the G_i-protein, which mediates inhibition of adenylate cyclase. Inhibition of adenylate cyclase is not, however, the mechanism by which epinephrine induces and potentiates platelet aggregation. Thus, the effector mechanism by which α_2-adrenoceptor stimulation and G_i-protein activation causes and potentiates platelet aggregation is unknown.

α_2-Adrenoceptor activation and protein kinase C-activation synergistically induce platelet aggregation and secretion. The effect on aggregation is more pronounced than the effect on secretion. The mechanism underlying this synergism involves the interaction of two different signal-transducting pathways: *protein kinase C* and *G_i-protein activation* (Siess and Lapetina 1989). Epinephrine and phorbol ester do not induce phospholipase C activation or Ca^{2+} mobilization. Interestingly, the small Ca^{2+} mobilization, observed after the addition of epinephrine, is abolished when phorbol ester is added before epinephrine. The synergistically induced platelet aggregation induced by epinephrine plus phorbol ester is not affected by high concentrations of the intracellular Ca^{2+}-chelator Quin-2. Therefore, platelet aggregation can occur by a mechanism completely independent of inositol 1,4,5-trisphosphate formation and intracellular Ca^{2+} mobilization.

Epinephrine shows also a synergism with the Ca^{2+} ionophore A23187 in inducing platelet aggregation. This synergism is observed at low concentrations of the Ca^{2+} ionophore A23187 and occurs without phosphorylation of the 47-kDa protein, the substrate of protein kinase C (Olbrich and Siess 1989). Therefore, platelet aggregation can occur in the absence of protein kinase C-activation. The exact mechanism underlying the synergism of *Ca^{2+} mobilization and G_i-protein activation* is obscure. Interestingly, epinephrine increased the small production of phosphatidic acid observed after A23187 stimulation, and the increased phosphatidic acid production correlated with the potentiation of platelet aggregation. It has been shown recently that the Ca^{2+} ionophore A23187 activates phospholipase D in human platelets (Huang et al. 1991) leading to a production of phosphatidic acid, and it could be that the G_i-protein activated by epinephrine enhanced the Ca^{2+}-dependent activation of phospholipase D.

Therefore, there might be at least three signal transduction pathways present in platelets that can evoke platelet aggregation and that synergize with each other. Synergism has been demonstrated for each pair of these pathways. Each pathway can synergize with the other independently of the third pathway (Siess 1991). The synergism of the Ca^{2+} pathway and the G_i-protein pathway, and the protein kinase C and the G_i-protein pathway concern mainly platelet aggregation and less secretion. Strong physiological stimuli like thrombin, activate all three pathways, and induce rapidly through their synergistic interaction, large platelet aggregates and secretion. Weaker stimuli activate only one or two pathways. For example, epinephrine activates only the G_i-protein pathway, ADP may activate the G_i-protein and Ca^{2+} pathway, but does not induce phospholipase C-activation (Vickers et al. 1990). All physiological stimuli that stimulate phospholipase C, activate the protein kinase C and Ca^{2+} pathways and their synergistic interaction.

3 Platelet Inhibition by Cyclic AMP

Platelet agonists such as the prostaglandins prostacyclin, prostaglandin E_1, and prostaglandin D_2, which raise cyclic AMP levels, are the most powerful inhibitors of platelet activation. Platelet shape change, secretion, and aggregation induced by all physiological stimuli including epinephrine are inhibited by pretreatment with these prostaglandins. These substances bind to specific receptors on the platelet surface that couple to the guanine nucleotide-binding protein G_s (Edwards et al. 1987; Tsai et al. 1989). Activation of G_s stimulates adenylate cyclase and leads to an increase in intracellular cyclic AMP, the activation of protein kinase A, and the subsequent phosphorylation of specific proteins. Cyclic AMP inhibits platelet activation at several steps. It is currently assumed that one of the most important steps is the inhibition of receptor-mediated phospholipase C activation (Fig. 1). The inhibitory effects of cyclic

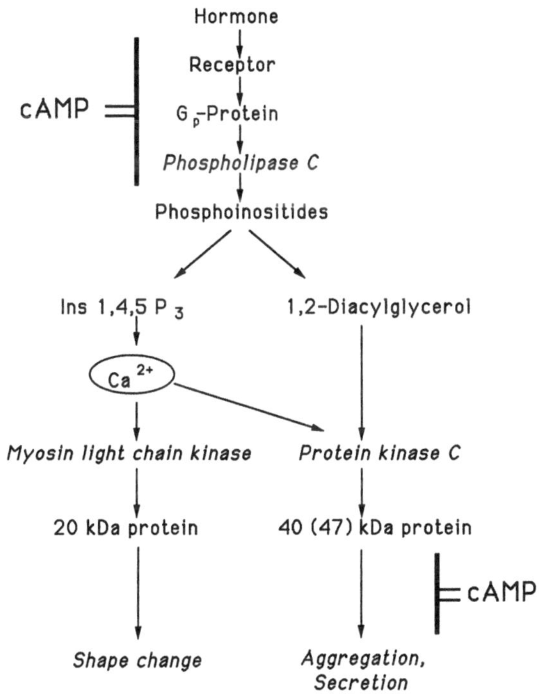

Fig. 1. Possible target sites for inhibition of platelet activation by cyclic AMP

AMP on protein kinase C activation, Ca^{2+} mobilization, fibrinogen-receptor exposure, myosin light chain phosphorylation, actin polymerization, and cytoskeletal assembly are believed to be consequences of the inhibition of receptor-mediated phospholipase C-activation (Siess 1989). However, there are also additional target sites for platelet inhibition by cyclic AMP. For example, cyclic AMP inhibits secretion and aggregation induced by Ca^{2+} ionophore at steps distal to Ca^{2+} mobilization (Pannocchia and Hardisty 1985). We have recently found that cyclic AMP inhibits platelet aggregation at steps distal to activation of protein kinase C and Ca^{2+}-dependent protein kinase (Fig. 1; Siess and Lapetina 1989).

Inhibition of platelet activation by cyclic AMP is associated with the phosphorylation of proteins with molecular masses of 22, 24, 38, 50, 60, 130, and 250 kDa (Fig. 2). The 22, 24, and the 250 kDa proteins have been identified (Table 4; Fox et al. 1987; Siess et al. 1990a). The 22 kDa protein, recently identified as *rap*1B (Siess et al. 1990a), a *ras*-like protein that belongs to the family of small GTP-binding proteins, is highly expressed in human platelets and is present in platelet membranes and cytosol. It was shown that phosphorylation of *rap*1B in the particulate fraction occurs at a C-terminal serine residue in close vicinity to cysteine-181 (Siess et al. 1990a). In *ras*P21 this amino acid is C-terminal, carboxymethylated, and farnesylated; the latter modification is essential for the membrane attachment of *ras*P21. *rap*1B is also identical to thrombolamban (White et al. 1990). Earlier studies indicated that cyclic AMP-dependent phosphorylation of thrombolamban stimulates Ca^{2+} uptake into platelet membrane vesicles thereby sequestering elevated cytosolic Ca^{2+}. That finding, however, was questioned in more recent studies.

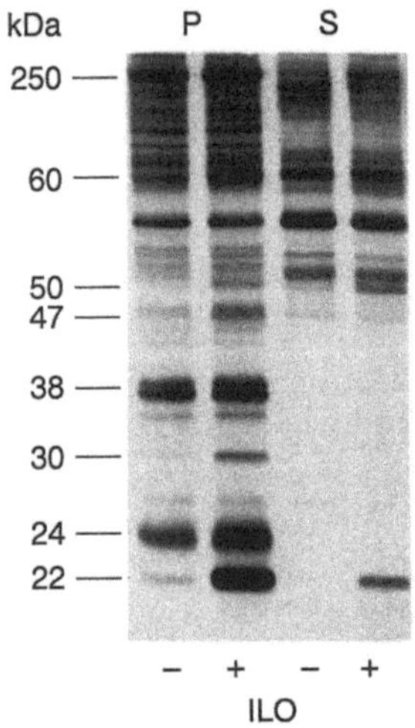

Fig. 2. Effect of the stable prostacyclin analog iloprost on protein phosphorylation. ^{32}P-labeled platelets were incubated with iloprost, 1 µM (ILO) for 60 min at 37 °C. Platelets were lysed by the addition of saponin, and pellet (P) and supernatant (S) proteins were separated on SDS/PAGE. An autoradiograph of the dried gel is shown.

We have recently explored the functional relationship between cyclic AMP-dependent phosphorylation of specific proteins and the inhibition of platelet activation (Siess and Lapetina 1990b). It was found that cyclic AMP-dependent phosphorylation of *rap*1B and the β-chain of glycoprotein Ib is *not* involved in the inhibition of thrombin-induced platelet activation. It was observed that platelet activation could be completely prevented by prostacyclin without significant phosphorylation of *rap1B*. Furthermore, platelets could be fully activated by thrombin despite maximal phosphorylation of *rap*1B and the β-chain of glycoprotein Ib. It was found that proteins such as p38 and p50 (Halbrügge et al. 1990) are more likely to be involved in platelet inhibition by cyclic AMP. Phosphorylation of these proteins correlated best with the inhibition of platelet activation by prostacyclin. It is to expect that further identification of such proteins may also help to understand better the pathways involved in platelet activation.

Table 4. Proteins phosphorylated by protein kinase A in human platelets

Mw (kDa)	Protein	Function
22	*rap* 1B	?
24	Glycoprotein Ib (β–chain)	Receptor for von Willebrand factor Binding site for thrombin
38	?	
50	?	
250	Actin-binding protein	Links glycoprotein Ia and Ib to actin filaments

References

Banno Y, Yada Y, Nozawa Y (1988) Purification and characterization of membrane-bound phospholipase C specific for phosphoinositides from human platelets. J Biol Chem 263:11459–11465

Brass L, Woolkalis MJ, Manning D (1988) Interactions in platelets between G proteins and the agonists that stimulate phospholipase C and inhibit adenylyl cyclase. J Biol Chem 263:5348–5355

Carlson KE, Brass LF, Manning DR (1989) Thrombin and phorbol esters cause the selective phosphorylation of a G protein other than G_i in human platelets. J Biol Chem 264:13298–13305

Cerione RA, Regan JW, Nakata H, Codina J, Benovic JL, Gierschik P, Somers RL, Spiegel AM, Birnbaumer L Lefkowitz RJ, Caron MG (1986) Functional reconstitution of the α_2-adrenergic receptor with guanine nucleotide regulatory proteins in phospholipid vesicles. J Biol Chem 261:3901–3909

Crouch MF, Winegar DA, Lapetina EG (1989) Epinephrine induces changes in the subcellular distribution of the inhibitory GTP-binding protein $G_{i\alpha-2}$ and a 38-kDa phosphorylated protein in the human platelet. Proc Natl Acad Sci USA 86:1776–1780

Crouch MF, Lapetina EG (1988) A role for G_i in control of thrombin receptor-phospholipase C coupling in human platelets. J Biol Chem 263:3363–3371

Edwards RJ, MacDermot J, Wilkins AJ (1987) Prostacylin analogues reduce ADP-ribosylation of the α-subunit of the regulatory G_s-protein and diminish adenosine (A_2) responsiveness of platelets. Br J Pharm 90:501–510

Ferrell JE, Martin GS (1989) Tyrosine-specific protein phosphorylation is regulated by glycoprotein IIb–IIIa in platelets. Proc Natl Acad Sci USA 86:2234–2238

Fox JE, Reynolds CC, Johnson MM (1987) Identification of glycoprotein Ib_β is one of the major proteins phosphorylated during exposure of intact platelets to agents that activate cyclic AMP-dependent protein kinase. J Biol Chem 262:12627–12631

Gennity JM, Siess W (1991) Thrombin inhibits the pertussis toxin-dependent ADP-ribosylation of a novel soluble G_i protein in human platelets. Biochem J 279:643–650

Halbrügge M, Friedrich Ch, Eigenthaler M, Schanzenbächer P, Walter U (1990) Stoichiometric and reversible phosphorylation of a 46-kDa protein in human platelets in response to cGMP- and cAMP-elevating vasodilators. J Biol Chem 265:3088–3093

Haslam RJ, Williams KA, Davis W, Sherwood J, van der Meulen J (1990) Roles of GTP-binding proteins and protein kinase C in signal transduction in the platelet. In: Nishizuka Y et al. (eds) The biology and medicine of signal transduction. Raven Press, New York

Huang R, Kucera GL, Rittenhouse SE (1991) Elevated cytosolic Ca^{2+} activates phospholipase D in human platelets. J Biol Chem 266:1652–1655

Jakobs KH, Saur W, Schultz G (1978) Inhibition of platelet adenylate cyclase by epinephrine requires GTP. FEBS Lett 85:167–170

Lapetina EG, Reep B, Chang KJ (1986) Treatment of human platelets with trypsin, thrombin, or collagen inhibits the pertussis toxin-induced ADP-ribosylation of a 41-kDa protein. Proc Natl Acad Sci USA 83:5880–5883

Manne V, Kung HF (1987) Characterization of phosphoinositide-specific phospholipase C from human platelets. Biochem J 243:763–771

Matsuoka M, Itoh H, Kozasa T, Kaziro Y (1988) Sequence analysis of cDNA and genomic DNA for a putative pertussis toxin-insensitive guanine nucleotide-binding regulatory protein α subunit. Proc Natl Acad Sci USA 85:5384–5388

Moriyama T, Narita H, Oki M, Matsuura T, Kito M (1990) Purification of polymeric phospholipase Cs from human platelets. J Biochem 108:414–419

Nagata K, Katada T, Tohkin M, Itoh H, Kaziro Y, Ui M, Nozawa Y (1988) GTP-binding proteins in human platelet membranes serving as the specific substrate of islet-activation protein, pertussis toxin. FEBS Lett 237:113–117

Nishizuka Y (1984) The role of protein kinase C in cell surface signal transduction and tumor promotion. Nature (London) 308:693–698

Olbrich Ch, Siess W (1989) Epinephrine and the Ca^{2+} ionophore A23187 synergistically induce platelet aggregation without protein kinase C activation. FEBS Lett 243:275–279

Pannocchia A, Hardisty RM (1985) Cyclic AMP inhibits platelet activation independently of its effect on cytosolic free calcium. Biochem Biophys Res Commun 127:339–345

Rink TJ, Sage SO (1990) Calcium signaling in human platelets. Annu Rev Physiol 52:431–449

Rittenhouse-Simmons S (1979) Production of diglyceride from phosphatidylinositol in activated human platelets. J Clin Invest 63:580–587
Siess W (1989) Molecular mechanisms of platelet activation. Physiol Rev 69:58–178
Siess W (1990) Mechanisms of platelet inhibition by prostaglandin: new aspects. Atherosclerosis Rev 21:159–164
Siess W (1991) Multiple signal transduction pathways synergize in platelet activation. News Physiol Sci 6:51–56
Siess W, Weber PC, Lapetina EG (1984) Activation of phospholipase C is dissociated from arachidonate metabolism during platelet shape change induced by thrombin or platelet activating factor. Epinephrine does not induce phospholipase C activation or platelet shape change. J Biol Chem 259:8286–8292
Siess W, Lapetina EG (1987) Phorbol esters sensitize platelets to activation by physiological agonists. Blood 70:1373–1381
Siess W, Lapetina EG (1988) Ca^{2+}-mobilization primes protein kinase C in human platelets. Ca^{2+} and phorbol esters stimulate platelet aggregation and secretion synergistically through protein kinase C. Biochem J 255:309–318
Siess W, Lapetina EG (1989a) Prostacyclin inhibits platelet aggregation induced by phorbol ester or Ca^{2+}-ionophore at steps distal to activation of protein kinase C and Ca^{2+}-dependent protein kinases. Biochem J 258:57–65
Siess W, Lapetina EG (1989b) Platelet aggregation induced by α_2-adrenoceptor and protein kinase C activation: A novel synergism. Biochem J 263:377–385
Siess W, Winegar DA, Lapetina EG (1990a) Rap1-B is phosphorylated by protein kinase A in intact human platelets. Biochem Biophys Res Commun 170:944–950
Siess W, Lapetina G (1990b) Functional relationship between cyclic AMP-dependent protein phosphorylation and platelet inhibition. Biochem J 271:815–819
Siess W, Wilhelm B, Aepfelbacher M, Gennity JM (1991) The Ca^{2+}-ionophore A23187 and thrombin inhibit the pertussis toxin-dependent ADP-ribosylation of $G_{i\alpha}$ and other proteins in human platelets. Eur J Biochem, in the press
Siffert W, Akkerman JWN (1989) Na^+/H^+-exchange and Ca^{2+}-influx. FEBS Lett 259:1–4
Simonds WF, Goldsmith PK, Codina J, Unson CG, Spiegel AM (1989) G_{i2} mediates α_2-adrenergic inhibition of andenylyl cyclase in platelet membranes: In situ identification with G_α C-terminal antibodies. Proc Natl Acad Sci USA 86:7809–7813
Smrcka AV, Hepler JR, Brown KO, Sternweis PC (1991) Regulation of polyphosphoinositide-specific phospholipase C activity by purified Gq. Science 251:804–807
Tsai A, Hsu MJ, Vijjeswarapu H, Wu KK (1989) Solubilization of prostacyclin membrane receptors from human platelets. J Biol Chem 264:61–67
Ui M (1984) Islet-activating protein, pertussis toxin: a probe for functions of the inhibitory guanine nucleotide regulatory component of adenylate cyclase. Trends Pharmacol Sci 5:277–279
Vickers JD, Kinlough-Rathbone RL, Packham MA, Mustard JF (1990) Inositol phospholipid metabolism in human platelets stimulated by ADP. Eur J Biochem 193:521–528
Williams AG, Woolkalis MJ, Poncz M, Manning DR, Gewirtz AM, Brass LF (1990) Identification of the pertussis toxin-sensitive G proteins in platelets, megakaryocytes, and human erythroleukemia cells. Blood 76:721–730
White TE, Lacal JC, Reep B, Fischer TH, Lapetina EG, White GC (1990) Thrombolamban, the 22-kDa platelet substrate of cyclic AMP-dependent protein kinase, is immunologically homologous with the *ras* family of GTP-binding proteins. Proc Natl Acad Sci USA 87:758–762

Platelet-Neutrophil Interactions

V. ULLRICH[1], G. HECKER[1], and M. SCHATZ-MUNDING[1]

Within an organ the various cell types have different tasks and require a communication system to guarantee the function of the organ and the whole organism. Chemical messengers maintain this intercellular communication which can be very sophisticated, so that the term *"cross-talk"* adequately describes this network of communication pathways.

There are many examples of cross-talking cells but the cells of the inflammatory response are an especially suitable example, although it seems that only a small part of the underlying network has been unraveled. The inherent difficulties are obvious: only when the cells involved are sufficiently characterized with regards to their intracellular metabolism, their regulatory systems, and physiological responses, can the field of intercellular messengers be approached, and then their large diversity, their chemical instability, and their low concentrations very often hamper detailed biochemical work.

Platelets and neutrophils are easily accessible and have been sufficiently characterized. They appear only at first sight to be components of blood and furthermore, they are not the predominant in the inflammatory response. Platelets are mainly involved in the clotting process which takes place in the vascular system. Neutrophils exert their functions only after emigration to the injured tissue and then take part in the host defense process. Thus, at first sight there seems to be little in common between the functions of platelets and neutrophils. However, an intercellular communication between them is not only proposed as a model, but is supported by clinical observations and clear-cut experiments.

In the adult respiratory distress syndrome (ARDS), which is considered to be caused by generalized neutrophil activation in the lung, the removal of platelets from the blood significantly improves the symptoms of this usually fatal disease (Ward et al. 1986). In contrast, the symptoms of the infarcted myocardium, which is essentially a process of platelet aggregation, are improved considerably after depletion of the granulocytes from blood (Bednar et al. 1985).

These in vivo findings ar corroborated by in vitro experiments with the FMLP-stimulated neutrophils, which in the presence of platelets will induce their aggregation (Del Maschio et al. 1990). If platelets are specifically stimulated by thrombin, they can cause secretion from coincubated granulocytes (Del Maschio et al. 1989). Hence, chemical messengers have to be released by the stimulated cell, which either by way of a surface receptor of by influencing an intracellular target, can cause a response in the adjacent cell.

[1] Faculty for Biology, University of Konstanz, Universitätsstr. 10, 7750 Konstanz, FRG

42. Colloquium Mosbach 1991
Molecular Aspects of Inflammation

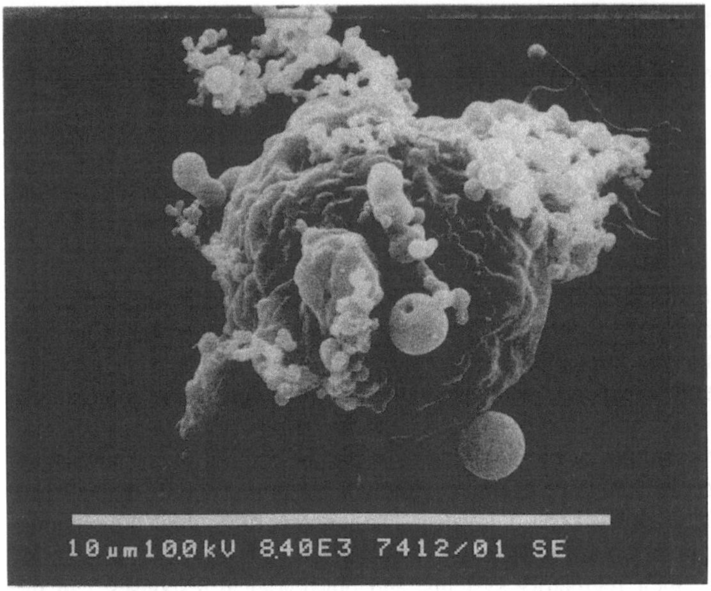

Fig. 1. Scanning electron microscopy of a PMN after a normal isolation procedure. Adhering platelets are either single or already aggregated, the spherical form probably originated from hypoosmolar treatment. *White bar* = 10 µm; magnification is 8400-fold. The photo was taken in collaboration with Prof. Plattner and Dr. Hentschel, Konstanz

Before elaborating the specific messengers involved in platelet- neutrophil interactions some general points should be discussed. One concerns the spatial requirements for such an interaction. Does it occur through a diffusion of independently migrating cells or is specific adherence or contact necessary? The latter would greatly increase the efficiency by which a certain amount of messenger, especially when it is unstable, can reach the communicating cell. Indeed, platelets are known to express a binding protein named PADGEM or GMP-140 after stimulation, thus linking with monocytes or granulocytes (McEver 1991) and resulting in mixed aggregates (Fig. 1).

If one investigates the nature of this interaction by electron microscopy of thin sections, one finds only few attachment points, indicating that the cells come in close proximity to each other but do not fuse so that each membrane surface remains fully functional for the exchange of messengers (unpubl. results).

Another point is the purpose of the communication process. It is obvious that this should occur in order to coordinate the response of cells within an organ, e.g. the conveyance of contracting or dilating stimuli from endothelial to smooth muscle cells of a vessel. In the case of platelets there appears to be either an enhancement of clot formation by the incorporation of granulocytes into the clot or the granulocytes are further stimulated by the responding platelets. However, it is not always a one-way stimulatory response but rather cross-talk, thus implying the exchange of information when a cell has fulfilled its function. This constitutes an important aspect of hemostasis.

A very general question deals with the economy of the information transfer. Since this has to take place during or before a cell is activated, only little energy can be directed toward the synthesis or release of chemical messengers. This is achieved impressively by using the energized state built up in the resting period of the cell. A slight triggering event can cause a downhill cascade of processes inducing cell activation with a concomitant release or synthesis of messengers. The most economical form is the storage of messengers in the resting state. If synthesis is required, as in the case of arachidonate-derived messengers, the biosynthesizing pathways usually consist of hydrolytic or dioxygenating reactions which do not consume energy.

As mentioned before the biosynthesizing and physiological properties of the communicating cells have to be known before the nature of their interaction can be elucidated. Although many details about platelets and neutrophils have been elaborated, the essential biochemical characteristics of both cell types should first be summarized.

Platelets lack a nucleus but otherwise possess most of the properties of normal cells. In their physiological action they have to respond to a variety of stimuli like thrombin, ADP, collagen, thromboxane A_2, or platelet-activating factor (PAF) by aggregation. This has to occur in a very rapid process which is supported autocatalytically through the biosynthesis of thromboxane A_2 and the secretion of ADP or its precursors (Fig. 2). Arachidonic acid, added from outside, can be efficiently converted to thromboxane A_2 via cyclooxygenase and thromboxane A_2 synthase but only in the absence of albumin (Broekman et al. 1989). In its presence or in serum, arachi-

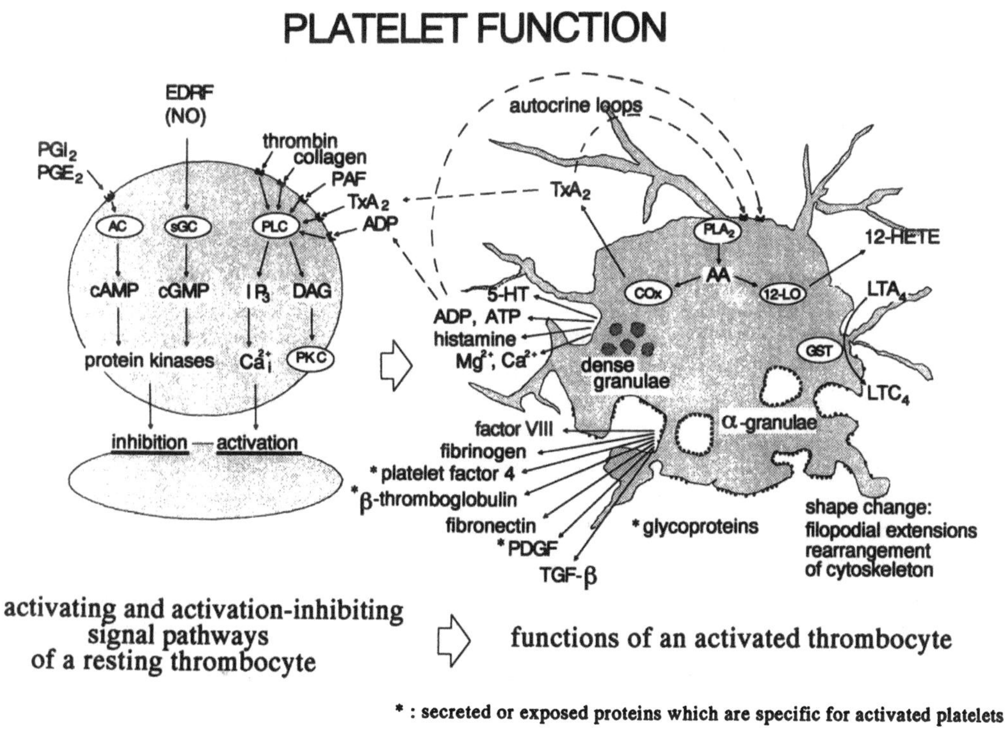

Fig. 2. Schematic representation of platelet function

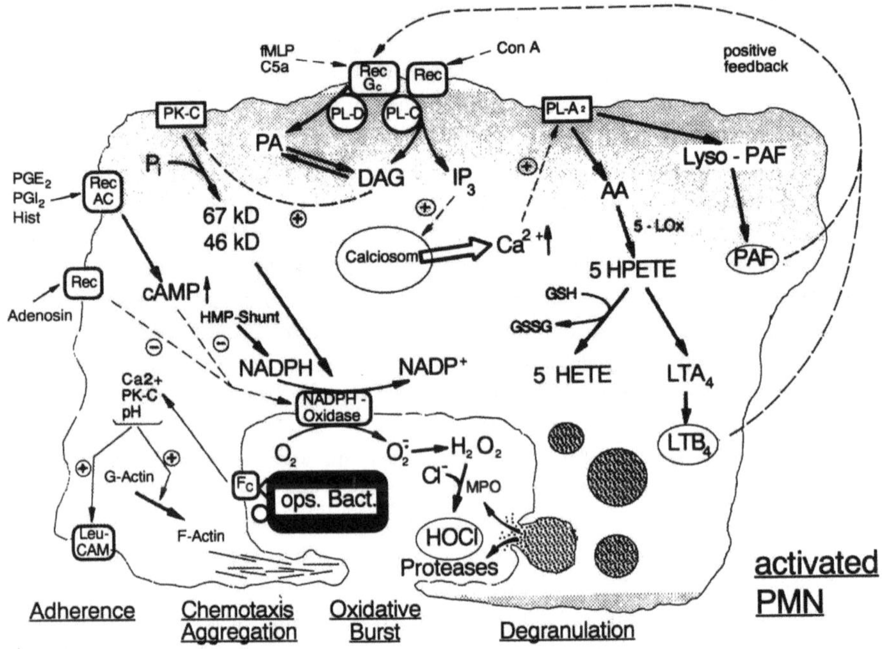

Fig. 3. For further explanation, see text

donate is largely bound to albumin so that about 100-fold higher concentrations are required to elicit a response. A second pathway for arachidonate exists in the action of 12-lipoxygenase, but no defined function for it is known. Patients with an inherited 12-lipoxygenase deficiency show some prolonged bleeding times, suggesting a supporting role in the aggregatory reaction (Kanaji et al. 1986). Interestingly, platelets contain a glutathione S-transferase, which allows exogenous LTA_4 to be converted to LTC_4, a potent vasoconstrictor (Maclouf et al. 1990). Since LTA_4 synthesis occurs mainly in granulocytes, a first intercellular communication pathway becomes apparent, i.e., "transcellular biosynthesis".

All platelet agonists increase intracellular Ca^{2+} either by the so-called PI response and/or by influx through Ca^{2+} channels (Lapetina 1986). The reversal of this process and hence an inactivation of platelets is achieved by increases in cAMP or cGMP, which through corresponding kinases act on various and, in part, still unknown steps of the signal transduction pathways. Hence, phosphorylation seems to induce inactivation rather than activation.

The activation of neutrophils can be visualized in an analogous way with the PI response as the signal transduction system for agonists like FMLP, a bacterial peptide mimic, complement C_{5a}, opsonized zymosan, Concanavalin A, PAF, or LTB_4. The latter two are formed by neutrophils after activation, so that like in platelets a second autocatalytic loop for activation seems to exist in neutrophils (Fig. 3).

However, it is important to note that with respect to arachidonic acid release and conversion to LTB_4 this enhancer loop does not operate as in platelets. The PI response, induced by the mentioned agonists, releases almost all intracellular Ca^{2+}, but

Table 1. Stimulation of human PMN with different agonists

Agonist		Σ 5-LOX metabolites (ng/10^7 cells)
Ionomycin	(2 μM)	597 ± 9
PAF	(0.2 μM)	108 ± 8
FMLP	(0.1 μM)	19 ± 4
LTB_4	(0.02 μM)	3 ± 1
X_m ± SEM, N = 4–5		

10^7 PMN/ml in PBS/1 mM $CaCl_2$ was prewarmed for 5 min at 37 °C and stimulated with the different agonists for a further 5 min. Metabolite formation was analyzed by HPLC (Rouzer and Samuelsson 1985).

the levels obtained only reach about 300 nmol/l, which we have shown to be insufficient for phospholipase A_2 activation in granulocytes (Schatz-Munding et al. 1991). Hence, the release of LTB_4 from granulocytes is only a small fraction of that obtained in the presence of a Ca-ionophore and extracellular Ca^{2+} (Table 1).

By analyzing the Ca-transients elicited by PAF, FMLP, and LTB_4 we could show further that the small amount of LTB_4 formed correlated with the second phase of the Ca-signal which is derived from an influx of extracellular Ca (Schatz-Munding et al. 1991) (Fig. 4).

From this we conclude that in platelets the amount of stored intracellular Ca^{2+} released by the agonists is sufficient to activate phospholipase A_2, but that in neutrophils the Ca^{2+} pools are too small and thus extracellular Ca^{2+} is required for activation. Alternately, external arachidonate can be used to supplement the substrate re

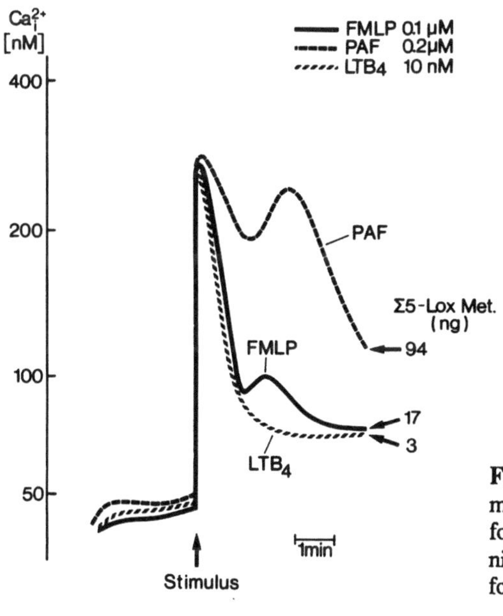

Fig. 4. Ca measurements by the Fura-2 method. 10^7 PMN/ml were preincubated for 3 min at 37 °C, stimulated by the agonists indicated, and after 5 min analyzed for 5-lipoxygenase metabolites by HPLC

Table 2. Arachidonic acid content in tissues and utilization by PMN

Tissue		PMN (0.25 g/ml)	PAF (10^7/ml)	Σ-5-LOX Met (0.2 μM)	Arachidonic acid (ng/10^7 PMN) nmol/g tissue)
	–	x	x	38 ± 6 (X_m ± SEM)	
Control:	Muscle	–	–	26	27
	Muscle	x	x	56	
	Skin	–	–	n. d.	20
	Skin	x	x	n. d.	
	Fat	–	–	n. d.	28
	Fat	x	x	9	
Inflamed:					
	Mamma carcinoma	–	–	9	138
	Mamma carcinoma	x	x	146	
	Mamma carcinoma	–	–	37	173
	Mamma carcinoma	x	x	312	
	Tumor (scrotum)	–	–	8	201
	Tumor (scrotum)	x	x	133	
	Tumor (scrotum)	–	–	23	78
	Tumor (scrotum)	x	x	152	
	Tumor (uterus)	–	–	405	97
	Tumor (uterus)	x	x	928	
	Appendix	–	–	22	40
	Appendix	x	x	56	
	Appendix	–	–	71	179
	Appendix	x	x	120	

Tissues were homogenized under nitrogen (– 80 °C) and dissolved in buffer. Arachidonic acid concentrations were measured by GC after derivation with α-bromo,2,3,4,5-pentafluorotoluene. 5-LOX activity was measured by HPLC.

quirement of 5-lipoxygenase in neutrophils. This enzyme is also Ca-dependent but the intracellular release is sufficient to keep it fully activated in human neutrophils (Schatz-Munding et al. 1991).

Following reports of arachidonate levels of more than 100 μmol/l in psoriatic skin (Hammarström et al. 1975), we analyzed inflamed human tissue samples and also found arachidonate levels ranging from 100–200 μM. Incubations of human PMN with such tissue homogenates in most cases should cause the synthesis of 5-lipoxygenase products after stimulation (Table 2).

The above-mentioned facts may correspond to a physiological situation in which arachidonate diffuses transcellularly into neutrophils, whereby platelets are likely candidates for such a role since their phospholipase A_2 can be maximally stimulated. This would give rise to transcellular interaction which allows neutrophils to synthesize more of the potent chemoattractant LTB_4.

Another point of control in the 5-lipoxygenase pathway is the peroxide tone required for the enzyme (Hatzelmann and Ullrich 1987). In the presence of glutathione and increasing concentrations of glutathione peroxidase, enzyme activity abruptly stops but can be stimulated again by the addition of fatty acid peroxides (Hatzelmann et al. 1989). This can be explained by the following mechanism in which an external electron donor, e.g., phenol or arachidonate itself, leads to a reduction of the ferric enzyme after a series of catalytic cycles (Fig. 5).

The product 5-hydroxyperoxyeicosatetraenoic acid (5-HPETE) or other lipophilic hydroperoxides can reoxidize the inactive ferrous enzyme, but in the cell this becomes effective only above certain concentrations. It seems reasonable to assume that the 12-hydroxyeicosatetraenoic acid (12-HPETE) produced in platelets can also contribute to the peroxide tonus in neutrophils, thus leading to the synergistic activation of 5-lipoxygenase if arachidonate is provided (Maclouf et al. 1986).

Of special importance in platelet-neutrophil interactions can be the influence of H_2O_2, formed during the oxidative burst of activated neutrophils. Adherent platelets

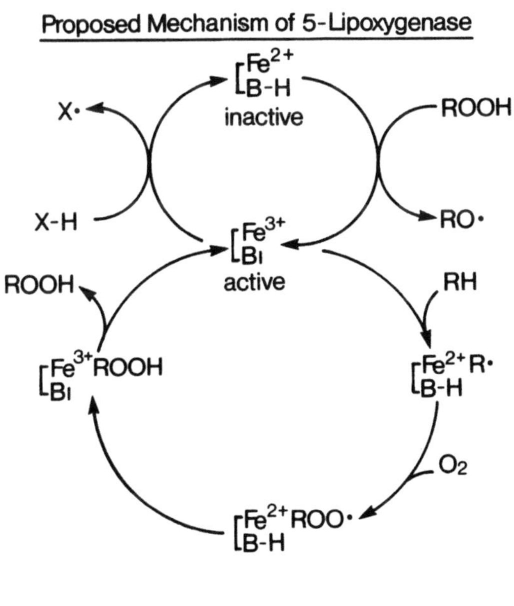

Fig. 5. Proposed Mechanism of 5-lipoxygenase

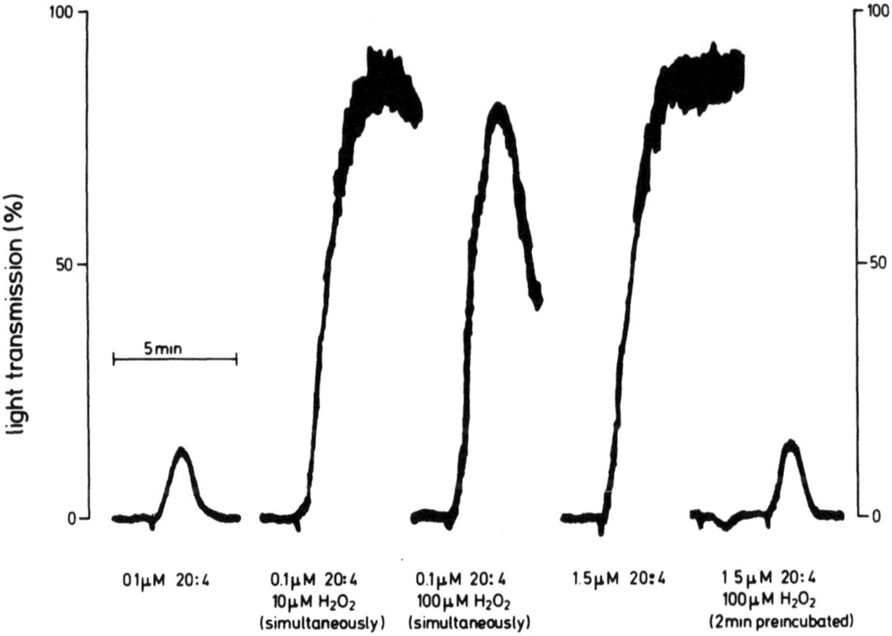

Fig. 6. Aggregation traces of washed human platelets. To *traces 2* and *3* H_2O_2 was added simultaneously to a subthreshold concentration of arachidonic acid (20:4). 15 µM 20:4 caused full aggregation *(trace 4)* which was abolished by preincubation with 100 µM H_2O_2 for 2 min *(trace 5)*

can be exposed to quite high concentrations of H_2O_2 and could respond to this active oxygen species.

In the literature stimulating and inhibiting effects of H_2O_2 on platelets have been reported (Clark and Klebanoff 1980; Del Principe et al. 1985). The apparent contradiction could perhaps be resolved by a more detailed analysis of the experimental conditions. When added together with a concentration of arachidonate, which itself does not yet cause aggregation, H_2O_2 can help to elicit a full aggregatory response, although H_2O_2 alone is ineffective (Fig. 6).

At higher arachidonate levels the synergistic action of H_2O_2 disappears, indicating that H_2O_2 affects a step in the arachidonic acid cascade. Indeed, whereas the levels of 12-HETE remained unchanged, the amounts of thromboxane A_2 (TxA_2) and heptadecatrienoic acid (HHT) greatly increased with H_2O_2 at 0.1 µM arachidonate. Since cyclooxygenase also requires a peroxide tone, for which is H_2O_2 also a mediator (Hemler and Lands 1980), an increased production of prostaglandinendoperoxide (PGH_2) is a plausible explanation. Obviously, peroxides determine the threshold level in platelets for arachidonate metabolism. The neighborhood of stimulated neutrophils by the release of H_2O_2 will therefore lead to an earlier response to arachidonate either from exogenous sources or from a normally subeffective PI response from agonists.

When using higher concentrations of H_2O_2 (50–200 µM), a second action of H_2O_2 on the thrombin-induced aggregation became apparent. The increase in aggregation

Table 3. Effect of preincubation with H_2O_2 on platelet aggregation, CGMP formation, and Ca^{2+} level

	Aggregation (%)	cGMP pmol/5 x 10^8 platelets	$[Ca^{2+}]$; nmol/l
Control	100	0.067 ± 0.012	859 ± 91
10 μM H_2O_2	85 ± 12	0.160 ± 0.024	787 ± 112
50 μM H_2O_2	44 ± 25	0.194 ± 0.027	426 ± 143
100 μM H_2O_2	16 ± 9	0.211 ± 0.039	361 ± 85
200 μM H_2O_2	4 ± 3	0.644 ± 0.133	203 ± 28

Washed human platelets were incubated for 2 min at 37 °C and 800 rpm in the presence of the indicated H_2O_2 concentrations. Stimulation was performed with 0.05 U/ml of thrombin and after 3 min the aggregation and the cGMP concentration were measured. For [Ca] the peak of the transient is given.

was not then influencing TxA_2 levels, but lead to an increased arachidonate release. Inhibitors of protein kinase C like H7 or staurosporin could block this effect and the 20 and 46 kD proteins showed increased phosphorylation (G. Hecker and V. Ullrich, submitted).

This finding is probably less relevant for a transcellular effect of neutrophil-derived H_2O_2 because of the high concentrations, but we have found that hydroperoxyeicosatetraenoic acids are already effective at micromolar levels. It remains to be shown whether 5-HPETE from neutrophils is able to reach platelets at these concentrations.

A different picture emerges again when H_2O_2 is preincubated with platelets for about 30 s or longer. Platelets then become refractory to agonists and this could be traced back to an increase in cellular cGMP levels (Table 3). It is known that guanylate cyclase can be activated by peroxides (Graff et al. 1978) but the mechanism by which this occurs is still unclear. The effect provides a means, however, to counter-regulate platelet aggregation.

The situation becomes even more complex when one considers the reported inhibitory action of superoxide anion on guanylate cyclase (Brüne and Ullrich 1987; Salvemini et al. 1989). Since O_2^- is the primary product of the oxidative burst in neutrophils, it remains to be elucidated whether its inhibitory action becomes more effective than the stimulatory properties of its dismutation product H_2O_2.

Besides these oxygen metabolites, several other factors released from neutrophils may also participate in their interaction with platelets. Cathepsin G, a serine protease, exhibits similar properties as thrombin and hence will activate platelets through a proteolytic process (Selak et al. 1988). On the other hand, neutrophils also secrete elastase, which can cleave v. Willebrand factor as well as the thrombin receptor from the platelet surface, and therefore may be involved in the downregulation of aggregation (Wicki and Clemetson 1985). This may not be paradox when one considers that elastase is primarily bound to its inhibitor and that an oxidative process via H_2O_2 and the secreted myeloperoxidase are necessary for the release of elastase from the inhibitory complex (Johnson and Travis 1979). Thus, the actions of cathepsin G and elastase could be coordinated by time-dependent processes.

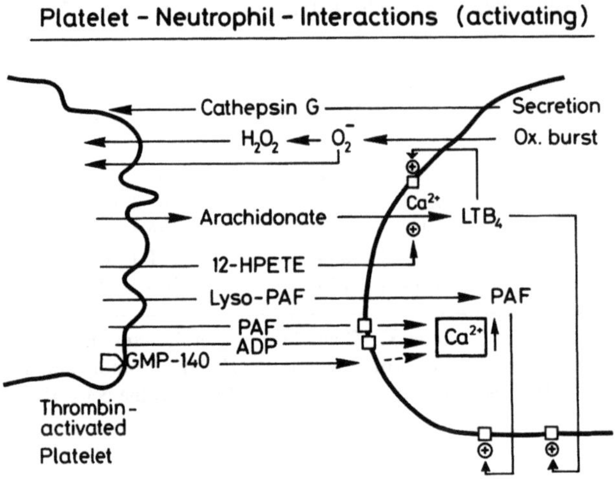

Fig. 7. Platelet-neutrophil interactions (activating)

Recently, it was reported that neutrophils in their resting state as well as upon stimulation synthesize nitric oxide (EDRF), a potent activator of guanylate cyclase. Since in the activated state superoxide is also released and rapidly reacts with NO (Gryglewski et al. 1986), this mediator may influence platelets mainly in their resting state, thus preventing spontaneous aggregations or diminishing responses to agonists.

Still unresolved is the role of 15-lipoxygenase in neutrophils in intra- and intercellular communication. The product 15-HPETE can be converted to lipoxin A_4 intracellularly by 5-lipoxygenase, but whether this new mediator inhibits protein kinase C of only neutrophils or also of neighboring cells remains to be tested (Serhan and Sheppard 1990).

Although the possible interactions starting from the platelets seem to be less in number, they are quite important, e.g. the already mentioned expression of GMP-140. Unlike granulocytes, platelets release an excessive amount of arachidonate which is not only required for the 12-lipoxygenase and the cyclooxygenase pathway. Together with ADP, which increases intracellular Ca^{2+} in neutrophils through purinergic receptors, arachidonate could enter the neutrophil and now serve as substrate for the synthesis of LTB_4. The intermediate LTA_4 in turn could be transformed in platelets to LTC_4 which is a strong agonist for smooth muscle contraction. Phospholipase A_2 in platelets would also act on O-alkyl-phosphatidylcholine, generating lyso-PAF which in platelets is only partly converted to PAF and in part is transferred to neutrophils, which also synthesize PAF from it. PAF is not only an agonist for platelets and neutrophils, but is also a very common messenger molecule between other cells of the immune system (Benveniste and Vargaftig 1983). A summarizing view of all mentioned platelet-neutrophil interactions is given in Fig. 7. For clarity the inhibitory pathways are separately depicted in Fig. 8.

It should be mentioned that platelet-neutrophil interactions are widely reported in the literature and many more or less established mechanisms could be added to both schemes e.g., the recently discussed synthesis of hepoxilins (Nigam et al. 1990). As

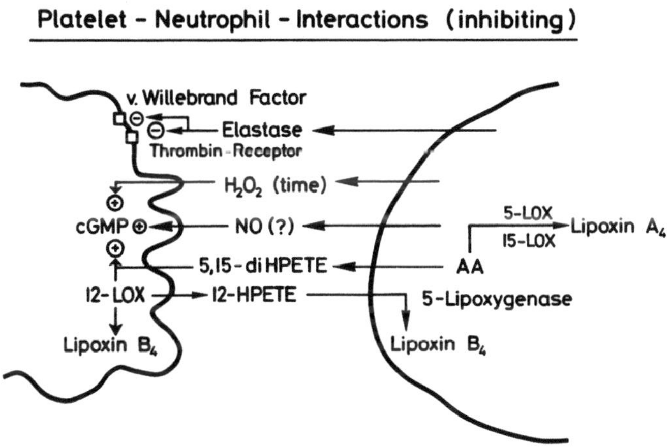

Fig. 8. Platelet-neutrophil interactions (inhibiting)

in all regulatory systems the antagonizing pathways not only provide stop signals for the agonist-induced physiological response of the cell, but they are present continuously to counteract the various stimuli that would invariably lead to autoactivation. Since a detailed picture of the signaling pathways in a single cell like platelets or granulocytes is not yet complete, the factors regulating the communication between cells are even less understood. The pathways discussed here represent probably only a few building blocks in a communication network that may become even more sophisticated when more closely cooperating cell types, e.g., granulocytes-endothelial cells, endothelial cells-smooth muscle cells, or immuncompetent cells among themselves, are considered. A further step in complexity would be the interaction of three or more cell types. Moreover, it is important that the time-dependent formation and degradation of the various messengers are achieved. Thus, cross-talk between cells may not only serve the same purpose as our languages, but may also reach a similar order of complexity.

References

Bednar M, Smith B, Pinto H, Mullane KM (1985) Neutrophil depletion suppresses ^{111}In-labeled platelet accumulation in infarcted myocardium. J Cardiovasc Pharmacol 7:906–912

Benveniste J, Vargaftig BB (1983) An ether lipid with biological activities: platelet-activating factor. In: Mangold HK, Paltauf F (eds) Ether-lipids: biomedical aspects. Academic Press, New York, p 355

Broekman MJ, Eiroa AM, Marcus AJ (1989) Albumin redirects platelet eicosanoid metabolism toward 12(*S*)-hydroxyeicosatetraenic acid. J Lipid Res 30:1925–1932

Brüne B, Ullrich V (1987) Inhibition of platelet aggregation by carbon monoxide is mediated by activation of guanylate cyclase. Mol Pharmacol 32:497–504

Clark RA, Klebanoff S (1980) Neutrophil-platelet interaction mediated by myeloperoxidase and hydrogen peroxide. J Immunol 124:399–405

Del Maschio A, Corvazier E, Maillet F, Kazatchkine MD, Maclouf J (1989) Platelet-dependent induction and amplification of polymorphonuclear leucocyte lysosomal enzyme release. Br J Haematol 72:329–335

Del Maschio A, Evangelista V, Rajtar G, Chen ZM, Carletti C, De Gaetano G (1990) Platelet activation by polymorphonuclear leukocytes exposed to chemotactic agents. Am J Physiol 258:H870–879

Del Principe D, Menichelli A, De Matteis W, Di Corpo ML, Di Giulio S, Finazzi-Agro A (1985) Hydrogen peroxide has a role in the aggregation of human platelets. FEBS Lett 185:142–145

Graff G, Stephenson JH, Glass DB, Haddox MK, Goldberg ND (1978) Activation of soluble splenic cell guanylate cyclase by prostaglandin endoperoxides and fatty acid hydroperoxides. J Biol Chem 253(21):7662–7676

Gryglewski RJ, Palmer RMJ, Moncada S (1986) Superoxide anion is involved in the breakdown of endothelium-derived relaxing factor. Nature (London) 320:454–456

Hammarström S, Hamberg M, Samuelsson B, Duell EA, Stawiski M, Voorhees JJ (1975) Increased concentrations of nonesterified arachidonic acid, 12*L*-hydroxy-5,8,20,14-eicosatetraenoic acid, prostaglandin E_2, and prostaglandin $F_{2\alpha}$ in epidermis of psoriasis. Proc Nat Acad Sci USA 72(12):5130–5134

Hatzelmann A, Ullrich V (1987) Regulation of 5-lipoxygenase activity by the glutathione status in human polymorphonuclear leukocytes. Eur J Biochem 169:175–184

Hatzelmann A, Schatz M, Ullrich V (1989) Involvement of glutathione peroxidase activity in the stimulation of 5-lipoxygenase activity by glutathione-depleting agents in human polymorphonuclear leukocytes. Eur J Biochem 180:527–533

Hemler ME, Lands WEM (1980) Evidence for a peroxide-initiated free radical mechanism of prostaglandin biosynthesis. J Biol Chem 255:6253–6261

Johnson D, Travis J (1979) The oxidative inactivation of human α_1-proteinase inhibitor. J Biol Chem 254:4022–4026

Kanaji K, Okuma M, Uchino H (1986) Deficient induction of leukotriene synthesis in human neutrophils by lipoxygenase-deficient platelets. Blood 67:903–908

Lapetina, EG (1986) Inositide-dependent and independent mechanisms in platelet activation. In: Phosphoinositol and receptor mechanisms. pp 272–286, Alan R. Liss

Levine PH, Weinger RS, Simon J, Scoon KL, Krinsky NI (1976) J Clin Invest 57:955–963

Maclouf J, Iliou JP, Del Maschio A (1986) Potentiation of 5-lipoxygenase metabolism by human neutrophils using a combination of activators. Prost Leukotr Med 23:105–108

Maclouf J, Murphy RC, Henson PM (1990) Transcellular biosynthesis of sulfidopeptide leukotrienes during receptor-mediated stimulation of human neutrophil/platelet mixtures. Blood 76:1838–1844

McEver RP (1991) GMP-140: a receptor for neutrophils and monocytes on activated platelets and endothelium. J Cell Biochem 45:156–161

Nigam S, Nodes S, Cichon G, Corey EJ, Pace-Asciak CR (1990) Receptor-mediated action of hepoxilin A_3 releases diacylglycerol and arachidonic acid from human neutrophils. Biochem Biophys Res Commun 171(3):944–948

Salvemini D, De Nucci G, Sneddon JM, Vane JR (1989) Superoxide anions enhance platelet adhesion and aggregation. Br J Pharmacol 97:1145–1150

Schatz-Munding M, Hatzelmann A, Ullrich V (1991) The involvement of extracellular calcium in the formation of 5-lipoxygenase metabolites by human polymorphonuclear leukocytes. Eur J Biochem 197:487–493

Selak MA, Chignord M, Smith JB (1988) Cathepsin G is a strong platelet agonist released by neutrophils. Biochem J 251:293–299

Serhan CN, Sheppard K-A (1990) Lipoxin formation during human neutrophil-platelet interactions. Evidence for the transformation of leukotriene A_4 by platelet 12-lipoxygenase in vitro. J Clin Invest 85:772

Ward PA, Till GO, Johnson KJ (1986) Role of platelets in neutrophil-dependent, oxygen radical mediated lung injury. Fed Proc 45:380

Wicki AN, Clemetson KJ (1985) Structure and function of platelet membrane glycoproteins Ib and V. Effects of leukocyte elastase and other proteases on platelet response to von Willebrand factor and thrombin. Eur J Biochem 153:1–11

Proteolysis-Induced Pathomechanisms in Acute Inflammation and Related Therapeutic Approaches

M. JOCHUM[1], W. MACHLEIDT[2], and H. FRITZ[1]

1 Introduction

1.1 Aspects of Proteolytic Pathomechanisms in Inflammation

Proteolysis-induced pathomechanisms seem to play a major role in the primary response of the organism to inflammatory stimuli such as tissue destruction due to multiple trauma and major surgery or invasive microbes and endotoxins in sepsis. Physiologically, the inflammation response is directed towards inactivating and eliminating the deleterious agents and to initiate the process of repair and healing. Yet, the activation of the complex interacting cellular and humoral defence mechanisms necessary for this purpose carries with it also the risk of damaging healthy tissue, thus perpetuating the inflammatory process. In this respect, the lysosomal serine proteinase elastase and the cysteine proteinase cathepsin B of the primary inflammatory cells, polymorphonuclear (PMN) granulocytes, and monocytes/macrophages are supposed to be potent effectors of proteolytic tissue damage if they are discharged extracellularly in high amounts during activation of the phagocytes. Moreover, proteinases of humoral origin (plasma kallikrein, thrombin, plasmin, complement esterases) and protein split products generated by their proteolytic action (fibrinopeptides and fibrin monomers, fibronectin peptides, complement-derived factors such as C3a, C4a, and C5a, etc.) have been proven to be of major importance as strong stimulators of the primary defence cells.

In addition to lysosomal proteinases, highly reactive oxygen species produced by the respiratory burst are also extracellularly liberated during phagocytosis. By this means, vital structural elements (basal membranes, cell receptors, fibronectin, elastin, collagens, proteoglycans, etc.) as well as humoral factors including plasma proteins in close vicinity to the phagocytizing cells may be impaired unless the lysosomal proteolytic enzymes and oxidants are inactivated by their physiological regulators, the proteinase inhibitors (α_1-proteinase inhibitor, α_2-macroglobulin, cysteine proteinase inhibitors) and antioxidants (superoxide dismutase, catalase, glutathione redox system, ceruloplasmin).

Yet, due to an overstressed phagocytic activity of PMN granulocytes and monocytes/macrophages during severe inflammation, the main antagonist of PMN elastase,

[1] Department of Clinical Chemistry and Clinical Biochemistry, Surgical Clinic City of the University of Munich, Nußbaumstraße 20, D-8000 Munich 2, FRG
[2] Institute of Physiological Chemistry, Physical Biochemistry and Cell Biology of the University of Munich, Goethestraße 33, D-8000 Munich 2, FRG

42. Colloquium Mosbach 1991
Molecular Aspects of Inflammation

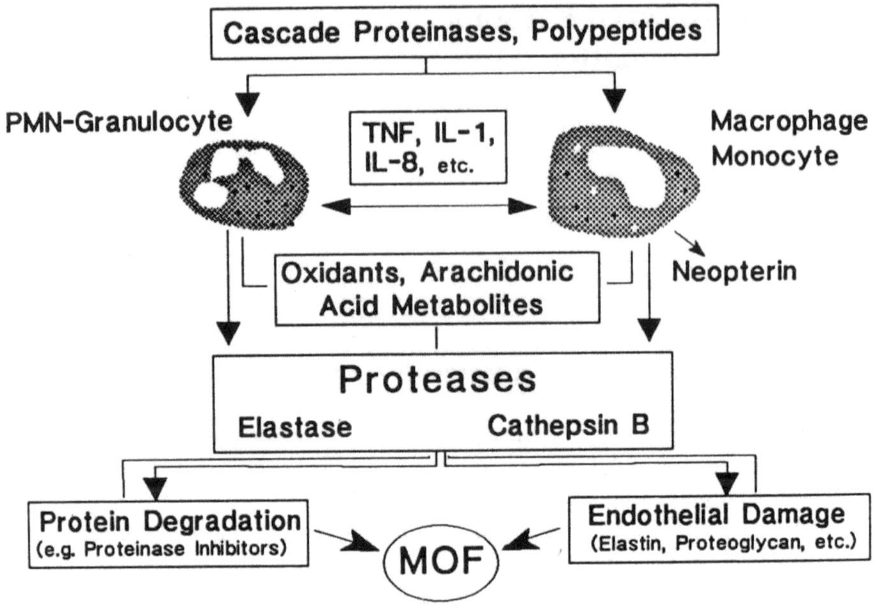

Fig. 1. Schematic depiction of humoral and cellular proteolytic pathomechanisms involved in the development of multiple organ failure (MOF). Destructive proteases (elastase, cathepsin B) and other cell constituents are released after activation of the phagocytes by blood cascade proteinases (plasma kallikrein, factor XIIa, thrombin, etc.) and polypeptides (fibrinogen and fibronectin split products, complement factors C3a, C4a, and C5a, etc.). For detailled explanation, see text

the α_1-proteinase inhibitor (α_1PI), is highly susceptible not only to proteolytic cleavage by cysteine and metallo proteinases released from the mononuclear cells, but also to oxidative denaturation of the reactive inhibitory site in the molecule. This enables an unrestricted local digestive activity by PMN elastase combined also with fatal consequences for the hemostasis system. As shown by several authors, PMN elastase can easily degrade and inactivate the principal inhibitors (antithrombin III, α_2-plasmin inhibitor, C1-inactivator) of proteinases of the blood cascade systems (clotting, fibrinolysis, complement), thus allowing the maintenance of life-threatening consumption of hemostasis factors and the additional production of potent stimulators of the phagocytes as mentioned above. Moreover, the proteolytic inactivation of the plasma inhibitors may be greatly enhanced by their oxidative denaturation in the surroundings of phagocytizing cells.

The interrelationship of cellular and humoral proteolytic pathomechanisms eventually contributing to the development of multiple organ failure (MOF) in the course of severe acute inflammation is depicted in a simplified version in Fig. 1. A comprehensive review of literature on this topic is given by Jochum (1988), Jochum and Fritz (1989), Assfalg-Machleidt et al. (1990), and Jochum and Fritz (1990).

1.2 Rationale for Therapeutic Use of Proteinase Inhibitors

To obtain fundamental information about the extent proteases may participate in the initiation and perpetuation of inflammatory processes, the following indications – as a modification of the Koch-Dale criteria – have to be proven in clinical and/or experimental studies:

1. The release of the lysosomal proteinases (e.g., PMN elastase, cathepsin B) and the activation of proteolytic blood cascade enzymes have to be verified in correlation to the severity of the inflammation.
2. The consumption of proteinase inhibitors and other plasma factors susceptible to proteolytic degradation should coincide with the occurrence of proteolytic activity.
3. Specific split products of functional proteins generated by the proteolytic action of lysosomal proteinases have to be shown in correlation to the extracellular release of these enzymes.
4. The use of specific exogenous proteinase inhibitors as therapeutic tools should prevent or at least reduce to some extent severe signs of inflammation.

In the following, characteristic results obtained from several clinical and experimental studies dealing with multiple trauma and/or septicemia are given. They demonstrate clearly the involvement of proteolysis-induced pathomechanisms in acute inflammation necessitating proteinase inhibition as a convenient therapeutic approach.

2 Methodology

Serially drawn citrated plasma, bronchoalveolar lavage fluid (BALF), and peritonitis exudate samples were used to quantify proteinases, inhibitors, and plasma protein split products.

Measurements of the lysosomal serine proteinase PMN elastase were carried out with a commercially available, highly specific two-site sandwich enzyme immunoassay kit (E. Merck, Darmstadt, Germany), which detects elastase only as an inactive complex with α_1PI (Neumann and Jochum 1984). Since, in contrast to plasma, BALF and peritonitis exudate samples may contain proteolytically active elastase in addition to the complex, parts of these specimens were incubated also with a surplus of α_1PI in vitro and reassayed for an increase in elastase-α_1PI complex as a measure of enzymatic elastase activity in vivo. The cysteine proteinase cathepsin B activity was quantified using a fluorometric assay as described by Assfalg-Machleidt et al. (1988). Caseinolytic activity of local body fluids due to elastase and/or cathepsin B was detected using resorufin-labeled casein (Boehringer, Mannheim, Germany) as a substrate. The method is outlined in detail by the manufacturer (Boehringer).

The coagulation proteinases, plasma kallikrein and thrombin, were estimated by their cleavage activity on the chromogenic peptide substrates S-2302 (Kabi, Stockholm, Sweden) and Chromozym TH (Boehringer, Mannheim, Germany), respectively, after the turnover of the proenzymes to the active proteinases.

The inhibitory activities of antithrombin III and α_1PI were measured with commercially available test systems (Coatest Antithrombin, Kabi, Stockholm, Sweden; α_1-Antitrypsin Farbtest, Boehringer, Mannheim, Germany) applying chromogenic peptide substrates for thrombin (S-2238) and trypsin (BAPA), respectively. The antigen levels of the inhibitors were quantified with commercial radial immunodiffusion plates (Nor-Partigen plates, Behringwerke, Marburg, Germany).

Opsonic activity, antigen levels, and split products of the opsonins IgG and complement factor C3 were determined as outlined in Billing et al. (1990, 1991) and Machleidt et al. (1991).

An elastase-specific split product of the Aα-chain of fibrinogen, called fibrino-elastase-peptide (FEP; equivalent to the Aα 1-21 peptide described by Weitz et al. 1986), was quantified with a two-step competitive immunoassay just recently developed in our laboratory (Gippner-Steppert 1991).

Clinical methods, treatment of patients, and experimental therapeutic studies are extensively described in the original publications cited in the following sections.

3 Results and Discussion

3.1 Proteolysis-Induced Pathomechanisms in Humans

3.1.1 Multiple Trauma

In an extended exploration 100 severely injured patients fulfilling previously defined study criteria were enrolled in a prospective study (directed by Drs. Nast-Kolb and Waydhas, Surgical Clinic City, University of Munich) on inflammatory mediators and multiorgan failure (MOF) associated with polytraumatic events (Nast-Kolb et al. 1990; Waydhas et al. 1991). Collection of blood samples and recording of clinical data were started within 30 minutes after arrival of the patients in the emergency department (about 1 h after the accident) and continued on a 6 h basis up to 48 h. Subsequently, the monitoring interval was extended to 24 h for a period of 14 days. Thereafter the clinical course was recorded until either transfer to a general ward or death of the patient.

Retrospectively, the patients could be assigned to three different groups: 16 of them died due to MOF between 3 and 28 days (mean survival time: 16 days) after the traumatic incident (group I), 47 patients survived the development of organ failure (group II), and 37 patients recovered from the accident without evident signs of organ dysfunctions (group III).

The extracellular release of PMN elastase and monocyte/macrophage-derived cathepsin B into plasma of the patients in the three outcome groups is depicted in Fig. 2. PMN elastase was elevated in all groups significantly above normal (upper range 120 ng/ml) already within 1 to 2 h after trauma and showed an additional increase up to the 12th posttraumatic hour. The differences between mean plasma levels in patients with (groups I and II) and without organ failure (group III) were highly significant ($p < 0.01$) throughout the whole observation period. Moreover, patients dying due to organ failure (group I) and those who survived organ dysfunctions (group II)

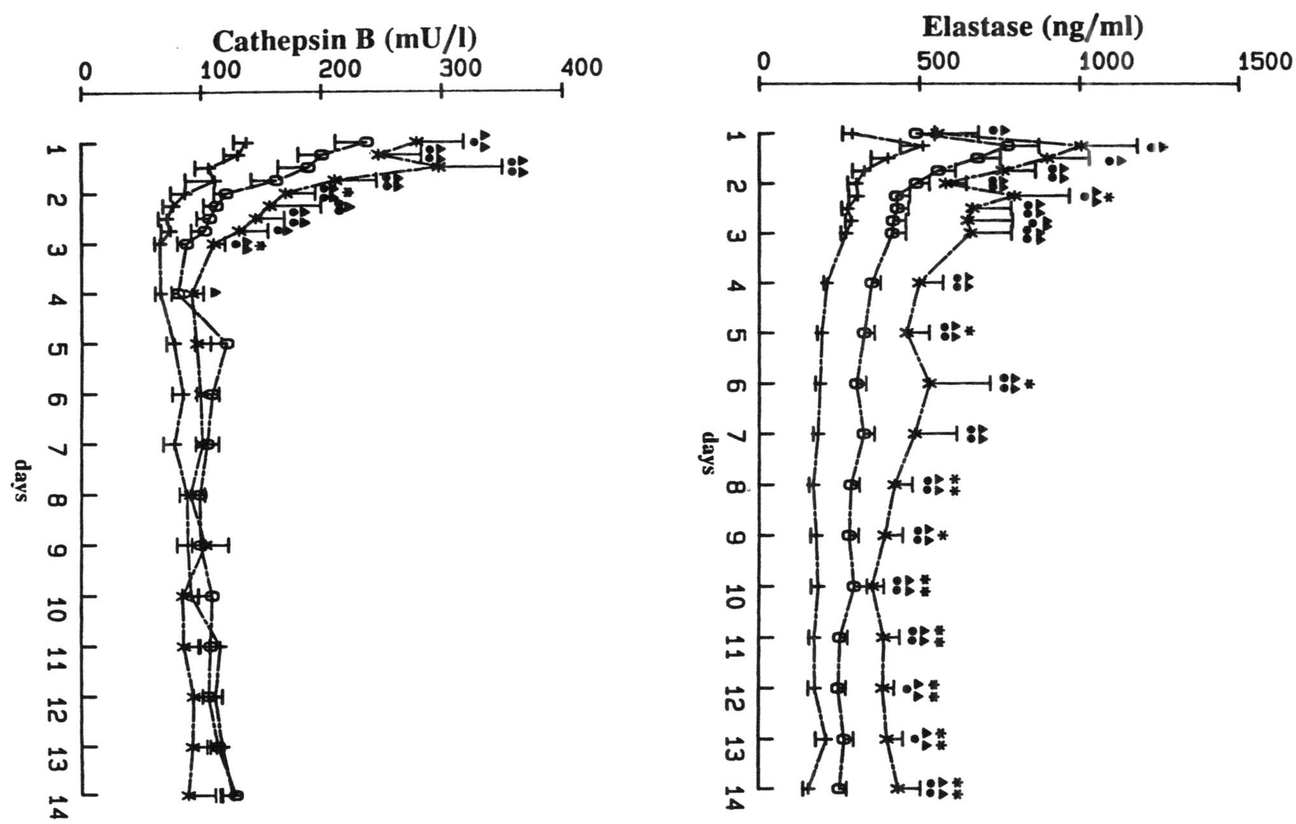

Fig. 2. Mean values (± SEM) of PMN elastase and cathepsin B in plasma of multiply injured patients assigned to three outcome groups: Group I (n = 16): nonsurvivors with multiple organ failure * --- *; group II (n = 47): survivors with reversible organ failure o --- o; group III (n = 37): survivors without organ failure + --- +. * $p < 0.05$, ** $p < 0.01$ group I vs group II; • $p < 0.05$, •• $p < 0.01$ group I vs group III; ▲ $p < 0.05$, ▲▲ $p < 0.01$ group II vs group III

were also significantly ($p < 0.05$) differentiated according to their mean PMN elastase plasma concentrations from the third posttraumatic day onwards. Cathepsin B plasma levels reached their maxima at least 6 h before complexed PMN elastase (Fig. 2). This early cathepsin B activity in circulating blood proved to be a sensitive and specific predictor of subsequent developing organ failure. Yet, it did not discriminate between lethal (group I) and reversible (group II) dysfunctions at that time. In contrast to PMN elastase, cathepsin B activity returned to only slightly elevated levels within 3 days in all patients (Fig. 2), thus not allowing a further differentiation of the three outcome groups. From these data the conclusion can be drawn, that the early release of lysosomal phagocyte proteinases after a severe traumatic event contributes to the occurrence of anticipated organ failure, and that at least the ongoing discharge of PMN elastase may be involved in the exacerbation of the clinical situation. The latter seems to be confirmed by the appearance of the elastase-induced split product FEP of fibrinogen which is correlated to the continuous release of elastase and the development of lethal multiple organ failure as outlined in Fig. 3 for a severely injured patient.

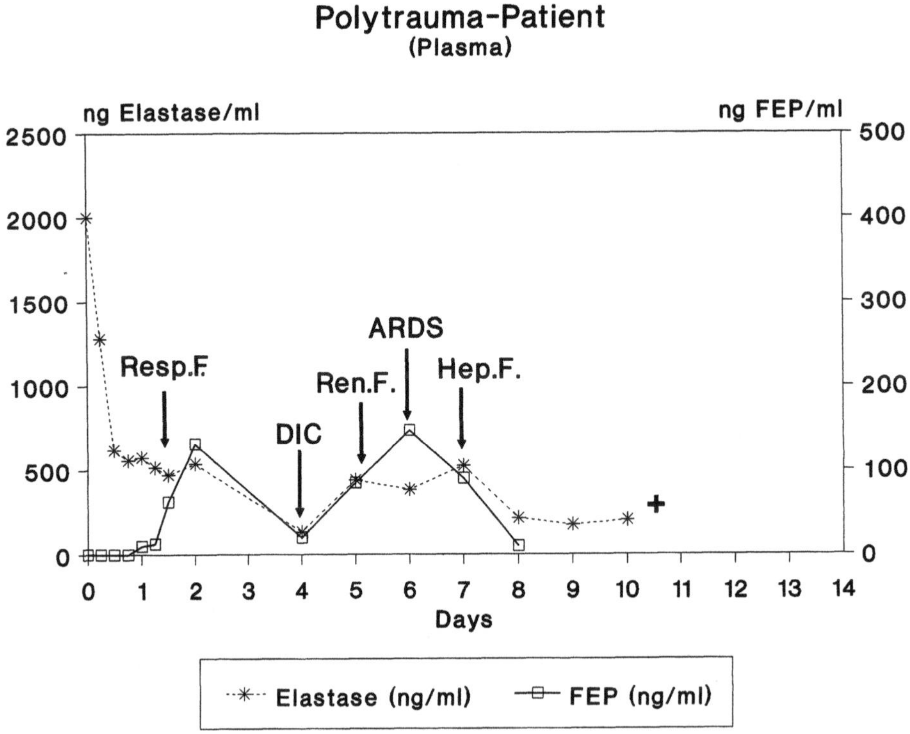

Fig. 3. Plasma levels of PMN elastase (in complex with α_1-proteinase inhibitor) and fibrino-elastase-peptide (FEP) in a polytraumatized patient with lethal multiple organ failure. *Resp. F* Respiratory failure; *DIC* disseminated intravascular coagulation; *Ren. F.* renal failure; *ARDS* acute respiratory distress syndrome; *Hep. F.* hepatic failure

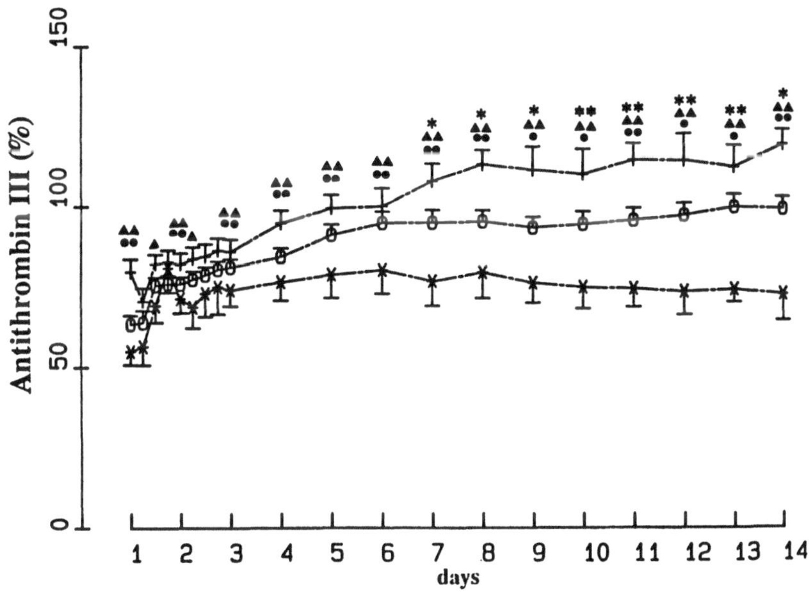

Fig. 4. Mean values (± SEM) of antithrombin III inhibitory activity in plasma of multiply injured patients assigned to three outcome groups. For further details, see Fig. 2

Interestingly, the main clotting inhibitor, antithrombin III (AT III), showed a significantly lower inhibitory activity ($p < 0.01$) in plasma of group I and II patients compared to group III during the whole observation period and in nonsurvivors (group I) compared to survivors with organ failure (group II) from the first posttraumatic week onwards (Fig. 4). Similar results were obtained concerning the rapid turnover of the coagulation proenzyme, prothrombin (data not shown) to thrombin, the essential target proteinase of AT III. Thus, plasma levels of AT III and prothrombin 60% below normal in the early posttraumatic phase were highly associated with the later appearance of severe organ dysfunctions, indicating that in addition to the release of lysosomal proteinases an overwhelming activation of the humoral proteolytic cascade systems is also conducive to the perpetuation of the posttraumatic inflammatory process.

In another clinical study on multiply injured patients (Sturm 1991) we were especially interested in local proteolysis-induced mechanisms which might be involved in the pathogenesis of the most severe lung dysfunction, the acute respiratory distress syndrome (ARDS). Daily drawn BALF (method described in Obertacke et al. 1991) samples allowed us to confirm a significantly increased local discharge of the phagocyte proteinases, elastase and cathepsin B, in subjects at high risk of developing ARDS (Jochum 1991). Although the α_1PI antigen levels in all BALF specimens of the traumatized patients were far above the normal values of healthy volunteers, these amounts were apparently insufficient in most cases to completely inhibit the PMN elastase released in the local epithelial milieu of the alveoli. Therefore, the obviously deficient inhibitory capacity of α_1PI may have been caused by proteolytic and oxidative denaturation as well. Applying special assay systems, both ways of α_1PI destruc-

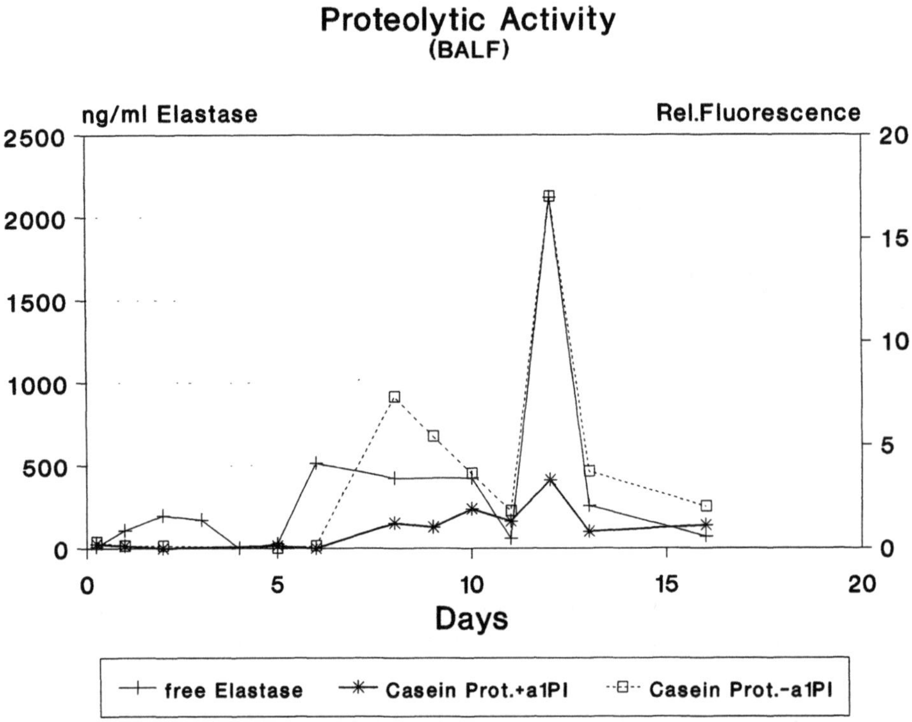

Fig. 5. Proteolytic activity (degradation of resorufin-casein with and without addition of α_1-proteinase inhibitor) and free PMN elastase in serial bronchoalveolar lavage fluids (BALF) of a severely traumatized patient. Acute respiratory distress syndrome and multiple organ failure persisted from the 5th to 12th posttraumatic day

tion were demonstrated by Schraufstatter et al. (1984), at least in some individual BALF samples of patients with manifest ARDS. Although other researchers could not detect elastase activity against high molecular weight protein substrates in BALF samples of ARDS patients (Idell et al. 1985; Wewers et al. 1988), we were recently able to show caseinolytic activity in a variety of BALF samples (kindly given to us by Dr. Obertacke, Surgical Clinic, University of Essen) from trauma patients with severe lung dysfunctions (Machleidt et al. 1991). In vitro incubation of these specimens with α_1PI abolished nearly completely this proteolytic activity (Fig. 5) and gave rise to an additional increase in the elastase-α_1PI complex, indicating that the enzymatic activity present in the alveolar environment was mostly due to PMN elastase.

Moreover, the destructive potency of elastase in vivo could be directly documented by the proof of augmented generation of FEP in close correlation to the rising amount of the extracellularly discharged PMN enzyme in BALF samples of patients with aggrevating respiratory failure (Fig. 6). Hence, the aggressive components of inflammatory cells, especially the lysosomal proteinases, may indeed contribute to the occurrence of lung dysfunction in severely traumatized patients.

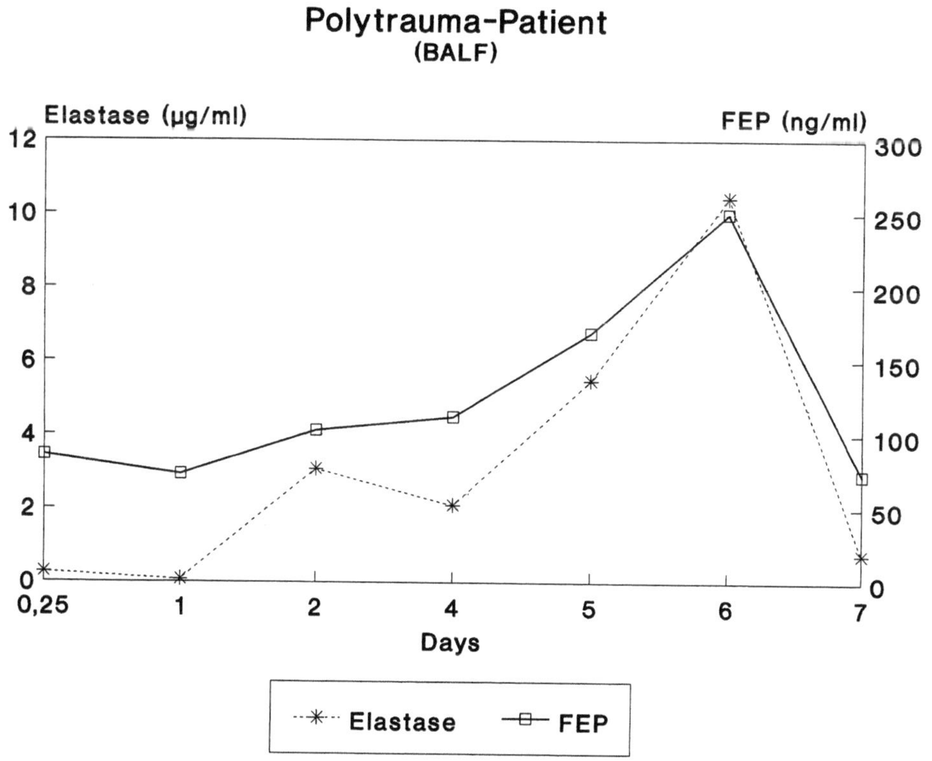

Fig. 6. Total amount of PMN elastase (measured as complex with α_1-proteinase inhibitor) and fibrino-elastase-peptide (FEP) in serial bronchoalveolar lavage fluids (BALF) of a severely traumatized patient with reversible respiratory dysfunction from the 2nd to the 7th posttraumatic day

3.1.2 Sepsis and Peritonitis

In several prospective clinical studies on patients suffering from bacterial infections after major surgical treatments, we could demonstrate an increasing release of PMN elastase into the circulation in accordance with the worsening of the inflammatory reaction (Duswald et al. 1985; Inthorn and Jochum 1988; Jochum et al. 1990) Patients without postoperative infections showed only moderate, transiently elevated plasma levels of complexed elastase (up to three times that of normal) after operation, whereas onset and course of sepsis were characterized by markedly increased concentrations, up to 20-fold in individual cases. In patients with persisting septicemia, the elastase plasma levels remained significantly elevated until death, while a recovery phase was accompanied by an obvious return to normal values. When complexed elastase increased in plasma, a prominent reduction of the humoral blood cascade proenzymes, prekallikrein and prothrombin, as well as of the AT III inhibitory activity was measurable. This indicates that the proteolytic mechanisms in this pathological situation are quite similar to those seen in the course of developing posttraumatic organ failure. In addition, we could demonstrate in some pilot measurements that

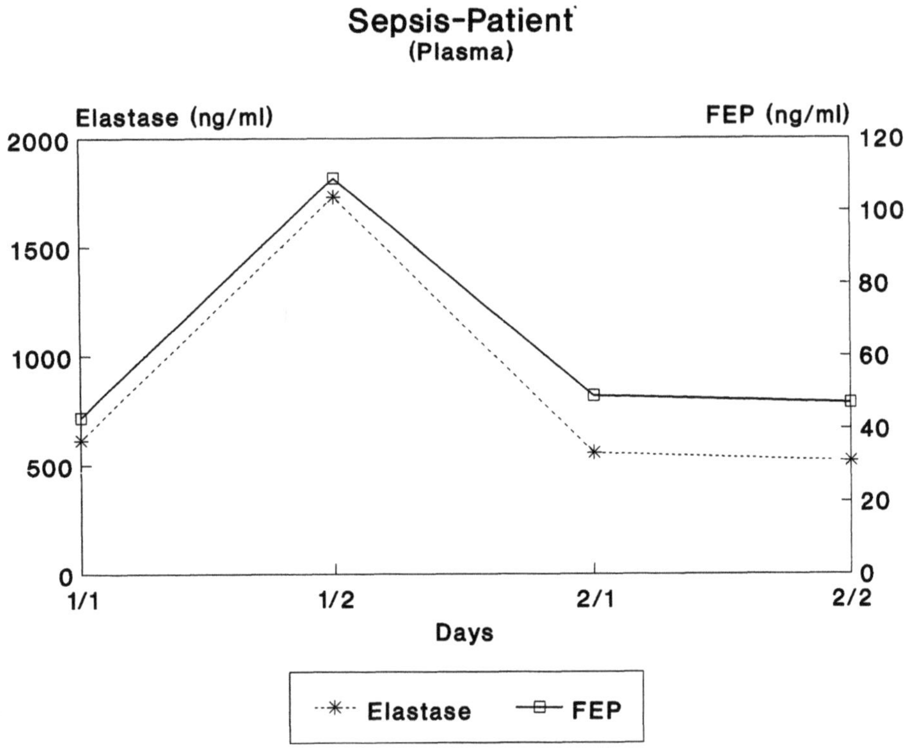

Fig. 7. Plasma levels of PMN elastase (in complex with α_1-proteinase inhibitor) and fibrino-elastase-peptide (FEP) of a patient with lethal outcome 2 days after diagnosis of sepsis due to bifurcation malacia of the trachea.

during a septic shock phase, the monocyte/macrophage-derived cathepsin B was released in a manner similar to the PMN elastase, thus intensifying the extracellular proteolytic capacity.

Just recently we were able to verify such a destructive proteolytic potency in a still ongoing prospective study (clinically directed by Dr. Inthorn, Surgical Clinic Großhadern, University of Munich) on septic patients. To be enrolled in this investigation, patients had to fulfill all four previously defined septic criteria (septic focus and/or positive blood culture; body temperature > 38.5 °C; leukocytes > 15 000/µl or < 5000/µl; platelets < 100 000/µl or drop of platelets > 20% within 24 h), indicating a severe inflammatory situation. Under these circumstances the generation of the elastase-induced fibrinogen split product FEP was highly correlated to the plasma level of PMN elastase as shown in Fig. 7 for a patient who died 2 days after diagnosis of sepsis. Moreover, the unspecific proteolytic degradation of AT III, which is also highly susceptible to PMN elastase cleavage (Jochum et al. 1981; Jordan et al. 1989), could be determined by an indirect method comparing the amount of AT III antigen with its inhibitory activity as depicted in Fig. 8 for a patient suffering from eventually lethal sepsis. Besides the fact that AT III was already highly consumed throughout the

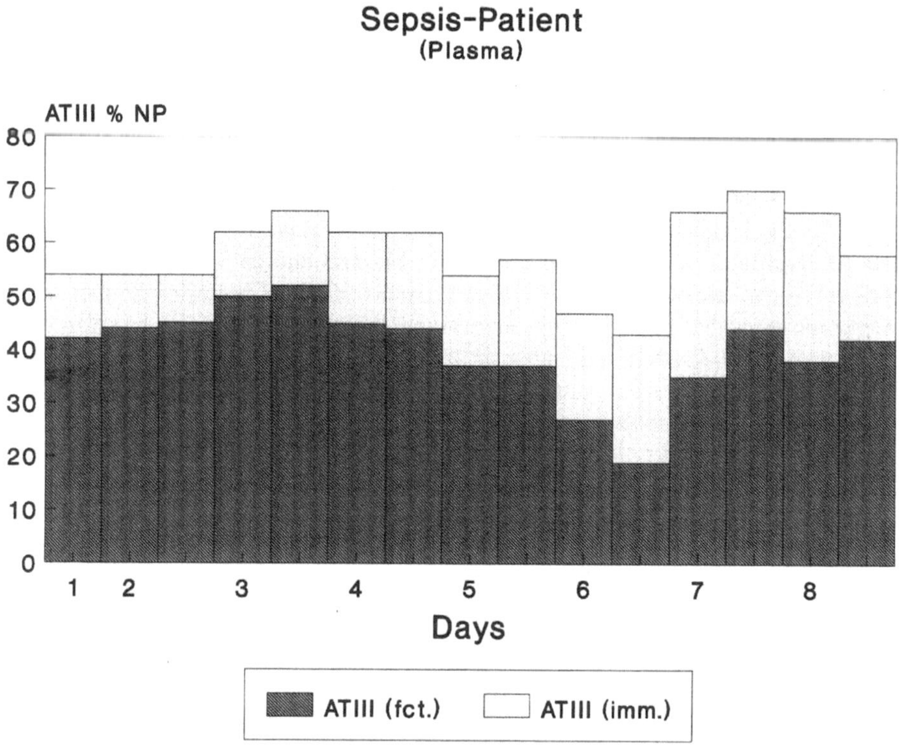

Fig. 8. Comparison of antithrombin III antigen (AT III) levels (*imm*) and inhibitory activity (*fct.*) in plasma samples of a patient with lethal sepsis due to stomach cancer. *NP %* of a norm plasma pool

whole sepsis period of 8 days, the clearly lower inhibitory activity indicates that part of the in antigen level may be due either to complexes of the inhibitor with clotting enzymes or to proteolytically degraded molecules. Since the latter have an only slightly lower molecular weight and nearly the same half-life as the intact molecules (Jordan et al. 1989), they cannot be distinguished from the native AT III when quantified by radial immunodiffusion. In our patient, however, we could definitely rule out the possibility that AT III-proteinase complexes contributed significantly to the higher antigen levels, because due to the very short half-life of only a few minutes, concentrations of AT III-thrombin complexes amounted to, at most, 0.01% of the overall AT III (data not shown in detail). Thus, elastase-inactivated AT III molecules should be present with high probability in the pathological situation of severe sepsis. This assumption is further confirmed by results of Jordan et al. (1989) which demonstrate that even catalytic amounts of heparin react with elastase in a way considerably accelerating the rate of AT III inactivation by this enzyme. Since heparin exists not only as a minor component on vascular endothelial cells, but is also used as a therapeutic drug in our patients, the positive regulation of hemostasis by heparin may be significantly counteracted by the adverse effect on AT III in the case of a high release of elastase from PMN granulocytes adhering to the endothelial layer of the blood vessel walls.

Diffuse peritonitis is often the source of a systemic spreading of local infectious complications, eventually leading to sepsis and multiorgan failure. In a previous work we were able to demonstrate that the impairment of opsonization in the peritoneal cavity allows the survival of huge numbers of bacteria despite the presence of intact phagocytes (Billing et al. 1988). Since the proteolytic breakdown of the main opsonins IgG and complement factor C3 in peritonitis exudates correlated well with the extracellularly released lysosomal proteinases elastase and cathepsin B in the local body fluids, we assumed that these enzymes are the principal causes of the deficiency in opsonic capacity. To assure this supposition, isolated IgG was incubated in vitro either with PMN elastase and cathepsin B or (in its isothiocyanate-labeled form = FITC IgG) with a cell-free, purulent peritonitis exudate (kindly supplied by Dr. Billing, Surgical Clinic Großhadern, University of Munich). The proteolytic degradation was followed by gel chromatography of the split products exhibiting the same type of IgG cleavage pattern under all conditions (Billing et al. 1991; Machleidt et al. 1991). Using resorufin-labeled casein as a substrate, similar results were obtained substantiating proteolytic elastase activity despite the presence of an up to 40-fold molar sur-

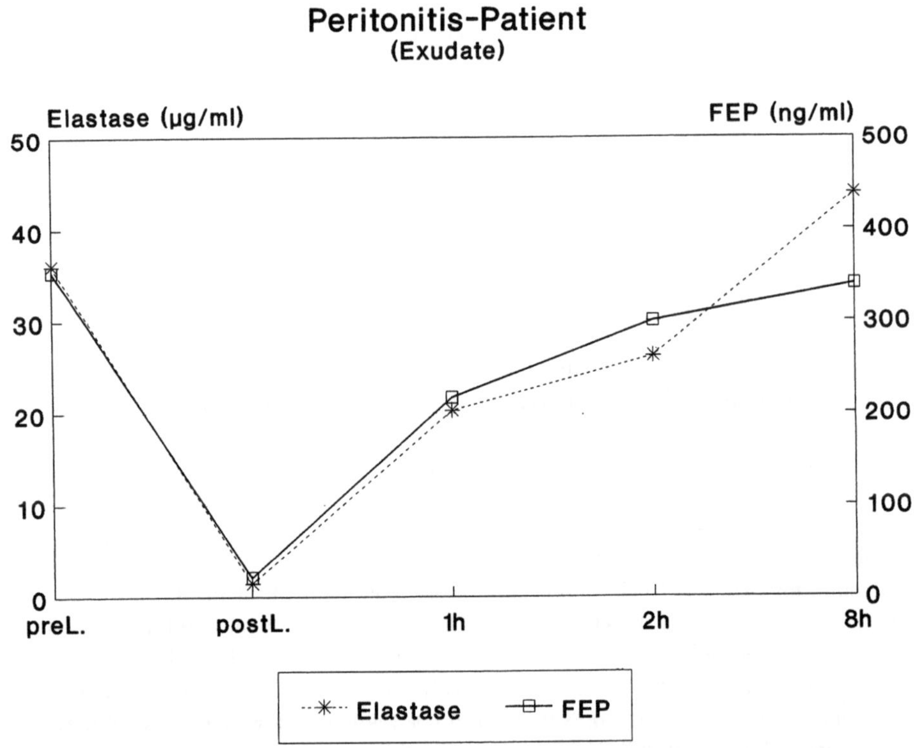

Fig. 9. PMN elastase (in complex with α_1-proteinase inhibitor) and fibrino-elastase-peptide (FEP) in exudate samples taken before (*preL.*) and immediately after (*postL.*) lavage of the abdomen of a peritonitis patient with Ringer lactate solution and in drainage fluids collected between 0 to 1 h, 1 to 2, and 2 to 8 h after lavage

plus of α_1PI antigen in peritonitis exudates (data not shown) as already discussed for BALF samples in trauma patients. In addition to the in vivo degradation of IgG and C3, we have successfully proved the generation of FEP in peritonitis exudates just recently. As depicted in Fig. 9, high amounts of complexed PMN elastase coincided with highly elevated FEP in the specimen drawn before surgical treatment of the abdomen of a patient with severe peritonitis. After rinsing the peritoneal cavity with 10 l of Ringer lactate solution both parameters decreased nearly to zero. Yet, as can be seen from the abdominal drainage fluids collected between 0 and 1, 1 and 2, and 2 and 8 h after operation, the release of PMN elastase started again, inducing also the production of FEP (Fig. 9). These observations may be taken as an indication of a still ongoing inflammatory reaction in this patient.

Summarizing the data obtained in our clinical studies on patients suffering from multiple trauma and/or septicemia, we could clearly show that excessive local (α_1PI) or systemic (AT III) consumption of proteinase inhibitors concomitant to the release of lysosomal phagocyte proteinases and the activation of proteolytic blood cascade enzymes during severe inflammatory reactions is a most critical event which may contribute to the propagation of (multiple) organ damage. Therefore, supplementation of the body's inhibitor potential by exogenous proteinase inhibitors – isolated from human material or produced by gene technology (Fritz et al. 1991) – seems to be a most promising therapeutic approach.

3.2 Proteinase Inhibition as a Suitable Therapeutic Approach in Acute Inflammation

3.2.1 Inhibitors of PMN Elastase or Thrombin in Experimental Endotoxin Shock

In a preliminary, controlled study on sepsis in young pigs the prophylactic application of the relatively specific recombinant elastase inhibitor, eglin c (originally isolated from the leech hirudo medicinalis), caused a significant reduction in the consumption of antithrombin III and other plasma proteins as well as in the formation of interstitial pulmonary edema (Jochum et al. 1987). As assessed by measurement of arterial blood pressure and total protein concentration in plasma, Siebeck et al. (1989a) demonstrated in a more extended study that eglin c can also reduce the overall capillary leakage induced by the infusion of live *E. coli* in pigs. Moreover, in a further controlled investigation, Siebeck et al. (1989b) and Hoffmann et al. (1990) could show that besides eglin c the thrombin-specific inhibitor hirudin – another recombinant inhibitor also formerly isolated from the medical leech – significantly improved endotoxin shock syndromes in minipigs. Fibrinogen consumption, formation of fibrin monomers, the occurrence of pulmonary vasoconstriction, and the release of PMN constituents were clearly lower in endotoxemic animals treated with hirudin as compared to those without continuous intravenous inhibitor infusion. Interestingly, the supplementation of a purified antithrombin III-heparin complex in another randomized porcine endotoxin shock model had only a slight, insignificant positive effect on the endotoxin-induced mortality and oxygen saturation in arterial blood (as an indication of pulmonary function), although the consumption of fibrinogen and the formation of soluble fibrin monomers were clearly prevented in the drug-treated animals

(Spannagl et al. 1991). Furthermore, prothrombin consumption was similar in the treatment and placebo groups, suggesting that local thrombin generation via factor Xa and binding of both enzymes to cell membrane receptors (e.g., thrombin to thrombomodulin) may protect them from inhibition by the AT III-heparin complex. On the other hand, the applied amount of the AT III-heparin complex may not have been sufficient to inhibit the activation of contact phase enzymes (plasma kallikrein, factor XIIa) which are also supposed to be potent stimulators of PMN granulocytes (Wachtfogel et al. 1983; 1985), thus maintaining the inflammatory process via the release of phagocyte proteinases and reactive oxygen metabolites. Moreover, since after an initial rise (up to about 130%) the AT III activity decreased during further drug application, while the antigen level still increased, the heparin in the complex may have also facilitated the inactivation of AT III by released PMN elastase (Jordan et al. 1989), thereby lowering the inhibitory capacity under an otherwise presumably effective threshold level.

The requirement of high levels of AT III inhibitory activity in the circulation is also confirmed by results of Emerson et al. (1989) concerning the efficacy of antithrombin III supplementation in several animal models (rat, sheep, baboon) of fulminant *E. coli* endotoxemia or bacteremia. Only very high dosage (up to threefold that of normal) and prophylactic administration of AT III prevented organ damage and increased permanent survival in the experimental animals. Interestingly, the combined application of AT III and α_1PI showed a significant synergistic improvement of the pulmonary function compared to the single drug treatment in the endotoxemic sheep model. This indicates again that a complex interaction of lysosomal and humoral blood cascade proteinases contribute to the perpetuation of a septic-like inflammation.

3.2.2 Antithrombin III Supplementation in Clinical Sepsis

From the therapeutic animal experiments and the apparent lack of sufficient efficacy of AT III supplementation in clinical studies (Blauhut et al. 1985; Vinazzer 1987), we draw the conclusion that only AT III levels well above the normal value may be able to improve organ dysfunctions in clinical sepsis. Therefore, we conducted a study (clinically directed by Dr. Inthorn, Surgical Clinic Großhadern) on septic patients with the aim to increase the AT III inhibitory activity above 120% that of normal. To achieve this, AT III concentrates were i.v. infused twice daily over 21 days according to a modified regimen originally described by Blauhut et al. (1985). Blood samples were taken twice daily throughout the whole observation period. Preliminary data and a detailled outline of the performance of the still ongoing study have been published (Jochum et al. 1991). Here, only the most important results will be presented. Up to now, 15 patients, each fulfilling the above-mentioned septic criteria, could be enrolled in the control and therapy groups. With our application scheme the AT III activity in the treatment group was elevated to mean plasma levels slightly below 120% during the first 9 days followed by an increase above 120% thereafter, whereas in the control group levels between 60% (early phase) and 80% (later phase) were measurable (Fig. 10). Although all AT III-treated patient received nearly the same amount of the inhibitor concentrate (between 8000 and 4000 U/day), those individuals who survived the septic events showed clearly higher AT III levels (up to a mean of 135% during

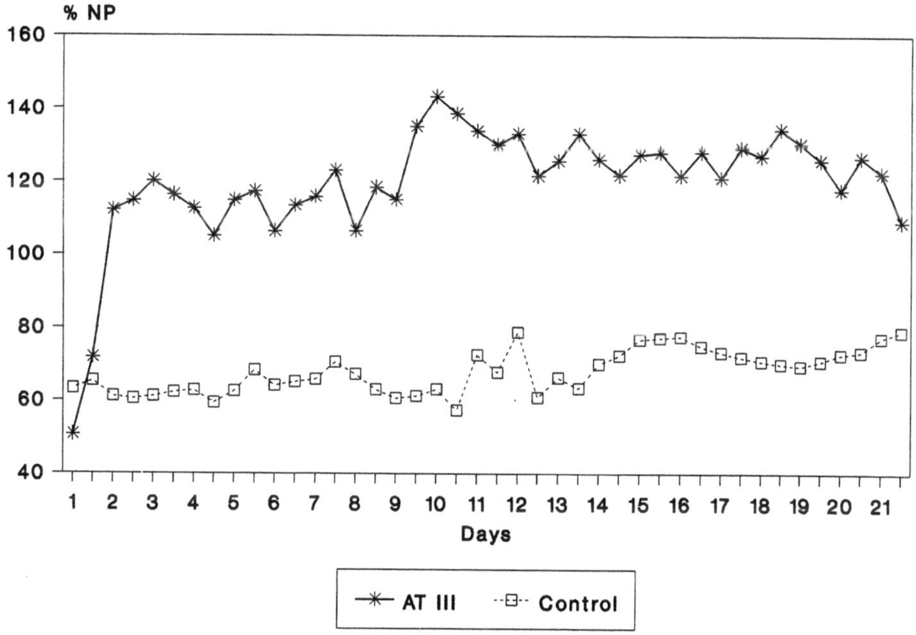

Fig. 10. Mean values of antithrombin III inhibitory activity in plasma of septic patients with (AT III; n = 15) or without (control; n = 15) antithrombin III supplementation

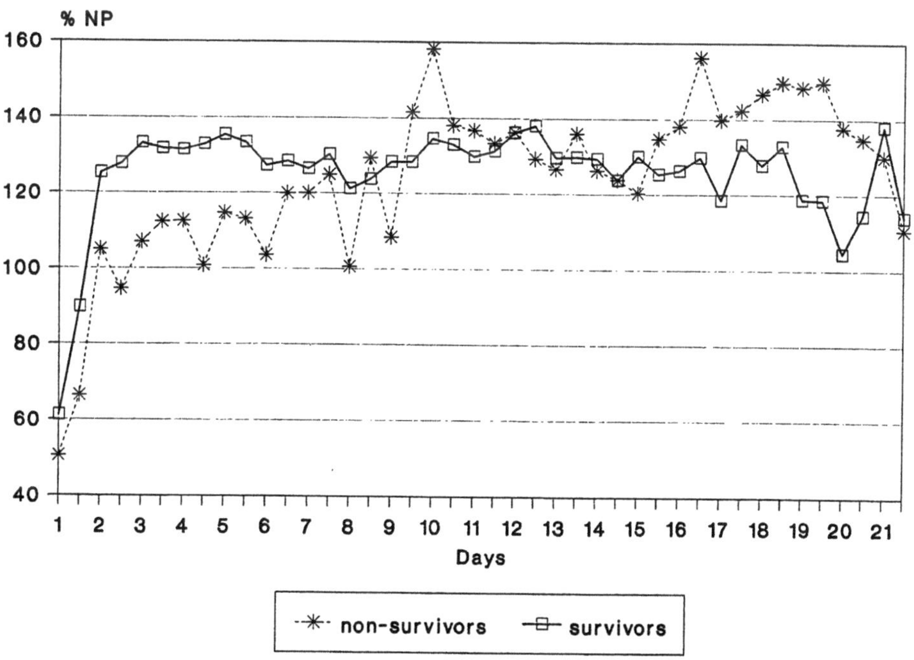

Fig. 11. Mean values of antithrombin III inhibitory activity in plasma of survivors (n = 6) and nonsurvivors (n = 9) of the AT III-treatment group

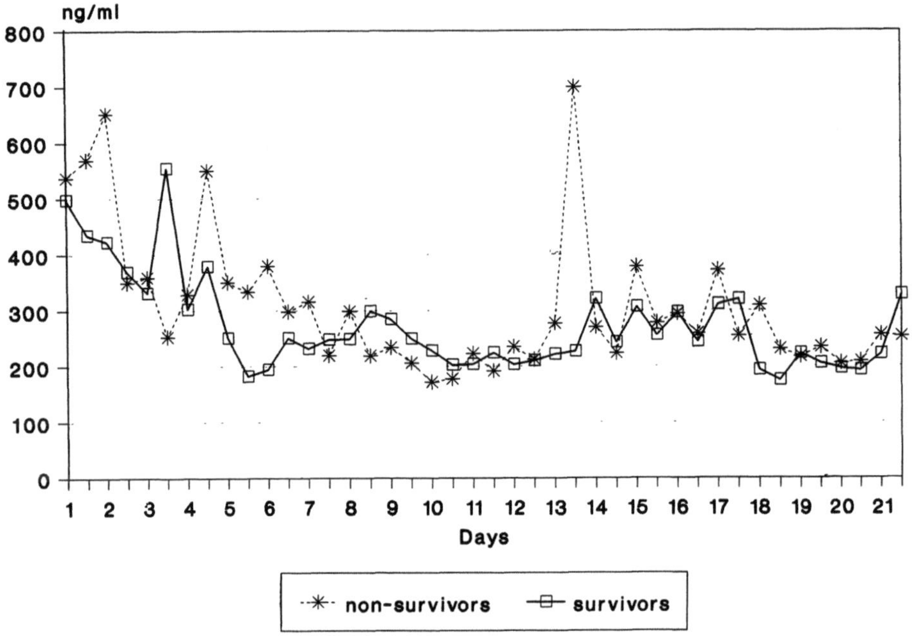

Fig. 12. Mean values of PMN elastase (in complex with α_1-proteinase inhibitor) in plasma of survivors (n = 6) and nonsurvivors (n = 9) of the AT III-treatment group

the first 5 days after sepsis diagnosis; Fig. 11) than patients who died despite AT III supplementation (mean AT III activity between 100% and 115% in the early septic phase). Thus, immediate AT III substitution in sufficiently high amounts after early diagnosis of a septic episode appears to be of great importance in improving the survival of the patients.

Probably due to a too late onset of the inhibition therapy and the application of still insufficiently high AT III dosages in some of our patients, the overall mortality could be reduced only from 87% in the control group to 60% in the AT III-treated collective. This diminution in lethality was statistically insignificant, yet a clear improvement of organ functions – especially lung, liver and kidney – in the treated patients as well as a further deterioration in the control group became evident. Mean plasma levels of complexed elastase were elevated up to sixfold that of normal upon admission and decreased gradually to about threefold in both groups until the end of the observation period. Although there were no statistically significant differences, a slight trend to lower plasma elastase levels appeared in the AT III group. Similar minute distinctions were seen in AT III-treated patients who survived compared to those who died (Fig. 12). Plasma prekallikrein and prothrombin levels were highly reduced up to 30 and 50%, respectively, upon admission and showed a more or less pronounced change to higher values (40 and 80% that of normal) later on without obvious differences in the placebo and treatment groups. Interestingly, however, prekallikrein levels rose steadily up to 50% in AT III-treated survivors, whereas in the non-

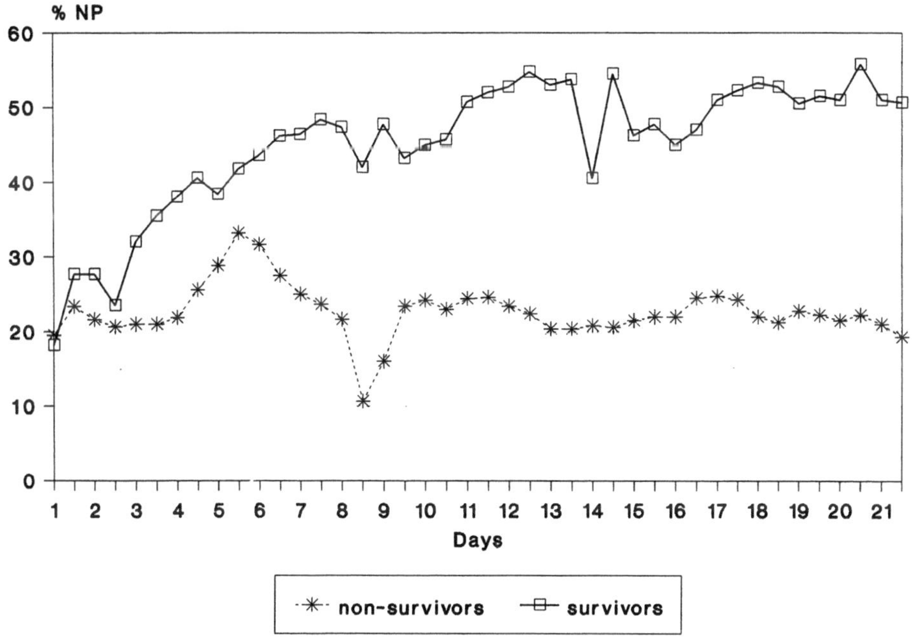

Fig. 13. Mean values of plasma prekallikrein (*PKK*) in plasma of survivors (n = 6) and nonsurvivors (n = 9) of the AT III-treatment group

survivors a further transient decrease up to 20% occurred (Fig. 13). Comparable behavior was observed for prothrombin which reached 80% of normal already 10 days after onset of sepsis in the AT III-treated survivors, while in the moribund patients plasma prothrombin levels as low as 60% were measured at that time. Thus, the rise in AT III inhibitory activity to nearly 140% in the early septic phase may have been beneficial enough to preserve clotting proenzymes from excessive activation throughout a septic period.

4 Conclusion

The given data, derived from clinical and experimental studies on acute inflammation, unequivocally indicate that proteolytic pathomechanisms play an important role in the onset and perpetuation of inflammatory processes like trauma- and sepsis-induced multiple organ dysfunctions. Therefore, the plasma levels of the regulatory proteinase inhibitors have to be elevated by suitable supplementation and kept well above the normal values to achieve significant improvement of the clinical situation. As the natural sources for the isolation of proteinase inhibitors from human material are very limited, the design of highly effective inhibitory proteins, on the basis of human inhibitor molecules by molecular modeling and their production by recombinant DNA technology, is the most promising approach at present to obtain the quantities necessary for proteinase inhibition therapy in the future (Fritz et al. 1991).

Acknowledgments. We are very grateful to the mentioned participating clinical colleagues for their intensive collaborations. Part of the work was financially supported by the Sonderforschungsbereich 207 of the University of Munich (grants G5 to M. J. and G1 to W. M.).

References

Assfalg-Machleidt I, Jochum M, Nast-Kolb D, Siebeck M, Billing A, Joka Th, Rothe G, Valet G, Zauner R, Scheuber HP, Machleidt W (1990) Cathepsin B – indicator for the release of lysosomal cysteine proteinases in severe trauma and inflammation. Biol Chem Hoppe-Seyler 371(Suppl):211–222

Billing A, Fröhlich D, Jochum M, Kortmann H (1988) Impaired phagocytosis in peritonitis exudate secondary to complement consumption. Surg Res Comm 3:335–345

Billing A, Fröhlich D, Jochum M, Kortmann H (1990) Deficient phagocytosis in abdominal sepsis: the influence of intraperitoneal substitution of opsonins – first results. Surg Res Comm 9:297–302

Billing A, Fröhlich D, Assfalg-Machleidt I, Machleidt W, Jochum M (1991) Proteolysis of defensive proteins in peritonitis exudate: pathobiochemic aspects and therapeutic approach. Biomed Biochim Acta 50 (in press)

Blauhut B, Kramer H, Vinazzer H, Bergmann H (1985) Substitution of antithrombin III in shock and DIC. A randomized study. Thromb Res 39:81–89

Duswald K-H, Jochum M, Schramm W, Fritz H (1985) Released granulocytic elastase: an indicator of pathobiochemical alterations in septicemia after abdominal surgery. Surgery 98:892–898

Emerson TE, Fournel MA, Redens TB, Taylor FB (1989) Efficacy of antithrombin III supplementation in animal models of fulminant *Escherichia coli* endotoxemia or bacteremia. Am J Med 87(Suppl. 3B):27–33

Fritz H, Collins J, Jochum M (1991) Proteinase inhibitor candidates for therapy of enzyme-inhibitor imbalances. In: Grassi C, Travis J, Casali L, Luisetti M (eds) Current concepts in the biochemistry of pulmonary emphysema. Springer, Berlin Heidelberg New York, Bi & Gi, Verona Publishers, London (in press)

Gippner-Steppert C (1991) Entwicklung eines spezifischen Testsystems für den Nachweis der Bildung eines proteolytischen Spaltproduktes des Fibrinogens durch lysosomale PMN-Elastase sowie Untersuchungen am Miniplasminogen, einem Elastase-spezifischen Spaltprodukt des Plasminogens. Dissertation, Fakultät für Chemie, Biologie und Geowissenschaften der Technischen Universität München

Hoffmann H, Siebeck M, Spannagl M, Weipert J, Geiger R, Jochum M, Fritz H (1990) Effect of recombinant hirudin, a specific inhibitor of thrombin, on endotoxin-induced intravascular coagulation and acute lung injury in pigs. Am Rev Respir Dis 142:782–788

Idell S, Kucich U, Fein A, Kueppers F, James HL, Walsch PN, Weinbaum G, Colman RW, Cohen AB (1985) Neutrophil elastase-releasing factors in bronchoalveolar lavage from patients with adult respiratory distress syndrome. Am Rev Respir Dis 132:1098–1105

Inthorn D, Jochum M (1988) Auswirkungen chirurgischer Infektionen auf die Stimulierbarkeit zur Chemilumineszenz von Granulozyten und die Freisetzung granulozytärer Elastase. In: Häring R (Hrsg) Risiko in der Chirurgie. Analyse und Kalkulation. Walter de Gruyter, Berlin, pp 219–224

Jochum M (1988) Lysosomale Faktoren aus polymorphkernigen Granulozyten: Pathobiochemische, diagnostische und therapeutische Aspekte. Habilitationsschrift an der Medizinischen Fakultät der Ludwig-Maximilians-Universität München

Jochum M (1991) Specific proteins of inflammatory cells and α_1-proteinase inhibitor in alveolar epithelial lining fluid of polytraumatized patients: do they indicate posttraumatic lung failure? In: Sturm JA (ed) Posttraumatic acute respiratory distress syndrome. Springer, Berlin Heidelberg New York, pp 193–211

Jochum M, Fritz H (1989) Pathobiochemical mechanisms in inflammation. In: Faist E, Ninnemann JL, Green DR (eds) Immune consequences of trauma, shock and sepsis. Springer, Berlin Heidelberg New York, pp 165–172

Jochum M, Fritz H (1990) Elastase and its inhibitors in intensive care medicine. Biomed Prog 3:55–59

Jochum M, Lander S, Heimburger N, Fritz H (1981) Effect of human granulocytic elastase on isolated human antithrombin III. Hoppe Seyler's Z Physiol Chem 362:103–112

Jochum M, Welter HF, Siebeck M, Fritz H (1987) Proteinase inhibitor therapy of severe inflammation in pigs. First results with eglin, a potent inhibitor of granulocyte elastase and cathepsin G. In: Taylor JC, Mittmann C (eds) Pulmonary emphysema and proteolysis. Academic Press, Orlando, pp 85–90

Jochum M, Assfalg-Machleidt I, Inthorn D, Nast-Kolb D, Waydhas Ch, Fritz H (1990) Leukozytäre Proteinasen und Hämostasestörung bei der Sepsis. In: Matthias R, Tilsner V (Hrsg) XXXII. Hamburger Symposium über Blutgerinnung: Infektion, Entzündung und Blutgerinnung. Editiones Roche, Basel, pp 241–254

Jochum M, Inthorn D, Nast-Kolb D, Fritz H (1991) AT III – ein neues therapeutisches Konzept bei der Behandlung der Sepsis und beim Organversagen? In: Henschel WF (Hrsg) Blut, Blutkomponenten und Blutersatzstoffe in der Intensivmedizin. Bericht über das 10. Bremer Interdisziplinäre Intensivtherapie-Colloquium. W. Zuckschwerdt-Verlag, München pp 46–58

Jordan RE, Nelson RM, Kilpatrick J, Newgren JO, Esmon PC, Fournel MA (1989) Antithrombin inactivation by neutrophil elastase requires heparin. Am J Med 87 (Suppl. 3B):19–22

Machleidt W, Assfalg-Machleidt I, Billing A, Fröhlich D, Joka Th, Nast-Kolb D (1991) The role of lysosomal cysteine proteinases as markers of macrophage activation and as non-specific mediators of inflammation. In: Proc 2nd Int. Congr on the Immune consequences of trauma, shock, and sepsis. Springer, Berlin Heidelberg New York (in press)

Nast-Kolb D, Waydhas Ch, Jochum M, Spannagl M, Duswald KH, Schweiberer L (1990) Günstigster Zeitpunkt für die Versorgung von Femurschaftfrakturen beim Polytrauma? Chirurg 61:259–265

Neumann S, Jochum M (1984) Elastase-α_1-proteinase inhibitor complex. In: Bergmeyer HU, Bergmeyer J, Graßl M (eds) Methods of enzymatic analysis, 3rd edn, vol 5. Verlag Chemie, Weinheim, pp 184–195

Obertacke U, Joka Th, Reuter M, Schmit-Neuerburg KP (1991) Bronchoalveolar lavage. In: Sturm, JA (ed) Adult respiratory distress syndrome. Springer, Berlin Heidelberg New York, pp 17–21

Schraufstatter I, Revak SD, Cochrane CG (1984) Biochemical factors in pulmonary inflammatory disease. Fed Proc 43:2807–2810

Siebeck M, Hoffmann H, Jochum M, Fritz H (1989a) Inhibition of proteinases with recombinant eglin c during experimental *Escherichia coli* septicemia in the pig. Eur Surg Res 21:11–17

Siebeck M, Hoffmann H, Weipert J, Spannagl M (1989b) Therapeutic effects of the combination of two proteinase inhibitors in endotoxin shock of the pig. In: Schlag G, Redl H (eds) Progress in clinical and biological research, vol 308. Second Vienna Shock Forum. Alan R. Liss, New York, pp 937–943

Spannagl M, Hoffmann H, Siebeck M, Weipert J, Schwartz HP, Schramm W (1991) A purified antithrombin III-heparin complex as a potent inhibitor of thrombin in porcine endotoxin shock. Thromb Res 61:1–10

Sturm JA (ed) (1991) Adult respiratory distress syndrome. An aspect of multiple organ failure. Results of a prospective clinical study. Springer, Berlin Heidelberg New York

Vinazzer H (1987) Clinical use of antithrombin III concentrates. Vox Sang 53:193–198

Wachtfogel YT, Kucich U, James HL, Scott CF, Schapira M, Zimmerman M, Cohen A, Colman RW (1983) Human plasma kallikrein releases neutrophil elastase during blood coagulation. J Clin Invest 72:1672–1677

Wachtfogel YT, Pixley RA, Kucich U, Abrams W, Weinbaum G, Schapira M, Colman RW (1985) Purified plasma factor XIIa aggregates human neutrophils and releases elastase. Circulation 70(Suppl. II):352

Waydhas Ch, Nast-Kolb D, Jochum M, Trupka A, Lenk S, Fritz H, Duswald KH, Schweiberer L (1991) Inflammatory mediators, infection, sepsis, and multiorgan failure after severe trauma. Arch Surg (in press)

Weitz JI, Landmann SL, Crowley KA, Birken S, Morgan FJ (1986) Development of an assay for in vivo human neutrophil elastase activity. J Clin Invest 78:155–162

Wewers MD, Herzyk DJ, Gadek JE (1988) Alveolar fluid neutrophil elastase activity in the adult respiratory distress syndrome is complexed to alpha-2-macroglobulin. J Clin Invest 82:1260–1264

Complement Activation

K. B. M. REID[1]

1 Introduction

The complement system is one of the major effector mechanisms in the blood which is involved with the generation of inflammation and the killing and clearance of pathogens. The classical and alternative pathways of the system can clearly be activated by both antibody-dependent and antibody-independent mechanisms. Immune aggregates containing antibody IgM, or certain subclasses of antibody IgG, are known to be efficient activators of the C1q $C1r_2$ $C1s_2$ complex of the classical pathway (via the Fc regions of these antibodies). Antibody IgG, while not a prerequisite for alternative pathway activation, is known to have an enhancing effect [via the $(Fab')_2$ region] on the activation of this pathway. Non-antibody-dependent mechanisms which may be of some importance in complement activation and the generation of inflammation include several acute phase proteins such as C-reactive protein (CRP), and mannan-binding protein (MBP), which can efficiently activate the classical pathway of complement, after interaction with appropriate ligands. The finding that the lectin MBP in involved in complement activation is a relatively recent observation (Ikeda et al. 1987; Ohata et al. 1989; Lu et al. 1990) and there is already evidence indicating that the level of MBP in the serum may be of importance in dealing with pathogenic organisms in the very young (Super et al. 1989) or immunodeficient individuals. MBP contains distinct globular and collagen-like domains and therefore has an overall structural similarity to C1q and it has been observed that MBP, and several other lectins containing collagenous domains, can all bind to the C1q receptor (C1qR) (Malhotra et al. 1990): This type of interaction could be of considerable physiological importance since the C1qR is very widespread, being found on fibroblasts, platelets, endothelial cells and most leukocytes (Ghebrehiwet 1989) and it is considered that binding of ligands to the receptor mediates a wide range of phenomena including phagocytosis, antibody-dependent cell-mediated cytotoxicity, modulation of cytokine and immunoglobulin secretion and polymorph-endothelium interaction.

As regards alternative pathway activation, the complement component C3 is known to be an acute phase protein and is involved in the initiation of non-antibody-dependent activation of the alternative pathway. The spontaneous hydrolysis at a low level of the intramolecular thiol-ester bond present in C3 induces a conformational change in the hydrolyzed C3 such that it adopts a C3b-like conformation and function (Pangburn et al. 1981). The hydrolyzed C3, designated $C3(H_2O)$, can, in the presence

[1] MRC Immunochemistry Unit, Department of Biochemistry, University of Oxford, South Parks Road, Oxford, OX1 3QU, UK

42. Colloquium Mosbach 1991
Molecular Aspects of Inflammation

of B and D, form the C3 convertase C3(H_2O) Bb. Splitting of C3 by this initial convertase allows freshly hydrolyzed C3b to bind to suitable sites on host cells or invading foregin cells. If C3b binds to host cells, then a variety of regulatory mechniams, as outlined below, prevent formation of the C3bBb complex. If the C3b binds to bacterial cell surfaces, it may escape the regulatory mechanisms and be utilized in the formation of C3bBb which, in turn, greatly amplifies the activation of C3. This result in the target becoming coated with a large number of C3b molecules, which can be utilized in phagocytic processes, and the generation of inflammation via C3a and the other anaphylatoxins.

Complement-mediated inflammation is due primarily to C5a, C3a and C4a, the peptide (approx. 77 residues long) activation fragments derived from the N-terminal ends of the α-chains of C5, C3 and C4. These so-called anaphylatoxins induce mast cell degranulation, smooth muscle contraction and vascular permeability with the triggering of the C5a receptor being especially important in the chemoattraction of neutrophils and macrophages. The recent molecular cloning of the human C5a receptor (Gerard and Gerard 1991) has shown that it is a member of the rhodopsin superfamily and, like the formyl peptide receptor, probably coupled to regulatory GTP-binding proteins to form a high affinity binding complex which leads to signal transduction. The large membrane- and particle-bound C4b and C3b fragments and their breakdown products, derived from the activation of C4 and C3, interact with a variety of receptors which include complement receptors CR1, CR2 and CR3. CR1 binds both C4b and C3B and is found on a wide variety of cells, including red blood cells. On binding to C4b or C3b, CR1 can act as a cofactor for the enzyme factor I which splits C3b to iC3b and then to C3d, g. CR1 acts as an adherence receptor but does not play a major role in phagocytosis. CR2, which is found mainly on B-lymphocytes, binds weakly to iC3b but more strongly to C3d, g and C3d, and appears to be involved in the regulation of the immune response (Cooper et al. 1988). CR1 and CR2 provide good examples of the regulators, or cell surface receptors, in the complement system which are composed of multiple copies of an independently folding molecule of 60 amino acids known as the short consensus repeat (SCR). CR1 contains 30 of these modules, while CR2 contains 15/16 modules. The determination of the three-dimensional structure of one SCR, derived from the control protein factor H, could allow assessment of the molecular interactions between the regulatory proteins and the enzyme complexes of the complement system. The availability of this information and also cDNA clones for all these regulatory proteins/receptors should allow the protein engineering of molecules with the potential to increase the breakdown of C3b and C4b by factor I, thus allowing control of undesirable inflammatory activity generated by overactivation of the system. The potential for this approach has clearly been demonstrated by the generation of a soluble form of CR1 which has been found to be a potent inhibitor of the C3 and C5 converting enzyme complexes in solution and to act, in vivo, in the suppression of complement-mediated post-ischemic myocardial inflammation and necrosis (Weisman et al. 1990) and also in the inhibition of the reversed passive Arthus reaction in rats (Yeh et al. 1991). Complement receptor type 3, CR3, is a member of the integrin superfamily and along with LFA-1 and gp150,95 forms the β2 subfamily of integrins which share a common β-chain and have homologous α-chains. CR3 and gp150,95 are found on monocytes, macrophages and granulocytes and bind iC3b. CR3 also binds β-glucan and the interaction of CR3 with zymosan stimulates the respiratory burst. The importance of CR3 as an inflammatory receptor is reflected in that individuals with a genetic deficiency of CR3 (and gp150,95 and LFA-1) suffer from recurrent stapylococcal septicaemias due to the lack migration of monocytes or polymorphs to the inflammatory site.

The main initial objective of the host inflammatory response is to draw neutrophil and monocytes, acute phase proteins and certain metal ions to a site of injury or infection. However, in certain pathological conditions, overactivation of the response can result in undesirable inflammatory conditions and in host tissue destruction, involving the components of the membrane attack complex (C5b–C9) overcoming the restriction factors found on the surface of the host cells.

2 Non-Antibody-Dependent Activation of the Classical Pathway of Complement Involving C-Reactive Protein and Mannan-Binding Protein

The roles of antibody IgG, or IgM, in the binding and activation of the C1 complex of the classical pathway of complement are well documented (Schumaker et al. 1987). The precise site in the C_H2 domain Fc region of IgG which appears to be involved in binding to the globular heads of C1q, prior to the activation of the C1q–C1r_2 C1s_2 complex, is a $Glu_{(318)}$-X-$Lys_{(320)}$-Y-$Lys_{(322)}$ motif (Duncan and Winter 1988). A simple motif in the C_H3 domain of IgM, perhaps only involving two charged amino acid residues, appears to be the site of IgM-C1q interaction. The initiation of efficient classical pathway activation, and the generation of inflammatory processes from C3 and C5 activation are clearly not restricted to involvement of antibody since the two acute phase proteins CRP and MBP can both bypass this requirement for antibody. However, CRP still utilizes C1q to bring about activation of the C1r_2 C1s_2 proenzyme complex, while MBP can mimic the action of C1q.

CRP is composed of five identical, non-covalently linked, non-glycosylated, subunits, each of 24 000 kDa. It has a Ca^{+2}-dependent binding specificity for phosphocholine which is found in the cell wall and membrane of *Streptococcus pneumoniae*. It was shown by in vitro studies that CRP can activate the classical pathway after indication with pneumoccal C-polysaccharide (Kaplan and Volanakis 1974) or certain polycations (Siegel et al. 1975). Recently, utilizing CRP trimers prepared by cross-linking, Jiang et al. (1991) identified the site of CRP-C1q interaction as being at the junction where the collagen-like region of C1q merge into the globular heads and, moreover, they demonstrated that this site was distinct from the IgG-C1q interaction site considered to be located within the globular heads of C1q. As well as being involved in classical pathway activation. CRP has been shown to promote phagocytosis by polymorphonuclear leukocytes and thus has been implicated as playing a general role in host defence against certain bacterial infections. This view is supported by the finding that human CRP can be used to passively protect mice from lethal infection with *Streptococcus pnenumoniae* and that this protective effect, in the absence of antibody, was complement-dependent (Horowitz et al. 1987). Although levels of CRP can rise several hundred-fold during pneumococcal infection, the protein has a short half-life which is consistent with the view that high CRP levels during the early stage of the disease may be of importance in preventing a fatal outcome. The subsequent production of anticapsular antibodies for the elimination of the pathogen would be followed by a concomitant decrease in inflammation and CRP levels.

Like CRP, mannan-binding protein (MBP) appears to be an acute phase protein which can play an important role in non-antibody defence mechanisms during an inflammation reaction period. MBP is a Ca^{+2}-dependent C-type lectin (Drickamer 1988) with specificity for terminal non-reducing N-acetyglucosamine, mannose, fucose and glucose residues (Childs et al. 1989). Thus any bacteria, viruses or yeasts which bear mannose rich oligosaccharides are potential ligands for MBP. Although MBP was first characterized by extraction from liver (Kawasaki et al. 1978), it is clearly a plasma protein which appears to circulate as a mixture of trimers, tetramers, pentamers and hexamers of approx. 90-kDa structural unit (Lu et al. 1990). The hexameric form of MBP has an overall similarity to C1q since both molecules are composed of six globular heads each joined by collagen-like strands to a fibril-like central core. It is therefore perhaps not too surprising that after binding to appropriate carbohydrate ligands, MBP isolated from serum can mimic C1q function in terms of bringing about activation of the classical pathway of complement (Ikeda et al. 1987; Kawasaki et al. 1989; Lu et al. 1990; Ohta et al. 1990). Efficient activation of C4 and C3 leads to the deposition of their activated forms onto the cell surface of pathogenic bacterial recognized by the MBP. The bound C4b and C3b (or iC3b) is, in turn, recognized by surface receptors on phagocytic cells and results in opsonization of the pathogenic material (Kuhlman et al. 1989; Super et al. 1989, 1990). Although MBP is normally present at relatively low levels (70 µg/l) in human serum compared to the level of C1q (80 mg/l), it could play an important role in situations where specific antibody levels are low, especially if the MBP level is raised by an acute phase reaction. The evidence for MBP being an acute phase reactant comes from the observations of raised liver mRNA level and serum levels for MBP in certain stress and disease situations (Ezekovitz et al. 1988) and the finding of heat shock consensus elements and glucocorticoid responsive elements in the 5' region of the human MBP gene (Taylor et al. 1989). There is now convincing evidence, from a variety of studies, that MBP is of importance in complement-mediated non-immune defence mechanisms: activation of mannose-rich, gram-negative bacteria for phagocytosis by monocytes and polymorphonuclear leukocytes (Kuhlman et al. 1989); the neutralization of viral infection of cells by the apparent masking of a site on the virus which is used to interact with receptors on the target cell – as seen in the prevention of human immunodeficiency virus infection lymphocytes (Ezekovitz et al. 1989) and the inhibition of infection cells by influenza A H1 and H3 subtypes (Anders et al. 1990). Further clinical evidence for the possible importance of MBP in non-antibody-dependent defence mechanisms comes from the finding that a common opsonic deficiency present in about 5% of the population is due to very low levels of MBP since there is a clear correlation between serum MBP levels, classical pathway activation (measured by C4 consumption) and the ability to generate C3b opsonin (Super et al. 1989, 1990). The defect can be corrected, in vitro, by the addition of purified MBP and clinical studies have shown that a group of children with this opsonic defect all had low MBP levels and suffered from frequent upper respiratory tract and intestinal infections (Super et al. 1989).

Conglutinin is another lectin, with an overall similarity to C1q, which exhibits a leukocyte- and complement-dependent bactericidal activity (Fris-Christiansen et al. 1990) and this may reflect the general importance which serum lectins may have in the early stages of infection and inflammation. However, in contrast to MBP, conglutinin does not appear to be directly involved in $C1r_2C1s_2$ activation. The opsonic effects seen with conglutinin are dependent upon complement activation taking place at the C3b stage, then the conglutinin binds to carbohydrate which becomes exposed by the degradation of bound C3b to iC3b. It is possible that some of the opsonic effects, and triggering of other effector functions, are mediated by the collagen-like regions present in MBP and conglutinin. MBP and conglutinin both appear to bind efficiently to the widespread C1q receptor (Malhotra et al. 1990) and this could trigger a number of biological responses similar to those seen by interaction of C1q with the receptor, i.e., an increase in oxidative metabolism, modulation of Ig synthesis, enhanced phagocytosis and induction of cell-mediated cytotoxicity (Ghebrehiwet 1989).

3 Control of Complement Activation

Control of overactivation of the complement system, and therefore of the potential inflammatory responses and tissue damage generated by the system, is achieved by a wide variety of soluble proteins and cell surface molecules (Table 1). Since there are two distinct pathways of activation, and many activators of these two pathways, it is necessary, if the system is to be regulated for therapeutic purposes, to identify a regulator which has the potential to effectively control both classical and alternative pathway activation at the C3/C5 activation stages. The cell surface CR1 and decay-acceleration factor (DAF) molecules both play an inhibitory role in control of the C3 and C5 converting enzyme complexes of both pathways. Both molecules have the capacity to disrupt the enzyme complexes but only CR1 acts as a cofactor for the serum enzyme factor I which specifically cleaves and inactivates C4b and C3b. Thus, CR1 has the greater potential as an overall inhibitor of activation. The genes for the cell surface proteins CR1, CR2, DAF, membrane cofactor protein (MCP), and the serum proteins, C4b-binding protein and factor H, are encoded within the regulation of complement activation (RCA) gene cluster. Most of the members of this cluster of genes can serve as cofactors for factor I-mediated cleavage of C3b and/or C4b (CR1, MCP, C4bp and factor H) and as accelerators of the decay of the C3 convertase (CR1, DAF, C4bp and factor H). Although CR2 is a member or this gene cluster, and shows a binding affinity for C3d, C3dg and iC3b, it is not considered to have a cofactor or to accelerate decay function. CR1, CR2, C4bp, MCP and DAF are almost entirely composed of varying numbers of distinct protein modules each 60 amino acids long. In these RCA proteins this module is known as the 'short consensus repeat' (SCR) and within the 60 amino acid long consensus sequence, four, almost invariant, cysteines are found along with other highly conserved residues (Klickstein et al. 1987; Campbell et al. 1988). Each SCR is generally encoded by a discrete exon at the gene level

Table 1. Control of the complement system

(i) Plasma proteins	Specificity	Role
C1-inhibitor	C1r, C1s	Formation of covalent complex with both C1r and C1s and their removal from the C1 complex
C4b-binding protein	C4b	Accelerates decay of C4b2a and is cofactor for cleavage of C4b by I
Factor H	C3b	Accelerates decay of C3bBb and is cofactor for cleavage of C3b by I
Factor I	C4b, C3b	Protease which inactivates C4b and C3b with the aid of cofactors C4bp, H, CR1 and MCP
Anaphylatoxin inactivator	C3a, C4a, C5a	Carboxypeptidase which inactivates C3a, C4a and C5a by removal of C-terminal arginine in each case
S-protein/vitronectin	C5b-7	Up to three molecules of S-protein bind to C5b-7, thus preventing the complex from binding to cell surfaces
SP-40, 40 (clusterin)	C5b-7	Fluid phase inactivation of C5b-7
Properdin	C3bBb	Positive regulatory protein of the alternative pathway which stabilizes the C3/C5 convertases

(ii)	*Membrane*	*Specificity*	*Role*
	Complement receptor 1 (CR1)-CD 35	C3b, C4b	Regulation of C3b breakdown as cofactor I. Accelerates decay of C3/C5 convertases. Binding of immune complexes to erythrocytes, phagocytosis
	Complement receptor 2 (CR2)-CD 21	C3d, C3dg, iC3b	Regulation of B cell functions Epstein-Barr virus receptor
	Membrane cofactor proteine (MCP)-CD 46	C3b, C4b	Regulation of C3b breakdown as cofactor for I
	Decay-accelerating factor (DAF)-CD 55	C4b2a, C3bBb	Accelerates decay of C3/C5 convertases
	Complement receptor 3 (CR3)-CD 11b/CD18	iC3b	Phagocytosis
	Complement receptor 4 (CR4) (p150,95)-CD 11*c*/CD 18)	iC3b	Monocyte migration
	Homologous restriction factor/C8bp	C8, C9	Prevention of formation of the membrane attack complex on homologous cells
	CD59/Protectin	C8, C9	Inhibits formation of the membrane attack complex by preventing C9 insertion into the membrane and polymerization of C9

and is considered to fold independently with the four conserved framework cysteines, forming two intradomain disulphide bridges. The SCR module is quite a common structural feature since it occurs over 140 times in more than 20 extracellular mosaic proteins which includes 12 proteins of the complement system. In addition to the complement regulatory proteins, two more SCR modules are found in C2, factor B, C1r, C1s, C6 and C7. All these complement proteins are involved in binding C3, C4 or C5 (or the activated forms C3b, C4b and C5b), thus implying that SCRs (or at least certain SCRs) are involved in the binding of C3b, C4b and C5b. However, in complement proteins containing large numbers of these SCR modules, such as factor H (20 SCRs) and CR1 (30 SCRs), it is clear that the C3b/C4b binding sites are restricted to a relatively small number of modules, e. g. the C3b binding site in factor H is likely to be localized within repeats 3 to 6, while in CR1 not more than 6 of the 30 SCRs appear to be involved in the three C3b/C4b binding sites found in the molecule – with 2SCRs at each site (Klickstein et al. 1988). The three-dimensional structure of a typical SCR has been determined by use of two-dimensional ^{1}H nuclear magnetic resonance spectroscopy of the 16th repeating SCR module, of factor H, after expression of the module in yeast (Barlow et al. 1990; Norman et al. 1991). This has shown the SCR structure to be approx. 3.8 nm long and to have a β-sandwich arrangement with one free face made up of three β-strands, hydrogen bonded to form a triple-stranded region at its centre, while the other face is formed from two separate strands. This study, along with the analysis of a two SCR-module structure, is providing data for accurate molecular modelling of the complement control proteins containing SCRs and the assessment of the crucial residues involved in binding to C3b or C4b. Although this could lead to the development of peptides/small molecules designed to inhibit the C3 and C5 convertases, another promising route to provide therapeutically useful inhibitors of complement activation lies in the expression of discrete portions of the various regulatory proteins.

Weisman et al. (1990) appear to have taken the first definitive steps in the protein engineering of a complement inhibitor which can efficiently control the soluble C3 and C5 convertases. They expressed a soluble form of CR1 (designated sCR1) by transfection of CHO cells with a construct containing a modified form of the cDNA sequence encoding the A allotype of CR1. The modified form of cDNA contained a translational stop codon at the junction of the extracellular and transmembrane regions. Therefore, an sCR1 molecule of 1998 amino acid residues (approx. 200 kDa) identical with the cell surface form of CR1, except for the absence of the 67 C-terminal residues, was secreted into the cell culture medium. Use of the sCR1 in in vitro assays showed that it was a potent inhibitor of both classical and alternative pathway activation in whole serum as monitored by the dose-dependent inhibition of the C3a-des Arg and C5a-des Arg cleavage fragments derived from C3 and C5 activation (Weisman et al. 1990). A remarkable feature of this inhibitory effect of sCR1 was that the inhibitory concentrations of sCr1 used were approx. 100 times lower than the concentrations of the serum inhibitors, C4b-binding protein and factor H, which are normally found in serum. In vivo assays to test the potential anti-inflammatory properties of CR1 were carried out in rats which were subjected to transient myocardial ischemia with subsequent reperfusion. The test animals were given 1 mg of sCR1 immediately before occlusion of the left coronary artery for a period of 35 min. Examination of the nature and size of the myocardial infarct, 7 days later, showed that the size of the infarction was reduced by 44% by the prior treatment with sCR1 and, as judged by the lack of deposition of the terminal components C5b-9 along the endothelium of capillaries and venules in the treated animals, and that complement activation had been markedly reduced. It has also been shown that sCR1 can inhibit immune complex-mediated inflammation in vivo as judged by studies using a rat reversed passive Arthus reaction model (Yeh et al. 1991). The administration sCR1 at dermal sites reduced Arthus vasculitis and also reduces the deposition of C3 and C5b-9, thus indicating that inhibition of complement activation was an important feature of the reduction of the in vivo immune complex-induced tissue damage. Further protein engineering studies on the CR1 molecule may allow the expression of a much smaller molecule than sCR1, which still fully retains the inhibitory activities displayed by the normal cell surface bound CR1, and which may be useful as an anti-inflammatory agent.

References

Anders EM, Hartley CA, Jackson DC (1990) Bovine and mouse serum beta inhibitors of influenza A viruses are mannose-binding lectins. Proc Natl Acad Sci USA 87:4485–4489

Barlow PN, Baron M, Norman DG, Day AJ, Willis AC, Sim RB, Campbell ID (1991) Secondary structure of the complement control protein module by two-dimensional 1H NMR. Biochemistry 30:997–1004

Campbell RD, Law SKA, Reid KBM, Sim RB (1988) Structure, organization and regulation of the complement genes. Annu Rev Immunol 6:161–195

Childs RA, Drickamer K, Kawasaki T, Thiel S, Mizuochi T, Feizi T (1989) Neoglycolipids as probes of oligosaccharide recognition by recombinant and natural mannose-binding proteins of the rat and man. Biochem J 262:131–138

Cooper NR, Moore MD, Nemerow GR (1988) Immunobiology of CR2, the B lymphocyte receptor for Epstein-Barr virus and the C3d complement fragment. Annu Rev Immunol 6:85–114

Drickamer K (1988) Two distinct classes of carbohydrate-recognition domain in animal lectin. J Biol Chem 263:9557–9560

Duncan AR, Winter G (1988) The binding site for C1q on IgG. Nature (London) 332:738–740

Ezekowitz RAB, Day LE, Herman GA (1988) A human mannose-binding protein is an acute-phase reactant that shares sequence homology with other vertebrate lectins. J Exp Med 167:1034

Ezekowitz RAB, Kuhlman M, Groopman JE, Byron RA (1989) A human serum mannose-binding protein inhibits in vitro infection by human immunodeficiency virus. J Exp Med 169:185–196

Friis-Christiansen P, Thiel S, Svehag SE, Dessau R, Svendsen P, Andersen O, Laursen SB, Jensenius JC (1990) In vivo and in vitro antibacterial activity of conglutinin, a mammalian plasma lectin. Scand J Immunol 31:453–460

Gerard NP, Gerard C (1991) The chemotactic receptor for human C5a anaphlatoxin. Nature (London) 349:614–617

Ghebrehiwet B (1989) Functions associated with the C1q receptor. Behring Inst Mitt 84:204–215

Horowitz J, Volanakis JE, Briles DE (1987) Blood clearance of *Streptococcus pneumoniae* by C-reactive protein. J Immunol 138:2598–2603

Ikeda K, Sannoh T, Kawasaki N, Kawasaki T, Yamashina I (1987) Serum lectin with known structure activates complement through the classical pathway. J Biol Chem 262:7451–7454

Jiang H, Siegel JN, Gewurz H (1991) Binding and complement activation by C-reactive protein via the collagen-like region of C1q and inhibition of these reactions by monoclonal antibodies to C-reactive protein and C1q. J Immunol 146:2324–2330

Kaplan MH, Vjolanakis JE (1974) Interaction of C-reactive protein complexes with the complement system. Consumption of human complement associated with the reaction of C-reactive protein with pneumococcal C-polysaccharide and with the choline phosphatides, lecithin and spingonyelin. J Immunol 112:2135–2141

Kawasaki N, Kawasaki T, Yamashina I (1989) A serum lectin (mannan-binding protein) has complement-dependent bactericidal activity. J Biochem (Tokyo) 106:483–489

Kawasaki T, Etoh R, Yamashina I (1978) Isolation and characterisation of mannan-binding protein from rabbit liver. Biochem Biophys Res Commun 81:1018

Klickstein LB, Wong WW, Smith JA, Weis JH, Wilson JG, Fearon DT (1987) Human C3b/C4b receptor (CR1). Demonstration of long homologous repeating domains that are composed of the short consensus repeats characteristic of C3/C4 binding proteins. J Exp Med 165:1095–1112

Klickstein LB, Bartow TJ, Miletic V, Rabson LD, Smith JA, Fearon DT (1988) Identification of distinct C3b and C4b recognition sites in the human C3b/C4b receptor (CR1, CD35) by deletion mutagenesis. J Exp Med 168:1699–1717

Kuhlman M, Joiner K, Ezekowitz RA (1989) The human mannose-binding protein functions as an opsonin. J Exp Med 169:1733–1745

Lu J, Thiel S, Wiedemann H, Timpl R, Reid KBM (1990) Binding of the pentamer/hexamer forms of mannan-binding protein to zymosan activates the proenzyme C1r2 C1s2 complex, of the classical pathway of complement, without involvement of C1q. J Immunol 144:2287–2294

Malhotra R, Thiel S, Reid KBM, Sim RB (1990) Human leukocyte C1q receptor binds other soluble proteins with collagen domains. J Exp Med 172:955–959

Norman DG, Barlow PN, Baron M, Day AJ, Sim RB, Campbell ID (1991) The three-dimensional structure of a complement control protein module in solution. J Mol Biol 219:717–725

Ohta M, Okada M, Yamashina I, Kawasaki T (1990) The mechanism of carbohydrate mediated complement activation by the serum mannan binding protein. J Biol Chem 265:1980–1984

Pangburn MK, Schreiber RD, Muller-Eberhard HJ (1981) Formation of the initial C3 convertase of the alternative complement pathway. Acquisition of C3b-like activities by spontaneous hydrolysis of the putative thiolester in native C3. J Exp Med 154:856–862

Schumaker VN, Zavodsky P, Poon PH (1987) Activation of the first component of complement. Annu Rev Immunol 5:21–42

Siegel J, Osmand AP, Wilson MF, Gewurz H (1978) Interactions of C-reactive protein with the complement system II. C-reactive protein mediated consumption of complement by polylysine polymers and other polycations. J Exp Med 142:709–714

Super M, Thiel S, Lu J, Levinsky J, Turner MW (1989) Association of low levels of mannan binding protein with a common defect of opsonisation. Lancet 2:1236–1239

Super M, Levinsky RJ, Turner MW (1990) The level of mannan-binding protein regulates the binding of complement-derived opsonins to mannan and zymosan at low serum concentrations. Clin Exp Immunol 79:144–150

Taylor ME, Brickell PM, Craig RK, Summerfield JA (1989) Structure and evolutionary origin of the gene encoding a human serum mannose-binding protein. Biochem J 262:763

Weisman HF, Bartow T, Leppo MK, Marsh HCJ, Carson GR, Concino MF, Boyle MP, Roux KH, Weisfeldt ML, Fearon DT (1990) Soluble human complement receptor type 1: in vivo inhibitor of complement suppressing post-ischemic myocardial inflammation and necrosis. Science 249:146–151

Yeh GC, Marsh HC, Carson GR, Berman L, Concino MF, Scesney SM, Kuestner RE, Skibbens R, Donahue KA, Ip SH (1991) Recombinant soluble human complement receptor type 1 inhibits inflammation in the reversed passive Arthus reaction in rats. J Immunol 146:250–256

Proteinase-Protein Inhibitor Interaction

W. BODE[1] and R. HUBER[1]

1 Introduction

Protein inhibitors are important in regulating the proteolytic activity of their target proteinases, in blocking these in emergency cases, or in signaling receptor interactions or clearance. Endogenous inhibitors appear to be always proteins; only in microorganisms are small inhibitors produced which block the proteolytic activity of host proteases.

The majority of protein inhibitors known and characterized so far are directed toward serine proteinases. Recently, a large number of protein inhibitors of cysteine proteinases have been discovered and characterized [1]. In contrast, only a few protein inhibitors directed toward metallo-proteinases or aspartyl proteinases are known to date. The α_2-macroglobulin family presents an exception, in that these proteins can inhibit each of these proteinases by virtue of a promiscuous "bait region".

Until recently, X-ray crystal structures were available of only a few serine proteinase inhibitors, one carboxypeptidase inhibitor, and some of their complexes with cognate proteinases, reviewed by Read and James [2]. Since 1986, further proteinase inhibitor-related crystal structures have been determined, in particular some serine proteinase inhibitors of hitherto unknown folding [3, 4, 5, 6, 7, 8], and the first two cysteine proteinase inhibitors [9, 10]. A new aspect is the appearance of inhibitor structures elucidated by two-dimensional NMR methods; until now, however, no NMR structures of an isolated protein inhibitor have been reported and no X-ray structures available.

In this chapter we shall attempt to elucidate the characteristic properties confering inhibitory activity to proteins. Several diverse approaches are available in nature to achieve proteinase inhibition. This is particularly illustrated by some recently published structures [7, 10]. In the interest of brevity, it is not possible to present all relevant results; thus, the selection of examples and emphasis on certain features must be to a certain extent subjective.

[1] Max-Planck-Institut für Biochemie, D-8033 Martinsried, FRG

42. Colloquium Mosbach 1991
Molecular Aspects of Inflammation

2 Protein Inhibitors of Serine Proteinases

These inhibitors can be grouped into various families [11] based on sequence homology, topological similarity, and the mechanism of binding. By far the largest group of inhibitors reacts with cognate enzymes according to a common, substrate-like "standard mechanism" [11, 12]. This group comprises relatively "small" proteins (or protein domains of multiheaded inhibitors) of between 29 and about 180 amino acid residues. They all possess an exposed binding loop of a characteristic "canonical" conformation, but are otherwise quite unrelated in structure.

The serpins (*Ser*ine *P*roteinase *In*hibitors) form a family of quite homologous, large (glyco)proteins comprising about 400 amino acid residues (see [13]). Most members exhibit inhibitory activity, exclusively toward serine proteinases; only a few (e. g., ovalbumin) apparently lack inhibitory properties and have other functions.

Hirudin binds according to a very different mechanism [7, 8]. This noncanonical interaction probably occurs more frequently in nature, but has been so far defined in hirudin only.

2.1 The "Small" Serine Proteinase Protein Inhibitors

The molecular frameworks of inhibitors belonging to different families do not possess a common folding motif. These inhibitors all exhibit, however, a relatively compact shape and an (often disulfide-interconnected) hydrophobic core. The stability of the native inhibitor domains toward unfolding is generally high. The domain segments form a supporting scaffold for the exposed proteinase binding loop. "Injuries" of the scaffold (such as main chain cleavage) can considerably reduce the melting temperature (M. Laskowski, pers. comm.). This loop, spanned between scaffold-anchored "pillars", has a flat shape that fits into the active-site clefts of cognate serine proteinases.

The residues of this proteinase binding loop flanking the scissile peptide bond exhibit a quite characteristic conformation from P3 to P3' [14]: the main chain conformation and angles of antiparallel β-strands at P3, of polyproline II at P2 and P1', of an approximate 3_{10}-helix at P1, and of parallel β-strands at P2' and P3'. The same inhibitors analyzed in different crystal environments [15, 16, 17, 18] and/or by two-dimensional NMR techniques [19, 20, 21] exhibit similar conformations. Thus, this common canonical conformation, presumed also to be attained by a productively bound substrate, is an inherent property of the inhibitors themselves.

The exposed (often remarkably hydrophobic) binding loop is further stabilized in all inhibitors by additional interactions between residues flanking the reactive site and the inhibitor core (see Fig. 1A). In most inhibitors a cysteine residue at P2 (small Kunitz inhibitors; mucous proteinase inhibitor [4]) or P3 (Kazal-, Bowman-Birk-, squash seed and potato-II inhibitors) is disulfide-connected with the hydrophobic core. In addition, side chain "spacers" and inter main chain hydrogen bonds clamp the loop to the main body. In some inhibitors (potato-I inhibitors; large Kunitz inhibitors) stabilization of the binding loop is achieved instead by an elaborate electrostatic/hydrogen bond network through side chains extending from the core to the binding loop (particularly well illustrated by the two parallel arginine side chains of potato-I inhibitors [14, 18]). Scaffold and loop together form a single cooperative unit; amino acid replacements, in particular at spacer residues, can considerably weaken the whole molecule [16, 22, 23].

2.2.1 The "Canonical" Interaction

These "small" inhibitors (I) interact with cognate enzymes (E) according to the following minimal scheme [24, 25]

$$E + I \underset{koff}{\overset{kon}{\rightleftharpoons}} EI \underset{kon^*}{\overset{koff^*}{\rightleftharpoons}} E + I^*$$

under rapid formation of (quite stable) complexes (EI), which can dissociate (mostly very slowly) into free enzymes and unmodified inhibitors or modified inhibitors (I^*); the latter are specifically cleaved at the scissile peptide bond P1-P1' and have similar free energy as the native inhibitors [26].

In complexes, the inhibitor reactive-site loops bind across the catalytic residues of their cognate proteinases, in a manner presumably similar to that of productively bound substrates (see Fig. 1A). In the case of trypsin-like proteinases, the segment amino-terminal to the scissile bond forms an antiparallel β-pleated sheet structure through main chain-main chain hydrogen bonds at P3 (a three-stranded antiparallel β-sheet is formed in subtilisin complexes [14, 18]). The reactive site of the inhibitor is close to the catalytic residue of the proteinase, with its P1 carbonyl carbon fixed in "sub-van der Waals" contact with Ser195 O_γ (typically around 2.7 Å [27]). The scissile peptide bond remains intact; these inhibitor complexes therefore resemble distorted Michaelis complexes (see Fig. 1A).

Several of the exposed side chains of the (8 to 12) contacting loop residues around the reactive site (between P9 and P4') have manifold (mostly hydrophobic) interactions with proteinase subsites (comprising 15 to 25 residues). In trypsin-like proteinase complexes, the interactions of the P1-residue with the "specificity pocket" (S1 subsite, see Fig. 1A) are most important; the side chain contacts of loop residues P2, P1', and P2' with their more shallow opposing subsites rank second. The almost independent arrangement of side chains explains the usually observed additivity of individual binding contributions to overall binding. This in turn allows reliable affinity predictions for optional loop sequences from the knowledge of only a few parameters (see the "sequence-function algorithm" of M. Laskowski [28]).

Besides these "primary" interactions, most inhibitors contact their cognate proteinases additionally by a "secondary" binding segment up to four residues long. These contacts (like those of more peripheral loop residues) are, however, in most cases not very specific, and are apparently not important for tightness of binding. In the cases of proteinases with quite narrow active-site clefts (e.g., thrombin [29]) these "secondary" elements can, however, cause severe steric hindrance and thus prevent complex formation.

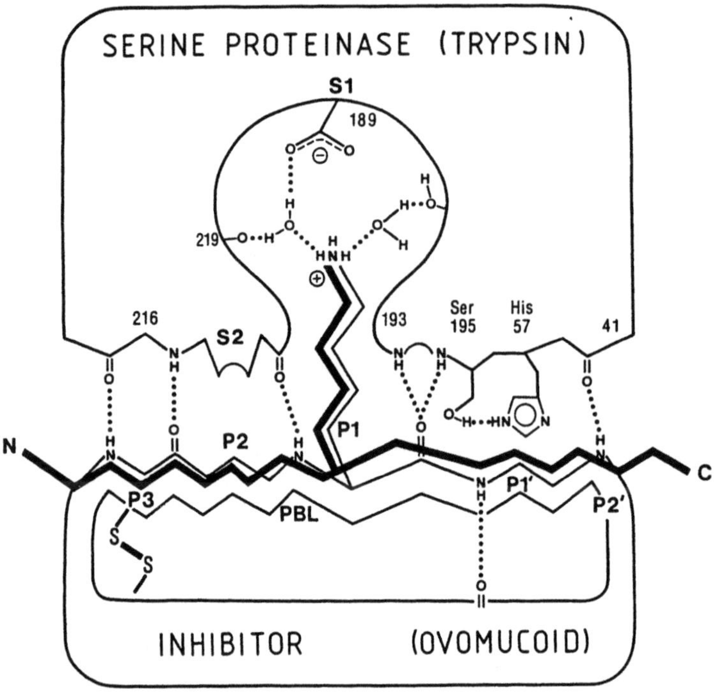
SERINE PROTEINASE (TRYPSIN)
S1
189
219
216
S2
193
Ser 195
His 57
41
N
P2
P1
C
P3
PBL
P1′
P2′
INHIBITOR (OVOMUCOID)

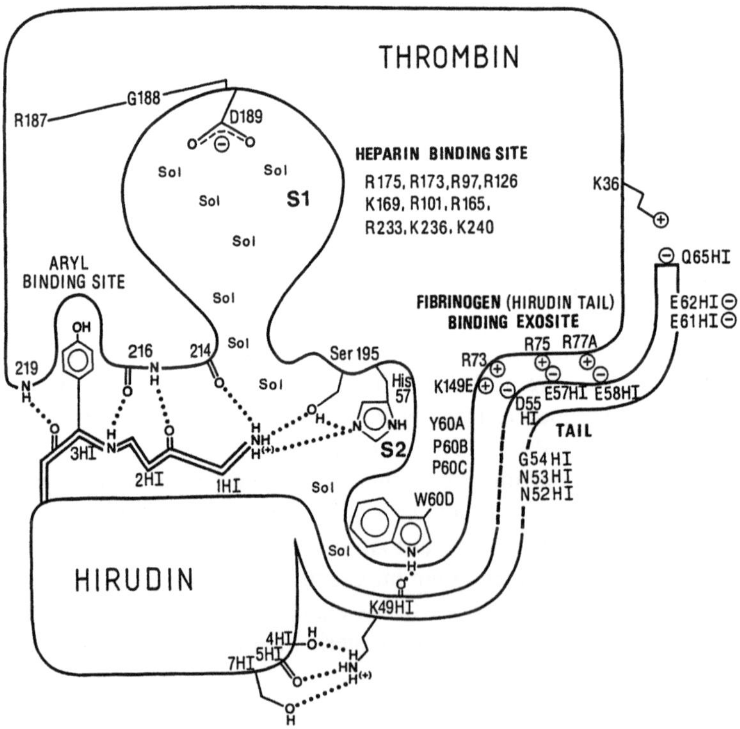
THROMBIN
R187
G188
D189
HEPARIN BINDING SITE
R175, R173, R97, R126
K169, R101, R165,
R233, K236, K240
K36
S1
ARYL BINDING SITE
Q65HI
FIBRINOGEN (HIRUDIN TAIL) BINDING EXOSITE
E62HI
E61HI
219
216
214
Ser 195
His 57
R75 R77A
R73
K149E
D55
E57HI E58HI
TAIL
Y60A
P60B
P60C
G54 HI
N53 HI
N52 HI
S2
3HI
2HI
1HI
W60D
HIRUDIN
K49HI
4HI
5HI
7HI

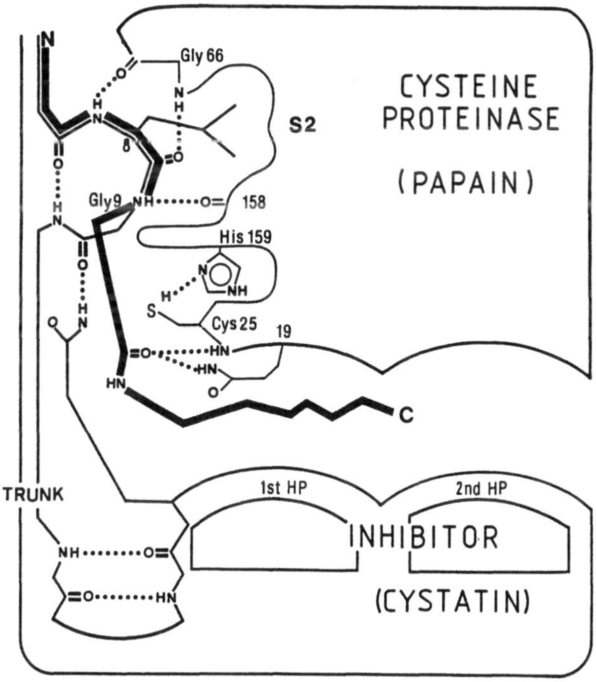

Fig. 1. Comparison of the canonical inhibition mode (**A**) with those of hirudin (**B**) and cystatin (C). *P1, P2* and *P1', P2'* designate substrate/inhibitor residues, amino- and carboxy-terminal to the scissile peptide bond, respectively; *S1, S2* and *S1', S2'* the corresponding subsites of the cognate proteinase. **A** Schematic representation of a trypsin-ovomucoid inhibitor complex. the canonical binding loop (*PBL*) of the inhibitor is spanned between scaffold-anchored pillars and further connected to its molecular core through a disulfide bridge and other spacer elements. It is bound in a similar manner to that of a substrate (*bold lines*) to the proteinase-binding site through several main chain-main chain hydrogen bonds. The catalytic Ser195 Oγ of trypsin is in "sub-van der Waals" contact to the P1 carbonyl carbon of the scissile peptide bond. **B** Schematic representation of the thrombin-hirudin complex. The three amino-terminal residues (*1HI – 3HI*) align with thrombin segment Ser214-Gly219 through parallel β-sheet formation. The reactive Ser195 is not blocked, and the specificity pocket is filled with water molecules. The carboxy-terminal tail loops around the characteristic thrombin insertion loop (Tyr60A-Trp60D) and binds to the fibrinogen-binding exosite of thrombin via several salt bridges. C Schematic representation of the cystatin-papain complex. Cystatin binds largly via two hairpin loops (*1stHP, 2ndHP*) to papain surface areas adjacent to the catalytic residues. The amino-terminal "trunk" areas are over the reactive Cys25, but interact via residues 8 and Gly9 with subsite S2 in a manner similar to a productively bound substrate, bold structure

In the interfaces of the complexes, the amino acid residues of both components are as densely packed as in the interior of proteins or in amino acid crystals [30]. The contact surfaces of *native* inhibitors and their cognate proteinases are essentially complementary; complex formation is accompanied by only slight conformational rearrangements (with rms deviations in the order of 0.35 Å), but (except for BPTI, the basic bovine pancreatic trypsin inhibitor) [15] result in considerable "freezing" of the binding loop. The intermolecular contacts are restricted to relatively small surface strips (of 2 x 600 up to 2 x 900 $Å^2$) on both components, i.e., inhibitor loop and binding cleft do not interdigitate in a tight "mortise and tenon" manner. The residual spaces left between inhibitor loop and cleft rims are filled with (often localized) solvent molecules. In various eglin c-complexes, a "rolling" motion of the inhibitor around its central binding loop "axis" relative to the enzyme's binding site has been observed [31], although binding loop interactions are maintained.

This surface complementarity in part explains the tightness of these complexes: the rigid inhibitor does not lose as much conformational freedom upon binding as a flexible substrate; it might also contribute to the fast association reaction believed to be necessary for many physiologically important blocking reactions. However, modified "small" inhibitors, specifically cleaved at their reactive site, bind with similar affinity as and sometimes even with an almost equivalent rate to their native compounds [2, 25, 26].

The structure analysis of two specifically cleaved ovomucoid inhibitors (Musil, Bode, Huber, and Laskowski, submitted) and preliminary NMR results [32] reveal that upon cleavage the residues adjacent to the cleaved bond become more flexible (resulting in a disorder of the binding site), whereas the remainder of the molecule maintains the same shape and conformation. The most stable complex formed upon association with a proteinase is, of course, also achieved with the uncleaved inhibitor: the packed binding loop of the inhibitor component is intact and quite rigid. The elaborate interactions with the subsites of the proteinase, and with the inhibitor's own core appear to be particularly tight. These two types of interaction mutually stabilize the binding loop in the complex, and confer thermodynamic stability to the complex; amino acid substitutions of inhibitor loop-core-spacers can seriously disturb the favourable interactions and weaken the complex (Wagner, Grütter and Heinz, pers. comm.).

These mutually stabilizing interactions apparently prevent the reactive site deformation presumed to be important for peptide bond cleavage, and thus slow down catalytic processing (a property which itself does not seem to be of great importance for inhibitor potency, but might rather be a side effect).

The inhibitor interaction with zymogens (in particular trypsinogen [12, 33]) provides evidence, that the mutual stabilization of inhibitor and proteinase allows complex formation even in the case of noncomplementarity of the reacting components. The substrate binding site of these proenzymes is organized quite differently to that of the activated proteinase and is often disordered. The reorganization of the "activation domain" [12, 33] upon inhibitor binding is similar to that found upon activation cleavage, presumably arising according to an induced-fit mechanism; the free energy needed to "force" the disordered trypsinogen segments into the well-fitting, ordered activation domain is provided in part by the free energy of binding with a concomitant reduction in affinity. This indicates again that not the structure of the single components, but rather that of the resulting complex confers stability [2]. This is further underlined by results showing that complexes with the same inhibitor (BPTI [25] or turkey ovomucoid [26]), but with different proteinases dissociate either slowly or rapidly (see [34]).

2.2 Serpins

To date, crystal structures of three specifically modified serpin inhibitors [α_1-proteinase inhibitor [35], antithrombin III [36], and α_1-antichymotrypsin (Baumann et al., submitted)] and of two ovalbumin species (modified ovalbumin – plakalbumin [37] and native ovalbumin [38]) are available. In the first three cleaved serpin species, the newly formed terminal segments S4A and S1C (see Fig. 2) are incorporated into two different β-pleated sheets A and C; the P1 and P1' residues are separated by a distance of 70 Å. The more compact structure explains the higher thermodynamic stability of the cleaved compared with native inhibitors [39].

In the two ovalbumin species, the peptide segment equivalent to the carboxy-terminal strand S4A does not insert to sheet A but is (in the native species) instead covalently connected to the incorporated strand S1C in a helical segment [38] (see Fig. 2; binding loops of serpins with inhibitory function are very probably nonhelical). In contrast to the above modified serpin structures, strands S1A, S2A, and S3A in both ovalbumin structures are annealed to the opposing strands S5A and S6A, forming a new contiguous five-stranded sheet A'. The main reason for the failure of strand S4A to incorporate into sheet A' in plakalbumin seems (besides some "slipping" restraint) to be steric hindrance through the bulky side chain of an arginine residue [37]; the equivalent Thr345 of α_1-proteinase inhibitor projects into the molecular body.

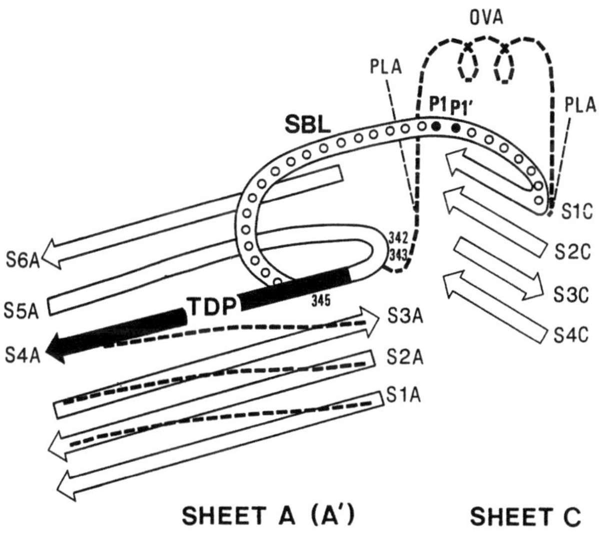

Fig. 2. Schematic representation of sheet *A* and *C* in serpins. *Open arrows* represent peptide strands as observed in cleaved inhibitors. In both ovalbumin structures strands S1A, S2A, and S3A (*dashed lines*) anneal with strand S5A; in plakalbumin the free ends (*PLA*) project from the surface, while in ovalbumin the connecting peptide (*OVA*) adopts a helical conformation. In functional native serpin inhibitors the serpin-binding loop (*SBL*) is presumed to be partially inserted in sheet A and this way spanned over the inhibitor surface with a canonical-like conformation. The functional native serpins would therefore appear to have structures midway between ovalbumin and the cleaved serpins. Upon insertion of an exogeneous tetradecapeptide (*black arrow*) the binding segment probably attains a more coiled conformation

Neither the loop structure of isolated inhibitory serpins nor the geometry of their complexes are yet known. Several lines of evidence (the susceptibility of residues P10 to P2' to proteolytic cleavage, see [13]; docking experiments with thrombin [40], which places the tightest constraints in serpin models [29]) suggest that serpins might interact with substrate-binding sites of cognate proteinases through a particularly flat, exposed binding loop of canonical conformation. To confer stability to the formed complex without ready cleavage, the reactive site of this binding loop requires support through side chains extending from the inhibitor core, possibly in a similar manner to that observed for eglin c [14]. In functional intact serpins, strand S4A (after passing the S5A-S4A tight turn, see Fig. 2) may reenter sheet A with a few (three to seven) residues before leaving the sheet and looping in an extended conformation over the inhibitor surface to merge with strand S1C (Fig. 2).

Evidence for this comes from very recent results in our laboratory [41], according to which intact human α_1-proteinase inhibitor is rendered inactive upon insertion of a tetradecapeptide sequentially identical to Thr345-Met358 (i.e., its strand S4A, see Fig. 2). In the latter binary complex, the integrated peptide obviously prevents the reentry or strand S4A; such a partial incorporation is presumably required for holding the loop segment in the canonical conformation and pulling it close to the inhibitor surface. The sequences around the active sites of serpins, in particular their P1 residues, match the requirements for canonical binding of their target proteinases. The detailed interaction, however, remains to be established.

2.4 Hirudin Interaction with Thrombin

Hirudin, a 65 or 66 amino acid residue protein [42], is an extremely tight binding and selective inhibitor of the coagulation protease thrombin. The structural analyses of two recombinant hirudins by two-dimensional NMR [43, 44] revealed that hirudin segment 3I-47I (except 31I–36I) forms a globular, compact domain, whereas the first two residues and the carboxy-terminal 18 residues are flexible in solution.

Recently, the X-ray crystal structure of two complexes formed between human α-thrombin and two slightly different recombinant hirudin variants have been solved at 2.3 Å [7] and 2.95 Å resolution [8]. In these complexes hirudin binds in an extended manner along the canyon-like active-site cleft of thrombin [29]. Its globular domain binds to characteristic thrombin surface patches adjacent to the thrombin active site (see Fig. 1B). The amino-terminal hirudin segment 1I–3I forms a parallel β-pleated sheet structure with thrombin segment Ser214-Gly219. This is in contrast to the antiparallel binding observed for the canonical proteinase inhibitors (cf. Fig. 1A, B). The reactive Ser195 of thrombin is not blocked; its specificity pocket is not used by hirudin residues, but filled with (structured) water molecules (see Fig. 1B).

The extended carboxy-terminal "tail" of hirudin (48I-65I, see Fig. 1B) runs in a long groove which extends from the active-site cleft of thrombin. Three nonpolar and several acidic side chains of this tail segment make numerous hydrophobic and a few surface salt-bridge contacts with this positively charged putative "fibrinogen secondary binding site". The major contribution to thrombin binding of these negatively charged residues of the hirudin tail has been demonstrated by systematic removal of charges using recombinant methods [45] and by recent electrostatic calculations in our laboratory (A. Karshikoff, unpubl. results). The surface area buried from bulk water by thrombin-hirudin complex formation (2 x 1800 Å^2) is considerably larger than that in case of the "small" serine proteinase inhibitors. The three hirudin regions (the amino-terminal segment, the globular domain, and the tail) participate in the intermolecular contact to similar degrees [7]. The globular domain (including the amino-terminal part) and the tail portion exhibit additive binding contributions, i.e., they bind independently of one another [46].

In approaching thrombin, hirudin might be preoriented by the complementary, strong electrostatic fields of each binding surface. This would then account for the almost diffusion-controlled association rate [45]. Kinetic data analyzed at varying ionic strengths indicate that in a first step the negatively charged hirudin tail associates (under rigidification) with the positively charged thrombin exosite, before the compact domain binds adjacent to thrombin's active site, through a favorable induced-fit of the amino-terminal segment (see Fig. 1B).

3 Cystatin – Cysteine Proteinase Interaction

Cystatins and stefins are tight, reversibly binding protein inhibitors of papain-like cysteine proteinases (see [1]). Recently, the X-ray crystal structures of two representatives, chicken egg white cystatin [9] and stefin B in complex with papain [10], have been elucidated. Cystatins/stefins consist of a long central α-helix, wrapped in a five-stranded antiparallel β-pleated sheet, with a subsidiary helix or strand, respectively. At one end of the sheet, an exposed "first" β-hairpin loop (comprising a highly conserved "QVVAG" or similar sequence) is flanked on both sides by the projecting amino-terminal segment and a second hairpin loop (see Fig. 1C). This wedge-shaped hydrophobic edge is complementary in shape to the active-site cleft of papain [9]. In the complex [9, 10] both cystatin loops interact with conserved primed subsites adjacent to the papain catalytic residues; the (initially flexible [47]) amino-terminal segment (the "trunk") loops over the catalytic Cys25 residue of papain and interacts via two more amino-terminal residues with the putative subsites S2 and S3. In contrast to bound substrates, this inhibitor "trunk" is removed from the catalytic residues in the complex and thus is not cleavable (see Fig. 1C).

The primarily hydrophobic side-chain interactions made by the rigid "first" hairpin loop (1stHP in Fig. 1C) confer most of the stability to the complex; in cystatin complexes, the interactions in the S2 subsite (see Fig. 1C) would appear to strengthen complexes with papain considerably [48].

4 Interaction of Carboxypeptidase A with Its Potato Inhibitor

No atomic spatial structure of any metallo-endoproteinase inhibitor is known to date. The X-ray crystal structure of a 39 amino acid residue protein inhibitor from potatoes has been determined in complex with carboxypeptidase A [49]. The projecting carboxy terminus of the potato inhibitor (the four-residue "primary contact region") is inserted into the active-site cleft of the enzyme; in the S1' subsite the carboxy-terminal residue, Gly39I, is split off, but remains buried by the rest of the residual inhibitor moiety; a few additional "secondary contacts" confer stabilization to the complex. This complex therefore represents a product stage in the catalytic mechanism.

In the isolated carboxypeptidase A-inhibitor determined by NMR [50], the projecting carboxy-terminal residue exhibits considerable flexibility. Upon binding to carboxypeptidase A it becomes ordered (presumably in a substrate-like manner), with Gly39I (which does not match the specificity requirements of carboxypeptidases) being slowly cleaved. The intermolecular contact made via its "primary" contact residues P1, P2, and P3 and the "secondary contact region" [49] keep the truncated inhibitor in position; they prevent fast dissociation as well as further intrusion into the active-site groove.

5 Conclusion

All protein inhibitors of proteinases prevent access of substrates to the proteinase's catalytic site through steric hindrance.

One class of inhibitors (the carboxypeptidase inhibitor, and most serine proteinase inhibitors) achieves this by binding with a peptide segment directly to the catalytic site in a substrate- or product-like manner. In the case of product-like binding the remaining interactions are strong enough to prevent fast dissociation; in the case of substrate-like binding, the intra- and intermolecular interactions of the inhibitor primary binding segment with the inhibitor core (through spacer elements) and with the enzyme binding site stabilize one another mutually. Inhibition selectivity is achieved simply through utilization of the proteinase substrate recognition subsites.

The second class of inhibitors (comprising the cysteine proteinase inhibitors and hirudin) binds mainly to surface sites adjacent to the catalytic residues of the cognate proteinase, utilizing substrate-binding facilities as well as other proteinase sites. If proteinase-specific surface patches are involved, the interaction may be very selective. Hirudin is an excellent example of this special case; obviously, nature has used this concept more often to achieve selectivity (see [51]).

Acknowledgements. We thank Dr. Milton Stubbs for carefully reviewing the manuscript, and Mrs. Karin Epp for help in preparing the figures. The financial support of the Sonderforschungsbereich 207 of the University of Munich (projects H-1 and H-2) and of the Fonds der Chemischen Industrie are acknowledged.

References

1. Barrett AJ, Rawlings ND, Davies ME, Machleidt W, Salvesen, G, Turk V (1986) Cysteine proteinase inhibitors of the cystatin superfamily. In: Barrett AJ, Salvesen G (eds) Proteinase inhibitors. Elsevier, Amsterdam, pp 515–569
2. Read RJ, James MNG (1986) Introduction to the protein inhibitors: X-ray crystallography. In: Barrett AJ, Salvesen G (eds) Proteinase inhibitors. Elsevier, Amsterdam, pp 301–336
3. Tsunogae Y, Tanaka I, Yamane T, Kikkawa J, Achida JT, Ishikawa C, Watanabe K, Nakamura S, Takahashi K (1986) Structure of the trypsin-binding domain of Bowman-Birk type protease inhibitor and its interaction with trypsin. J Biochem (Tokyo) 100:1737–1646
4. Grütter MG, Fendrich G, Huber R, Bode W (1988) The 2.5 Å X-ray crystal structure of the acid-stable proteinase inhibitor from human mucous secretions analyzed in its complex with bovine α–χηψμοτρψπσιν. EMBO J 7:345–351
5. Greenblatt HM, Ryan CA, James MNG (1989) Structure of the complex of *Streptomyces griseus* proteinase B and polypeptide chymotrypsin inhibitor-1 from Russet Burbank potato tubers at 2.1 Å resolution. J Mol Biol 205:201–225
6. Bode W, Greyling HJ, Huber R, Otlewski J, Wilusz T (1989) The refined 2.0 Å X-ray crystal structure of the complex formed between bovine β-trypsin and CMTI-I, a trypsin inhibitor from squash seeds *(Cucurbita maxima)*. Topological similarity of the squash seed inhibitors with the carboxypeptidase A inhibitor from potatoes. FEBS Lett. 242:285–292
7. Rydel TJ, Ravichandran KG, Tulinsky A, Bode W, Huber R, Roitsch C, Fenton JW (1990) The structure of a complex of recombinant hirudin and human α-thrombin. Science 249:277–280
8. Grütter MG, Priestle JP, Rahuel J, Grossenbacher H, Bode W, Hofsteenge J, Stone SR (1990) Crystal structure of the thrombin-hirudin complex: a novel mode of serine proteinase inhibition. EMBO J 9:2361–2365
9. Bode W, Engh R, Musil D, Thiele U, Huber R, Karshikov A, Brzin J, Kos J, Turk V (1988) The 2.0 Å X-ray crystal structure of chicken egg white cystatin and its possible mode of interaction with cysteine proteinases. EMBO J 7:2593–2599
10. Stubbs MT, Laber B, Bode W, Huber R, Jerala R, Lenarcic B, Turk V (1990) The refined 2.4 Å X-ray crystal structure of recombinant human stefin B in complex with the cysteine proteinase papain: a novel type of proteinase inhibitor interaction. EMBO J 9:1939–1947
11. Laskowski M, Kato I (1980) Protein Inhibitors of proteinases. Annu Rev Biochem 49:593–626
12. Huber R, Bode W (1978) Structural basis of the activation and action of trypsin. Acc Chem Res 11:114–122
13. Huber R, Carrell RW (1989) Implications of the three-dimensional structure of α_1-antitrypsin for structure and functions of serpins. Biochemistry 28:8951–8966
14. Bode W, Papamokos E, Musil D, Seemüller ?, Fritz H (1986) Refined 1.2 Å crystal structure of the complex formed between subtilisin Carlsberg and the inhibitor eglin c. Molecular structure of eglin and its detailed interaction with subtilisin. EMBO J 5:813–818
15. Wlodawer A, Deisenhofer J, Huber R (1987) Comparison of two highly refined structures of bovine pancreatic trypsin inhibitor. J Mol Biol 193:145–156
16. Eigenbrot C, Randal M, Kossiakoff AA (1990) Structural effects induced by removal of a disulfide-bridge. The X-ray structure of the C30A/C51A mutant of basic pancreatic trypsin inhibitor at 1.6 Å. Prot Eng 3:591–598
17. Papamokos E, Weber E, Bode W, Huber R, Empie MW, Kato I, Laskowski M (1983) Crystallographic refinement of Japanese quail ovomucoid, a Kazal-type inhibitor, and model building of complexes with serine Proteinases. J Mol Biol 158:515–537
18. McPhalen CA, Svendsen I, Jonassen I, James MNG (1985) Crystal and molecular structure of chymotrypsin inhibitor 2 from barley seeds in complex with subtilisin Novo. Proc Natl Acad Sci USA 82:7242–7246

19. Wagner G, Braun W, Havel TF, Schaumann T, Go N, Wüthrich K (1987) Protein structures in solution by nuclear magnetic resonance and distance geometry; the polypeptide fold of the basic pancreatic trypsin inhibitor determined using two different algorithms, DISGEO and DISMAN. J Mol Biol 196:611–639
20. Clore GM, Grönenborn AM, James MNG, Kjaer M, McPhalen CA, Poulsen FM (1987) Comparison of the solution and X-ray structures of barley serine proteinase inhibitor 2. Prot Eng 1:313–318
21. Holak TA, Bode W, Huber R, Otlewski J, Wilusz T (1989) Nuclear magnetic resonance solution and X-ray structures of squash trypsin inhibitor exhibit the same conformation of the proteinase binding loop. J Mol Biol 210:649–654
22. Jandu SK, Ray S, Brooks L, Leatherbarrow RJ (1990) Role of arginine 67 in the stabilization of chymotrypsin inhibitor 2: examination of amide proton exchange rates and denaturation thermodynamics of an engineered protein. Biochemistry 29:6264–6269
23. Goldenberg DP, Frieden RW, Haack JA, Morrison TB (1989) Mutational analysis of a protein-folding pathway. Nature (London) 338:127–132
24. Finkenstadt WR, Hamid MA, Mattis IA, Schrode I, Sealock RW, Wang D, Laskowski MJr (1974) Kinetics and thermodynamics of the interaction of proteinases with protein inhibitors. In: Fritz H, Tschesche H, Greene LJ, Truscheit E (eds) Proteinase inhibitors. Proc Int Res Conf – Bayer Symp V. Springer, Berlin Heidelberg New York, pp 389–411
25. Quast V, Engel J, Steffen E, Tschesche H, Kupfer S (1978) Stopped-flow kinetics of the resynthesis of the reactive site peptide bond in kallikrein inhibitor (Kunitz) by β–τρψπσιν. Biochemistry 17:1675–1682
26. Ardelt W, Laskowski M Jr (1985) Turkey ovomucoid third domain inhibits eight different serine proteinases of varied specificity on the same ... Leu^{18} – Glu^{19} ... reactive site. Biochemistry 24:5313–5320
27. Marquart M, Walter J, Deisenhofer J, Bode W, Huber R (1983) The geometry of the reactive site and of the peptide groups in trypsin, trypsinogen and its complexes with inhibitors. Acta Crystallogr Sect B39:480–490
28. Laskowski MJr, Kato I, Ardelt W, Cook J, Denton A, Empie MW, Kohr WJ, Park SJ, Parks K, Schatzley BL, Schoenberger OL, Tashiro M, Vichot G, Wheatley HE, Wieczorek A, Wieczorek M (1987) Ovomucoid third domains from 100 avian species: isolation, sequences, and hypervariability of enzyme-inhibitor contact residues. Biochemistry 26:202–221
29. Bode W, Mayr I, Baumann U, Huber R, Stone SR, Hofsteenge J (1989) The refined 1.9 Å crystal structure of human α-thrombin: interaction with D-Phr-Pro-Arg chloromethylketone and significance of the Tyr-Pro-Pro-Trp insertion segment. EMBO J 8:3467–3475
30. Janin J, Chothia C (1990) The structure of protein-protein recognition sites. J Biol Chem 265:16027–16030
31. Gros P, Betzel C, Dauter Z, Wilson KS, Hol WGJ (1989) Molecular dynamics refinement of a thermitase-eglin c complex at 1.98 Å resolution and comparison of two crystal forms that differ in calcium content. J Mol Biol 210:347–367
32. Rhyn GI, Markley JL (1988) Two-dimensional NMR studies of Kazal proteinase inhibitors 2. Sequence-specific assignments and secondary structure of reactive site modified turkey ovomucoid third domain. Biochemistry 27:2529–2539
33. Bode W (1979) The transition of bovine trypsinogen to a trypsin-like state upon strong ligand binding. II. The binding of the pancreatic trypsin inhibitor and of isoleucine-valine and of sequentially related peptides to trypsinogen and to p-guanidinobenzoate-trypsinogen. J Mol Biol 127:357–374
34. Estell DA, Laskowski MJr (1980) *Dermasterias imbricata* Trypsin 1: an enzyme which rapidly hydrolyzes the reactive-site peptide bonds of protein trypsin inhibitors. Biochemistry 19:124–131
35. Löbermann H, Tokuoka R, Deisenhofer J, Huber R (1984) Human α_1-proteinase inhibitor. Crystal structure analysis of two crystal modifications, molecular model and preliminary analysis of the implication for function. J Mol Biol 177:531–556

36. Delarue M, Samama JP, Moras D (1990) Crystal structure of bovine antithrombin III. Acta Crystallogr B46:550–556
37. Wright HT, Qian HX, Huber R (1990) Crystal structure of plakalbumin, a proteolytically nicked form of ovalbumin. Its relationship to the structure of cleaved α_1-proteinase inhibitor. J Mol Biol 213:513–528
38. Stein PE, Leslie AGW, Finch JT, Turnell WG, McLaughlin PJ, Carrell RW (1990) Crystal structure of ovalbumin as a model for the reactive centre of serpins. Nature (London) 347:99–102
39. Bruch M, Weiss V, Engel J (1988) Plasma serine proteinase inhibitors (serpins) exhibit major conformational changes and a large increase in conformational stability upon cleavage at their reactive sites. J Biol Chem. 263:16626–16630
40. Engh RA, Wright HT, Huber R (1990) Modeling the intact form of the α_1-proteinase inhibitor. Prot Eng 3:469©477
41. Schulze AJ, Baumann U, Knof S, Jaeger E, Huber R, Laurell CB (1991) Structural transition of α_1-antitrypsin by a peptide sequentially similar to β-strand S4A. Eur J Biochem 194:51–56
42. Dodt J, Seemüller U, Maschler R, Fritz H (1985) The complete covalent structure of hirudin. Localization of the disulfide bonds. Biol Chem Hoppe-Seyler 366:379–385
43. Folkers PJM, Clore GM; Driscoll PC, Dodt J, Köhler S, Grönenborn AM (1989) Solution structure of recombinant hirudin and the Lys47 --> Glu mutant: a nuclear magnetic resonance and hybrid distance geometry-dynamical simulated annealing study. Biochemistry 28:2601–2617
44. Haruyama H, Wüthrich K (1989) Conformation for recombinant desulfatohirudin in aquous solution determined by nuclear magnetic resonance. Biochemistry 28:4301–4312
45. Stone SR, Dennis S, Hofsteenge J (1989) Quantitative evaluation of the contribution of ionic interactions to the formation of the thrombin-hirudin complex. Biochemistry 28:6857–6863
46. Dennis S, Wallace A, Hofsteenge J, Stone SR (1990) Use of fragments of hirudin to investigate thrombin-hirudin interaction. Eur J Biochem 188:61–66
47. Bode W, Engh R, Musil D, Laber B, Stubbs M, Huber R, Turk V (1990) Mechanism of interaction of cysteine proteinases and their protein inhibitors as compared to the serine proteinase-inhibitor interaction. Biol Chem Hoppe-Seyler 371:111–118
48. Machleidt W, Thiele U, Laber B, Assfalg-Machleidt I, Esterl A, Wiegand G, Kos J, Turk V, Bode W (1989) Mechanism of inhibition of papain by chicken egg white cystatin. Inhibition constants of N-terminally truncated forms and cyanogen bromide fragments of the inhibitor. FEBS Lett. 243:234–238
49. Rees DC, Lipscomb WN (1982) Refined crystal structure of the potato inhibitor complex of carboxypeptidase A at 2.5 Å resolution. J Mol Biol 160:475–498
50. Clore GM, Grönenborn AM, Nilges M, Ryan CA (1987) Three-dimensional structure of potato carboxypeptidase inhibitor in solution. A study using unclear magnetic resonance, distance geometry, and restrained molecular dynamics. Biochemistry 26:8012–8023
51. Waxman L, Smith DE, Arcuri KE, Vlasuk GG (1990) Tick anticoagulant peptide (TAP) is a novel inhibitor of blood coagulation Factor Xa. Science 248:593–596

Attenuation of Inflammatory Disease by Reduction of Interleukin-1 Production or Receptor Antagonism

C. A. DINARELLO[1]

1 Introduction

Interleukin-1 (IL-1) is a 17-kDa pro-inflammatory cytokine synthesized from a variety of cell types primarily in association with disease states or during host perturbation such as immune responses. At pM or even fM concentrations, IL-1 triggers various responses in nearly all cells. It appears that there is little or no major role for IL-1 in homeostatic mechanisms. There are two IL-1's (α and β) each with its distinct sequence; there are two IL-1 receptors. Disease states such as local and systemic infection, septic shock, degenerative arthritis, and autoimmune diseases such as nephritis, vasculitis, and inflammatory bowel disease appear to be mediated, in part, by IL-1. Organ failure, capillary leak, and death occur in animals after a combination of tumor necrosis factor (TNF) and IL-1 which is more effective in inducing these changes than either cytokine alone. IL-1 is also a potent inducer of the endothelial cell adhesion molecules, IL-6, and IL-8, a neutrophil chemotactic and activating factor. Strategies for reducing the effects of IL-1 have been based on suppression of transcription, translation, or secretion; more recently, receptor blockade has been a new approach. A naturally occurring IL-1-specific receptor antagonist (IL-1ra), which shares 40% conserved amino acid homology with IL-1b, binds to IL-1 surface receptors with the same affinity as IL-1 but does not possess agonist activity and acts as a competitive inhibitor of IL-1. Studies using the IL-1ra to block endogenous IL-1 in a variety of animal disease models suggest that IL-1 plays a key role in triggering the cascade of inflammatory responses. In addition, the IL-1ra reduces the spontaneous production of growth factors and proliferation of leukemic cells. The IL-1ra may be an advantageous therapy in patients with sepsis, diabetes, inflammatory bowel disease, arthritis, and cancer.

2 Background

Interleukin-1 (IL-1) is a polypeptide produced as a result of infection, toxic injury, trauma, or antigenic challenge [1]. With the exception of the skin keratinocytes and certain cells in the central nervous system, mRNA coding for IL-1 is not observed in health in most other cells. However, within 15 min of stimulation by a variety of

[1] Department of Medicine, Tufts University School of Medicine and New England Medical Center, Boston, MA 02111, USA

42. Colloquium Mosbach 1991
Molecular Aspects of Inflammation

agents, the genes for IL-1 are transcribed. Although the macrophage monocyte is a primary source of IL-1, B-lymphocytes, endothelial, epithelial, mesangial, smooth muscle cells, and fibroblasts also synthesize IL-1. Two distinct IL-1 cDNA's were cloned in 1984 and represent two separate gene products; IL-1β [2] codes for a neutral polypeptide, whereas IL-1α [3] is an acidic protein. At the mRNA level IL-1β is more abundant than IL-1α [4]. As much as 2% of the polyadenylated RNA codes for IL-1β 4 h after human monocytes are activated by a few picograms of endotoxin. Neither IL-1α nor IL-1β contain a signal peptide; therefore, a considerable amount of IL-1 remains cell-associated. As much as 90% of IL-1α produced by human monocytes stimulated with bacterial endotoxin is found intracellularly, whereas 70% of IL-1β synthesized by these same cells is secreted [5, 6]. The second form of IL-1 is a 17000 kDa polypeptide cleaved from a 31-kDa precursor. It remains unclear how the precursor of IL-1 is cleaved or secreted but specific proteases (serine) appear to be involved. Cell-associated IL-1 is found primarily in the cytosolic rather than the endoplasmic reticulum and has been localized to "protective" vesicles [7].

Biologically active "cell surface membrane" IL-1α has been described by a number of investigators [8, 9, 10]. Although there was some evidence that membrane IL-1 was a phenomenon of "leaky" cells [11], recent studies have confirmed the original observation that with proper fixation of macrophages with paraformaldehyde, cell membrane IL-1α is biologically active and likely plays a role in paracrine functions [12]. Ovalbumin and fibroblast growth factor (α and β) are two other examples of proteins without signal peptides. Tumor necrosis factor (TNF) contains a typical cleavage sequence and 90% is secreted rapidly from cells, although membrane forms of TNF have been reported.

3 Biological Properties of IL-1

In considering strategies for anti-IL-1 treatment, the multiple biological effects of IL-1 reveal that this cytokine possesses both beneficial as well as detrimental activities. Some of these differences are clearly due to the amounts given in animal studies or added to in vitro experiments. For example, in recent clinical trials in patients with suppressed hematopoiesis, low doses of IL-1 stimulate bone marrow stem cells and the production of colony-stimulating factors leading to increased neutrophil and platelet production, but high doses cause hyperpyrexia, general malaise, gastrointestinal disturbances, and hypotension. Low doses of IL-1 stimulate insulin production, whereas high concentrations inhibit insulin transcription and are cytotoxic for the beta islet cells [13, 14]. Indeed, IL-1 is thought to play a role in the pathogenesis of type I diabetes mellitus. Other biological properties of IL-1 are clearly beneficial and there is justification for the use of IL-1 in the treatment of certain diseases, such as bone marrow transplants [15], protection from radiation [13, 14], and increase in nonspecific resistance to infection [16]. However, in this review, only anti-IL-1 strategies will be discussed.

4 IL-1 as a Prototype Inflammatory Cytokine

There is little question that IL-1 is a potent inflammatory molecule. Moreover, its inflammatory properties are dramatically enhanced by its biologically related cytokine, TNF [17]. IL-1 and TNF share many biological properties [18] and are highly synergistic in both animal and in vitro experiments. IL-1 induces fever; in fact, IL-1 was likely the first endogenous pyrogen described and purified [19]. Other cytokines, TNF, IL-6, and interferon (IFNα) also cause fever and are "endogenous pyrogens". In addition to triggering events in the central nervous system leading to fever, IL-1 also triggers the release of a variety of neuropeptides, most importantly ACTH, corticotropin releasing factor, and somatostatin. Although IL-1 is a potent appetite suppressant [20], its mechanism of action as an anorectic agent is thought to be due to its peripheral action on hepatic metabolism rather than on the central nervous system. IL-1 is a very potent sleep factor inducing a rapid increase in slow wave sleep, even before the onset of fever [21]. The somnogenic property of IL-1 is not, like fever, blocked by inhibition of cyclooxygenase.

A considerable number of biological responses to IL-1, particularly its inflammatory activity, are due to its ability to increase gene expression for cyclooxygenase [22]. Within minutes of exposing cells to IL-1, there is a dramatic increase in new mRNA coding for cyclooxygenase and phospholipase A_2. In isolated tissues perfused with IL-1, PGE_2 increases rapidly in the perfusate. The synthesis of PGE_2 and other products of arachidonic acid cyclooxygenation by vascular [23] and intestinal tissue [24] continues for several hours after cessation of the IL-1 perfusion. In vascular tissue perfused with IL-1, there is an associated increased pain response to bradykinin [23]. The increased synthesis of lipid metabolites such as the prostaglandins and also of platelet-activating factor by IL-1 contribute to the shock syndrome observed in several animal models [25]. The induction of hypotension, neutropenia, thromocytopenia, lactic acidosis, and decreased systemic vascular resistance is observed following an intravenous injection of IL-1 [26]; however, this is dramatically potentiated by co-injection with TNF. IL-1 and combinations of IL-1 plus TNF enhance endothelial pro-coagulant activity, the expression of leukocyte adhesion molecules, and the synthesis of a plasminogen activator inhibitor. In recent phase I studies in humans injected with recombinant IL-1β, the dose-limiting toxicity has been hypotension.

Synovial tissues removed from patients with rheumatoid arthritis actively produce IL-1 and PGE_2 and collagenase production by these cells is thought to be due to cells activated in vivo to produce IL-1 [27, 28]. It is therefore not surprising that inhibitors of cyclooxygenase such as aspirin or nonsteroidal antiinflammatory agents are used in diseases thought to be mediated by IL-1. However, agents which block the ability of IL-1 to induce cyclooxygenase and phospholipase A_2 (PLA_2) gene expression would have a higher therapeutic index since such IL-1 antagonists would not interfere with cyclooxygenation in gastric tissue or platelets. For example, agents that blocked IL-1 binding to its cellular receptors are highly specific for IL-1-mediated inflammatory responses.

Not all properties of IL-1 are mediated through its ability to increase cyclooxygenase and PLA_2 synthesis. Signal transduction is likely to be similar in most cells after binding of IL-1 to its receptor(s). The most consistent property of IL-1 is upregulation of cellular metabolism and increased expression of several genes. The expression of some genes, particularly household genes, are reduced by IL-1; for example, those for albumin, lipoprotein lipase, cytochrome P450, and aromatase. With these and other exceptions, IL-1 predominantly activates transcription of several genes and also stabilizes mRNA for others, such as GM-CSF [29, 30]. In both events, activation by IL-1 leads to increased new cytokine synthesis.

In the liver, IL-1 upregulates the transcription of several hepatic acute phase proteins such as alpha$_2$-acid glycoprotein and alpha$_2$-macroglobulin. IL-1 also initiates the transcription of serum amyloid A. This is a direct effect of IL-1 on hepatocytes but other effects of IL-1 on hepatic protein synthesis are due to IL-1 induced IL-6 production. IL-1 is a potent inducer of IL-6 in fibroblasts and other cells [31]; in fact, in animals given an inflammatory agent such as turpentine, pretreatment with an antibody to the IL-1 p80 receptor blocks IL-6 synthesis and reduces the acute phase response [32]. Even appetite suppression induced by IL-1 is blocked by the use of an anti-IL-1 receptor antibody, supporting the claim that IL-1 induced anorexia is an hepatic effect of IL-1 rather than an action on the central nervous system [20].

5 Antiinflammatory and Immunosuppressive Strategies Based on Inhibition of IL-1 Gene Expression

IL-1 is an inducible gene in most tissues following exposure to microbes, toxins, or inflammatory agents. The tissue macrophages or blood monocytes are the major sources of IL-1 in a variety of inflammatory diseases and in the inflamed joint, meninges, lung, and intestinal tract. One approach to antiinflammatory diseases has been the use of agents which block the induction of the IL-1 gene. The transcriptional activation of IL-1 by agents such as endotoxin begins within 15 min and mRNA for IL-1β reaches peak concentrations 3–4 h later [33, 34]. In human blood monocytes, there is a steady and rapid reduction of this RNA over the next 12 h and by 24 h, there is little or no IL-1β present. Cycloheximide blocks the synthesis of a putative transcriptional repressor in cell lines. Using IL-1 as a stimulant of its own gene expression, mRNA levels remain elevated in human blood monocytes for over 30 h [35]. Clearly, microbial agents trigger the synthesis of a transcriptional repressor or alternatively activate enzymes increasing the breakdown of IL-1 mRNA. What is most interesting for chronic inflammatory diseases is that the IL-1-induced IL-1 gene expression seems to escape rapid reduction in steady state mRNA.

5.1 Corticosteroids as Inhibitors of IL-1 Gene Expression

When added just prior to stimulation, corticosteroids block the transcriptional activation of IL-1 by a variety of exogenous stimulants [36]. When transcription has already taken place (for example, after monocytes have adhered to a surface), corticosteroids are less effective in reducing transcription. Other agents, such as some lipoxygenase inhibitors will also reduce IL-1 at the transcriptional level. We have taken the position that the dramatic improvement in several inflammatory and autoimmune diseases which takes place with corticosteroid therapy is, in part, due to the suppression of cytokine transcription. In addition, corticosteroids also inhibit cytokine action. Because IL-1 stimulates gene expression for other cytokines (IL-2 through IL-9, including TNF), reduction of IL-1 synthesis may provide an important approach to antiinflammatory or immunosuppressive therapy.

5.2 Interferons as Suppressors of IL-1 Gene Expression

IFNγ and also IFNα have two effects on IL-1 synthesis; when added to cells together with exogenous stimulants such as endotoxin or staphylococcal exotoxins, the interferons *increase* IL-1 production and this can be observed at the transcriptional level [37]. On the other hand, using IL-1 itself as a stimulator of IL-1 synthesis, both IFNγ and IFNα *reduce* IL-1 production by suppressing its transcription [35]. The decrease in transcription of IL-1 by interferons is not due to increased degradation of IL-1 mRNA since the half-life of mRNA in cells incubated with IFNγ was nearly the same as that in cells stimulated with endotoxin. Unlike corticosteroids, IL-1-induced IL-1 synthesis can be reduced when IFN is added as late as 6 h following stimulation.

Are there clinical examples where the suppression of IL-1-induced genes may be functioning? IL-1 is a growth factor for smooth muscle cells and there may be an IL-1 autocrine effect in which IL-1 induction of IL-1 contributes to the proliferation of smooth muscle cells and fibroblasts [38, 39]. Since IL-1 induces the genes for platelet-derived growth factor in endothelial cells, the efficacy of IFNα in Kaposi's sarcoma and in patients with angiogenic pulmonary fibrosis may be via IFNα suppression of IL-1-induced genes in these cases. Another example exists in synovial cells which when removed from the joint space of patients with rheumatoid arthritis are "pre-activated" to produce IL-1 spontaneously, whereas the peripheral blood leukocytes from these same patients do not spontaneously produce IL-1 upon culture in vitro [40]. However, the effect of IFNγ on cultured synovial cell production of IL-1 is one of inhibition, whereas the peripheral cells typically produce more IL-1 in the presence of IFNγ when stimulated by endotoxin. We conclude that the synovial cells have been stimulated by an IL-1 autocrine mechanism in the rheumatoid joint and that IFNγ suppresses IL-1 synthesis by mechanisms similar to those we observed in monocytes from healthy individuals stimulated by IL-1.

6 Preventing the Synthesis of IL-1

During cell activation by a variety of agents, arachidonic acid (AA) is liberated and undergoes either cyclooxygenation into protaglandins or lipoxygenation into leukotrienes (LT). Formation of LT's appears to provide a positive signal for cytokine synthesis. LTB_4 has been shown to stimulate IL-1 gene expression and synthesis. In addition, agents that inhibit the lipoxygenase pathway can also inhibit cytokine synthesis. Although the lipoxygenase product 13-hydroxyoctadecadienoic acid (13-

HODD) has been identified as a likely lipoxygenase metabolite for TNF synthesis [41], other lipoxygenase products may also be involved. On the other hand, there is ample evidence that metabolism of AA into prostaglandins and prostacyclin provides a negative signal for cytokine formation, particularly at higher concentrations. (Interestingly, low concentrations of PGE_2 have been shown to stimulate TNF and IL-6 synthesis.)

We speculated that a qualitative or quantitative change in lipoxygenase product synthesis during the induction of IL-1 gene expression would result in a corresponding decrease in the synthesis of IL-1 and related cytokines. We were aware of the studies of Lee et al. [42] and Payan et al. [43] that neutrophil LTB_4 formation was suppressed in human volunteers taking dietary supplements of the N-3 polyunsaturated fatty acids [eicosapentaenoic acid (EPA) in the form of fish oil concentrates] for 6 weeks. To evaluate the effect of reduced LT formation on cytokine synthesis, the ex vivo production of IL-1α, IL-1β, and TNF was measured before, after, and during two washout phases in human volunteers. A dramatic (50–60%) decrease in total cytokine synthesis in nine volunteers was reported [44]. The maximum effect occurred 10 weeks after stopping the 6-week supplement. The ex vivo production of these cytokines returned to pre-EPA therapy 20 weeks later. We have also observed similar decreases in ex vivo cytokine production after 2 and 3 month in women taking 1,5 g/day of EPA supplements [45].

The prolonged effect of EPA supplements in the volunteers has also been observed in clinical studies [46]. This may be related to the ratio of AA to EPA in the membrane of blood mononuclear cells. In humans consuming the usual Western diet this ratio is in the order of 20:1. After 6 weeks of therapy the ratio changes to 3:1 and 10 weeks after stopping the supplements it remains significantly ($p < 0.05$) altered at 11:1 [44]. The ratio returns to the presupplement level after 20 weeks. The altered ratio of AA to EPA results in a change in the concentration and type of AA metabolite. Instead of PGE_2 and LTB_4, the altered ratio results in the synthesis of PGE_3 and LTB_5. The LTB_5 may serve to block receptors for LT's involved in initiating cytokine synthesis.

The presence of cyclooxygenase inhibitors in vitro increases cytokine synthesis. Moreover, human volunteers given oral cyclooxygenase inhibitors increase ex vivo [5] and circulating cytokine production following challenge by endotoxin (G. Spinas, pers. comm.). To support these studies, a decrease in cytokine synthesis is observed by the addition of PGE_2 or PGI_2 to in vitro leukocyte cultures. The suppression of cytokine synthesis by the action of EPA via altered lipoxygenation prevails over that of reduced PGE_2 production.

The suppressive effect of PGE_2 and PGI_2 on cytokine synthesis appears to be via increase cAMP formation [47]. The methylxanthine drug pentoxifylline, which like PGE_2 increases cAMP formation, also reduces TNF synthesis [48]. We have recently observed similar suppression of cytokine synthesis using theophylline which blocks the cyclic nucleotide. hydrolyzing enzyme, phosphodiesterase; the addition of histamine, also an inducer of cAMP *via* the H-2 receptor, similarly reduces cytokine synthesis [49]. Thus, in addition to changes in lipoxygenase products, one can also reduce cytokine synthesis employing agents which increase cAMP levels in cells. It will be interesting to observe the effect of drugs such as pentoxifylline or the H-2 receptor blockers on in vivo and ex vivo cytokine production.

7 IL-1 Receptors

One approach to anti-IL-1 strategies is the use of antibodies to the IL-1 receptor(s) (IL-1R). There are at least two IL-1R's and each is a separate gene product [50, 51]. A p80 glycoprotein (termed IL-1RtI), a member of the immunoglobulin superfamily, is found on macrophages, T-cells, fibroblasts, endothelial cells, keratinocytes, and hepatocytes, whereas a p68 IL-1R (termed IL-1RtII) is found on B-cells and neutrophils. The p80 IL-1RtI contains an extracellular, a single transmembrane, and intracellular portion; the intracellular portion has no intrinsic tyrosine kinase activity, although it resembles substrates for protein kinase C. Monoclonal antibodies produced to the murine IL-1RtI block a variety of host responses including synthesis of hepatic proteins and IL-6 induction [32]. Antibodies to the IL-1RtI reduce endotoxin-induced fever in mice. In addition to these effects, anti-IL-1RtI prevents IL-1-induced protection to lethal radiation. Experiments using anti-IL-1RtII have not been reported. In addition to the 68-kDa IL-1RtII on B-cells, there is a 30-kDa binding protein observed on T-cells, macrophages, and mesangial cells, but it is unclear wether this is a receptor or an associated protein. The p68 and p80 IL-1R are separate gene products but the molecular nature of the 30-kDa binding protein remains unclear [52]. One hypothesis for the IL-1R is that, similar to other cytokine receptors, it is a heterocomplex of different polypeptide chains. However, a complete signal transduction event following exposure of cells to IL-1 leading to gene expression appears to be linked to the p80 IL-1RtI in some cells, whereas in other cells apparently lacking the p80 IL-1R chain, IL-1 itself causes only phospholipid (phosphatidylcholine) hydrolysis [53] without evidence of new gene expression. In those cells, a second signal is required for full IL-1 activity.

8 The IL-1 Receptor Antagonist

The IL-1 receptor antagonist (IL-1ra) molecule was originally called the IL-1 inhibitor and was isolated and purified from the urine of patients with myelomonocytic leukemia [54, 55]. A similar IL-1 inhibitory activity was found in the urine of patients with fever [56] and in the circulation of human subjects injected intravenously with endotoxin [57]. In these reports, the biological characteristic of the IL-1 inhibitory activity was its ability to block IL-1 but not IL-2 or mitogen-induced T-cell proliferation. The "IL-1 inhibitor" purified from the urine had a molecular weight of 23–25 kDa. It blocked the binding of IL-1 to receptors on T-cells but did not block the binding of TNF or IL-2 [54, 55, 58]. The IL-1 inhibitor did not bind to IL-1 itself, unlike the TNF inhibitor which was also isolated from urine [59]. Thus, the IL-1 inhibitor was a competitive inhibitor of IL-1/IL-1R interaction.

A similar IL-1 inhibitory activity was released from human monocytes adhering to IgG-coated surfaces [60]. Using the IL-1 inhibitor purified from adherent monocytes, an N-terminal sequence was obtained and the molecule was cloned [61]. The cDNA sequence deducted from the cDNA revealed 40% homology conserved amino acid to IL-1β and 39% to IL-1α. Similar to studies reported from the purified naturally occurring IL-1 urinary inhibitor [55], the recombinant IL-1 inhibitor competes with the binding of IL-1 to its cell surface receptors. Because of its sequence homology and mode of activity, the IL-1 inhibitor was renamed IL-1ra. Antibodies against recombinant human IL-1ra recognize the natural urinary IL-1 inhibitor [62].

9 Biological Effects of IL-1ra

It is not surprising that recombinant IL-1ra blocks the activity of IL-1 in a variety of biological responses. The efficacy of blockade may differ depending of cell types or the animal employed. Rabbits or baboons given recombinant IL-1 develop hypotension which is reversed by prior administration of the IL-1ra [63]. However, a more important question remains: during acute or chronic disease many cytokines are produced but an essential role for any cytokine has yet to be established. For example, when baboons given *E. coli* sepsis are pretreated with a monoclonal antibody to TNF, attenuation of severity of the hemodynamic parameters is observed. Although this would suggest an essential role for TNF in the pathogenesis of septic shock, baboons pretreated with anti-TNF demonstrated markedly diminished circulating IL-1 levels [64]. Thus, a role for TNF and IL-1 appears to be necessary for septic shock.

9.1 Effect of IL-1ra on Septic Shock

Injection of endotoxin or killed *E. coli* suspensions to rabbits intravenously produces several parameters of the septic shock syndrome, namely, hypotension, leukopenia, thrombocytopenia, tissue damage, and death which are significantly ameliorated by the IL-1ra [65, 66]. When rabbits were pretreated with the IL-1ra, only a transient hypotensive episode and a decrease in the leukopenia were observed. There is also reduced numbers of tissue-infiltrating neutrophils. In these studies the circulating levels of TNF and IL-1β (as determined by specific radioimmunoassay) were unchanged [66]. The interpretation of these results suggests that TNF may be responsible for the initial fall in blood pressure, but that IL-1 plays an essential role in the progression of the shock state.

9.2 The IL-1ra in Immune Complex-Induced Colitis

A role for IL-1 has been proposed in the pathogenesis of inflammatory bowel disease. Three studies support this conclusion: (1) rabbit colonic tissue releases large amounts of PGE_2 and LTB_4 for several hours following a brief period of perfusion with IL-1

[24]; (2) there is a reduction in the severity of colonic inflammation in rabbits pretreated with a single, low dose of IL-1 24 h before the induction of colitis [67]; and (3) the degree of inflammation, edema, and necrosis in colonic tissue correlates with the tissue levels of IL-1 in these tissues [68]. Although IL-1 levels and tissue injury correlate, these data do not necessarily support an essential role for IL-1 in the pathogenesis of colitis in this model as other inflammatory cytokines may also be involved. However, when rabbits were pretreated with the IL-1ra, a marked decrease in tissue inflammatory cell infiltration, edema, and necrosis was observed [68]. In addition, decreased PGE_2 was measured in the rectal lumen despite the fact that IL-1 tissue levels were unchanged [69]. Together, these data demonstrate that blockade of IL-1 prevents the onset and development of the inflammatory lesion in this model of immune complex-induced colitis.

9.3 Effect of IL-1ra on Spontaneous Colony-Stimulating Factor Production by Acute Myelogenous Leukemia Cells

Several studies have reported that IL-1 induces colony-stimulating factors from fibroblasts, endothelial cells, lymphocytes, and blood monocytes [30, 70, 71, 72, 73]. Moreover, antibodies to human IL-1β completely reduced the spontaneous proliferation and colony-stimulating activity of granulocytic and myelogenous leukemia cells [71, 74]. Therefore, it was hypothesized that growth factor production is under the control of IL-1 in these cells. Recent studies by Cozzolino, Rambaldi, and their coworkers have shown that the IL-1ra blocks the spontaneous proliferation as well as spontaneous production of GM-CSF, IL-1, and IL-6 colony-stimulating production in peripheral blood myelogenous leukemia cells in vitro [75]. These studies suggest that IL-1 controls the production of GM-CSF in these cells and that treatment with the IL-1ra may be highly effective in reducing the proliferation of these leukemia cells.

10 Other Models for Studying the Role of IL-1 in Disease Processes

The success of the IL-ra in reducing the severity of systemic and local disease in a variety of pathological events suggests that the IL-1ra should be tested in other models where IL-1 has been indirectly implicated in the process. For example, IL-1 is cytotoxic for the insulin-producing β-cells in the islets of Langerhans [13]. Recent studies in which the IL-1ra has been administered daily to the BB rat, there is a more prolonged time for the onset of diabetes in these diabetic prone rats (JM Dayer, pers. comm.). It appears that blocking IL-1 activity using the IL-1ra may have reduced the destruction of the islet cells in this model of type I insulin-dependent diabetes mellitus. In another model of acute inflammation, the IL-1ra reduces (70%) the infiltration of neutrophils into the peritoneal cavity following injection of either endotoxin or peptone broth [76].

Acknowledgment. These studies were supported by NIH grant AI15614.

References

1. Dinarello CA (1991) Blood 77:1627–1652
2. Auron PE, Webb AC, Rosenwasser LJ, Mucci SF, Rich A, Wolff SM, Dinarello CA (1984) Proc Natl Acad Sci USA 81:7907–7911
3. Lomedico PT, Gubler R, Hellmann CP, Dukovich M, Giri JG, Pan YE, Collier K, Semionow R, Chua AO, Mizel SB (1984) Nature (London) 312:458–62
4. Demczuk S, Baumberger C, Mach B, Dayer JM (1987) J Mol Cell Immunol 255–265
5. Endres S, Cannon JG, Ghorbani R, Dempsey RA, Sisson SD, Lonnemann G, van der Meer JWM, Wolff SM, Dinarello CA (1989) Eur J Immunol 19:2327–2333
6. Lonnemann G, Endres S, van der Meer JW, Cannon JG, Koch KM, Dinarello CA (1989) Eur J Immunol 19:1531–1536
7. Rubartelli A, Cozzolino F, Talio M, Sitia R (1990) EMBO J 9:1503–1510
8. Kurt-Jones EA, Beller DI, Mizel SB, Unanue ER (1985) Proc Natl Acad Sci USA 82:1204
9. Beuscher HU, Colten HR (1988) Mol Immunol 25:1189
10. Brody DT, Durum SK (1989) J Immunol 143:1183
11. Minnich-Carruth LL, Suttles J, Mizel SB (1989) J Immunol 142:526
12. Bailly S, Ferrua B, Fay M, Gougerot-Pocidalo MA (1990) Eur Cytokine Net 1:47–51
13. Bendtzen K, Mandrup-Poulsen T, Nerup J, Nielsen JH, Dinarello CA, Svenson M (1986) Science 232:1545–1547
14. Spinas GA, Mandrup-Poulsen T, Mlvig J, Baek L, Bendtzen K, Dinarello CA, Nerup J (1986) Acta Endocrinol (Copenh) 113:551–558
15. Fibbe WE, van der Meer JWM, Falkenburg JHF, Hamilton MS, Kluin PM, Dinarello CA (1989) Esp. Hematol 17:805–808
16. van der Meer JW, Barza M, Wolff SM, Dinarello CA (1988) Proc Natl Acad Sci USA 85:1620–1623
17. Beutler B, Krochin N, Milsark IW, Luedke C, Cerami A (1986) Science 232:977–980
18. Dinarello CA (1989) Adv Immunol 44:153–205
19. Dinarello CA, Cannon JG, Wolff SM (1988) Rev Infect Dis 10:168–189
20. Hellerstein MK, Meydani SM, Meydani M, Wu K, Dinarello CA (1989) J Clin Invest 84:228–235
21. Walter J, Davenne D, Shoham S, Dinarello CA, Krueger JM (1986) Am J Physiol 86:R96–R103
22. Raz A, Wyche A, Siegel N, Needleman P (1988) J Biol Chem 263:3022–3028
23. Schweizer A, Feige U, Fontana A, Muller K, Dinarello CA (1988) Agents Actions 25:246–251
24. Cominelli F, Nast CC, Dinarello CA, Gentilini P, Zipser RD (1989) Gastroenterology 97:1400–1405
25. Dejana E, Breviario F, Erroi A, Bussolino F, Mussoni L, Gramse M, Pintucci G, Casali B, Dinarello CA, Van Damme J, Mantovani A (1987) Blood 69:695–9
26. Okusawa S, Gelfand JA, Ikejima T, Connolly RJ, Dinarello CA (1988) J Clin Invest 81:1162–1172
27. Dayer JM, de Rochemonteix B, Burrus B, Demczuk S, Dinarello CA (1986) J Clin Invest 77:645–648
28. Schnyder J, Payne T, Dinarello CA (1987) J Immunol 138:496–503
29. Ernst TJ, Ritchie AR, Demetri GD, Griffin JD (1989) J Biol Chem 264:5700–3
30. Griffin JD, Cannistra SA, Sullivan R, Demetri GD, Ernst TJ, Kanakura Y (1990) Int J Cell Cloning 1:35–44
31. Kohase M, May LT, Tamm I, Vilcek J, Sehgal PB (1987) Mol Cell Biol 7:273–280
32. Gershenwald JE, Fong Y, Fahey TJI, Calvano SE, Chizzonite R, Kilian PL, Lowry SF, Moldawer LL (1990) Proc Natl Acad Sci USA 87:4966–4970
33. Fenton MJ, Clark BD, Collins KL, Webb AC, Rich A, Auron PE (1987) J Immunol 138:3972–3979

34. Schindler R, Clark BD, Dinarello CA (1990) J Biol Chem 265:10232–10237
35. Schindler R, Ghezzi P, Dinarello CA (1990) J Immunol 144:2216–2222
36. Knudsen PJ, Dinarello CA, Strom TB (1987) J Immunol 139:4129–4134
37. Ucla C, Roux-Lombard P, Fey S, Dayer JM, Mach B (1990) J Clin Invest 85:185–191
38. Warner SJC, Auger KR, Libby P (1987) J Exp Med 165:1316–1331
39. Warner SJC, Auger KR, Libby P (1987) J Immunol 139:1911–1917
40. Ruschen S, Lemm G, Warnatz H (1989) Clin Exp. Immunol 76:246
41. Schade UF, Burmeister I, Engel R (1987) Biochem Biophys Res Comm 147:695–700
42. Lee TH, Hoover RL, Williams JD et al. (1985) N Engl J Med 312:1217–1224
43. Payan DG, Wong MY, Chernov-Rogan T et al. (1986) J Clin Immunol 6:402–410
44. Endres S, Ghorbani R, Kelley VE, Georgilis K, Lonnemann G, van der Meer JW, Cannon JG, Rogers TS, Klempner MS, Weber PC, Wolff SM, Dinarello CA (1989) N Engl J Med 320:265–271
45. Meydani SN, Endres S, Woods MM, Goldin BR, Soo C, Morrill-Labrode A, Dinarello CA, Gorbach SL (1990) J Nutr (in press)
46. Kremer JM, Lawrence DA, Jubiz W, DiGiacomo R, Rynes R, Bartholomew LE, Sherman M (1990) Arthritis Rheum 33:810–820
47. Knudsen PJ, Dinarello CA, Strom TB (1986) J Immunol 137:3189–3194
48. Streiter RM, Remick PA, Ward PA, Spengler RN; Lynch JP, Larrick J, Kunkel SL (1988) Biochem Biophys Res Comm 155:1230–1236
49. Vannier E, Miller LC, Schindler R, Terlain B, Dinarello CA (1989) Cytokine 1:123
50. Bomsztyk K, Sims JE, Stanton TH, Slack J, McMahan CJ, Valentine MA, Dower SK (1989) Proc Natl Acad Sci USA 86:8034–8038
51. Chizzonite R, Truitt T, Kilian PL, Stern AS, Nunes P, Parker KP, Kaffka KL, Chua AO, Lugg DK, Gluber U (1989) Proc Natl Acad Sci USA 86:8029–8033
52. Savage N, Puren AJ, Orencole SF, Ikejima T, Clark BD, Dinarello CA (1989) Cytokine 1:23–25
53. Rosoff PM, Savage N, Dinarello CA (1988) Cell 54:73–81
54. Seckinger P, Dayer JM (1987) Ann Inst Pasteur/Immunol 138:461–516
55. Seckinger P, Lowenthal JW, Williamson K, Dayer JM, MacDonald HR (1987) J Immunol 139:1546–1549
56. Liao Z, Grimshaw RS, Rosenstreich DL (1984) J Exp Med 159:125–136
57. Dinarello CA, Rosenwasser LJ, Wolff SM (1981) J Immunol 127:2517–2519
58. Balavoine JF, de RB, Williamson K, Seckinger P, Cruchaud A, Dayer JM (1986) J Clin Invest 78:1120–1124
59. Engelmann H, Aderka D, Rubinstein M, Rotman D, Wallach D (1989) J Biol Chem 264:11974–11980
60. Arend WP, Joslin FG, Thompson RC, Hannum CH (1989) J Immunol 143:1851–1858
61. Eisenberg SP, Evans RJ, Arend WP, Verderber E, Brewer MT, Hannum CH, Thompson RC (1990) Nature (London) 343:341–346
62. Seckinger P, Klein-Nulend J, Alander C, Thompson RC, Dayer JM, Raisz LG (1990) J Immunol 145:4181–4184
63. Ohlsson K, Bjork P, Bergenfeldt M, Eisenberg SP, Childs J, Hannum CH, Thompson RC (1989) Cytokine 1:131a
64. Fong Y, Tracey KJ, Moldawer LL, Hesse DG, Manogue KB, Kenney JS, Lee AT, Kuo GC, Allison AC, Lowry SF, Cerami A (1989) J Exp Med 170:1627–1633
65. Ohlsson K, Bjork P, Bergenfeldt M, Hageman R, Thompson RC (1990) Nature (London) 348:550–552
66. Wakabayashi G, Gelfand JA, Burke JF, Thompson RC, Dinarello CA (1991) FASEB J 5:338–343
67. Cominelli F, Nast CC, Llerena R, Dinarello CA, Zipser RD (1990) J Clin Invest 85:582–586
68. Cominelli F, Nast CC, Clark BD, Schindler R, Llerena R, Eysselein VE, Thompson RC, Dinarello CE (1990) J Clin Invest 86:972–980

69. Cominelli F, Llerena F, Clark BD, Nast CC, Thompson RC, Dinarello CA (1990) Lymph Res 9:597
70. Bagby GCJ, Dinarello CA, Wallace P, Wagner C, Hefeneider S, McCall E (1986) J Clin Invest 78:1316–1323
71. Cozzolino F, Rubartelli A, Aldinucci D, Sitia R, Torcia M, Shaw A, Di Guglielmo R (1989) Proc Natl Acad Sci USA 86:2369–2373
72. Ernst TJ, Ritchie AR, Demetri GD, Griffin JD (1989) J Biol Chem 264:5700–3
73. Zucali JR, Dinarello CA, Oblon DJ, Gross MA, Anderson L, Weiner RS (1986) J Clin Invest 77:1857–1863
74. Bagby GCJ, Dinarello CA, Neerhout RC, Ridgway D, McCall E (1988) J Clin Invest 82:1430–1436
75. Rambaldi A, Torcia M, Bettoni S, Barbui T, Vannier E, Dinarello CA, Cozzolino F (1990) Blood 76:114a
76. McIntrye KW, Stepan GJ, Kolinsky KD, Benjamin WR, Plocinski JM, Kaffka KL, Campen CA, Chizzonite RA, Kilian PL (1991) J Exp Med 173:931–939

Interleukin-6, Its Hepatic Receptor and the Acute Phase Response of the Liver

P. C. Heinrich[1], G. Dufhues[1], S. Flohe[1], F. Horn[1], E. Krause[1], A. Krüttgen[1], L. Legres[1], D. Lenz[1], C. Lütticken[1], H. Schooltink[1], T. Stoyan[1], H. S. Conradt[2], and S. Rose-John[1]

1 The Acute Phase Response

The acute phase response is the answer of the organism to disturbances of its homeostasis due to infection, tissue injury, neoplastic growth, or immunological disorders. It is thought to be beneficial to the injured organism and its aim is the restoration of the disturbed physiological homeostasis. There is first a local reaction at the site of injury characterized by the aggregation of platelets, clot formation, dilation and leakage of blood vessels, accumulation of granulocytes and mononuclear cells and the activation of endothelial cells, fibroblasts, monocytes, and macrophages, which in turn release inflammatory cytokines such as interleukin-1, interleukin-6, interleukin-8, tumor necrosis factor α, and interferons. These mediators act via specific surface receptors on different target cells and lead to a systemic reaction of the organism characterized by fever, leukocytosis, increases in the erythrocyte sedimentation rate, in the secretion of ACTH and glucocorticoids, the activation of complement and clotting cascades, and in dramatic changes in the concentrations of some plasma proteins designated as acute phase proteins [reviews 1–3].

2 Acute Phase Proteins

The acute phase protein pattern varies from one species to another. In man, C-reactive protein, serum amyloid A, fibrinogen, haptoglobin, and $\alpha_1$antichymotrypsin are the major acute phase proteins, while α_2-macroglobulin, α_1-acid glycoprotein, and cysteine proteinase inhibitor represent the major acute phase proteins in the rat [1, 2]. Many acute phase proteins are proteinase inhibitors, others function in blood clotting and fibrinolysis, in the removal of foreign material from the organism (complement factors B and C3), and in the transport and inactivation of biologically active compounds.

Our major interest for several years has been the problem of the regulation of gene expression of rat α_2-macroglobulin, the major acute phase protein of the rat. Alpha$_2$-macroglobulin is a high molecular weight proteinase inhibitor (750 kDa) consisting of four identical subunits (reviewed in [4]). Two subunits are linked by two disulfide bridges resulting in a structure, which is a dimer of dimers. In most cases one

[1] Institut für Biochemie der RWTH Aachen, Pauwelsstraße 30, D-5100 Aachen, FRG
[2] Department of Genetics, GBF, Mascheroder Weg 1, D-3300 Braunschweig, FRG

42. Colloquium Mosbach 1991
Molecular Aspects of Inflammation

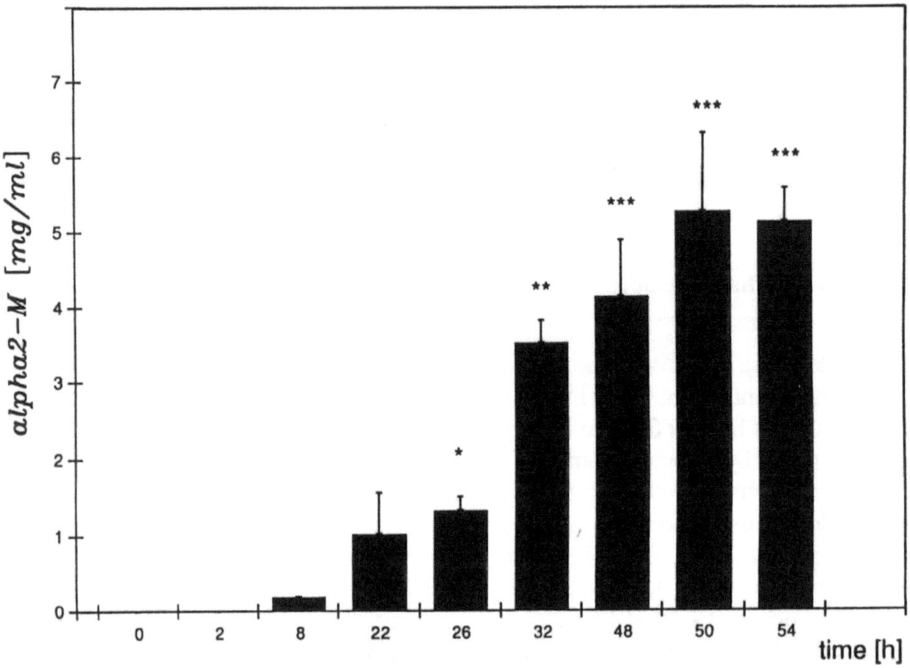

Fig. 1. Levels of α_2-macroglobulin in the blood of rats injected with 1, 2.5, and 50 mg/kg endotoxin at 0, 24, and 48 h, respectively. Alpha$_2$-macroglobulin concentrations were determined by rocket immunoelectrophoresis. Six animals per group were investigated [6]. *, **, and *** mean $P < 0.05$, < 0.01, and < 0.001, respectively

molecule of α_2-macroglobulin is capable of inhibiting two proteinase molecules. The proteinases are trapped within the interior of the α_2-macroglobulin molecule [5].

After intramuscular injection of turpentine or bacterial lipopolysaccharide dramatic increases in serum α_2-macroglobulin levels (up to 5 mg/ml) are observed (Fig. 1) [6]. In studies with adrenalectomized or hypophysectomized rats [7] it became clear that the expression of α_2-macroglobulin as well as the other acute phase proteins in the rat requires glucocorticoids and inflammatory mediators. Interleukin-6 (IL-6) was recognized as the major inducer of acute phase protein synthesis in rat [8, 9, 10] and human hepatocyte primary cultures [11, 12, 13] as well as in the rat in vivo [14, 15].

3 Properties and Pleiotropic Action of Interleukin-6

After appropriate stimulation many cells synthesize and secrete IL-6 [3]. The major IL-6-producing cells in the organism are monocytes, fibroblasts, and endothelial cells. As a secretory protein, IL-6 is synthesized with an extra amino terminal extension of 28 amino acids [16]. The molecule contains two sequential S-S bridges, two potential

N- and 27 possible O-glycosylation sites [16]. Since previous studies [17, 18, 19] have clearly shown that IL-6 from stimulated human monocytes, fibroblasts, and endothelial cells is N- and O-glycosylated, we asked whether glycosylation had an effect on biological activity, receptor recognition, or plasma half-life. In order to obtain large quantities of glycosylated IL-6, we have cloned the cDNA coding for human IL-6 behind a metallothionein promoter in the expression vector pBMGNeo and stably transfected mouse fibroblasts (NIH/3T3 cells) [19]. In the presence of cadmium chloride we were able to induce IL-6 synthesis and secretion in these cells. Thus, 10^6 of the transfected fibroblasts produced up to 10^6 units of IL-6 within 24 h [19]. The ana-

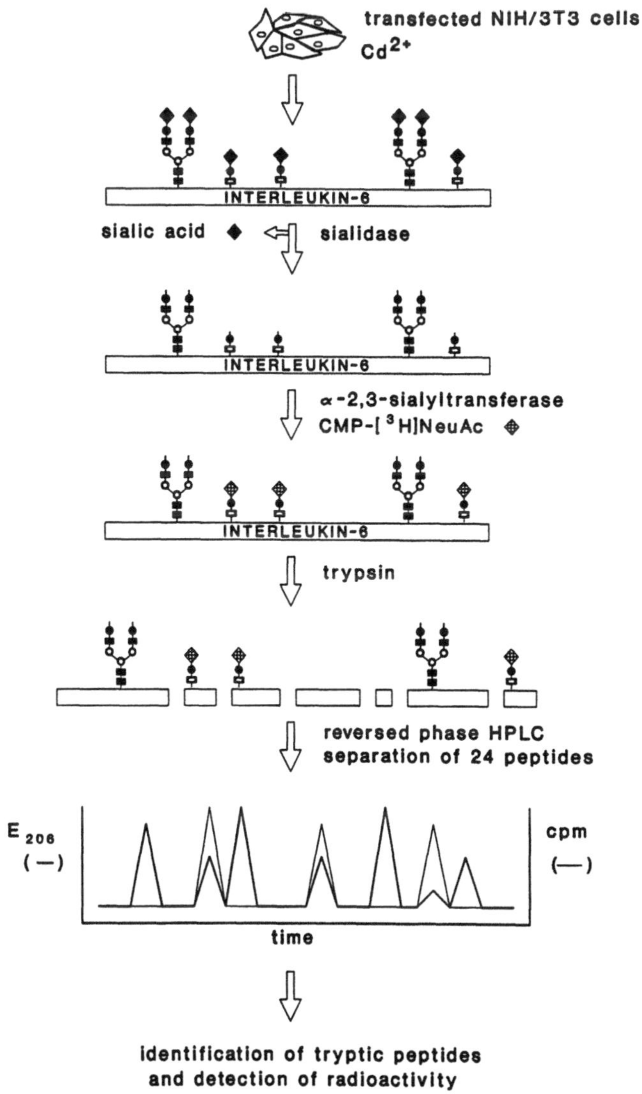

Fig. 2. Scheme for the identification of O-glycosylation sites in human IL-6 synthesized and secreted by transfected NIH/3T3 cells

lysis of the secreted cytokine showed the presence of < 5% N- and O-glycosylated IL-6, 60% O-glycosylated, and 35% nonglycosylated IL-6. As schematically shown in Fig. 2, we have identified the O-glycosylation site(s) of IL-6 produced in transfected NIH/3T3 cells after sialidase and O-glycan-specific sialyltransferase treatment, trypsin, and HPLC chromatography (Fig. 3A). Figure 3B shows that only one radioactively labeled tryptic peptide (#19) was detected. By radioactive Edman degradation and FAB MS sequencing of peptide 19 we identified threonine 138 as the only O-glycosylation site of IL-6. By an analogous procedure asparagine 45 was identified as the N-glycosylation site in peptide #8 (Fig. 3C, Table 1). Thus, only one (asn 45) of the two potential N-glycosylation sites and only one O-glycosylation site threonine 138 out of 27 potential sites are used in glycosylated IL-6. The analysis of the O-linked carbohydrate side chains resulted in the structures shown in Fig. 4. The comparison of biological activities of nonglycosylated and glycosylated IL-6 in the γ-fibrinogen induction assay in HepG2 cells as well as in the B9 cell proliferation test showed glycosylated IL-6 to be three to four times as active as the nonglycosylated cytokine.

4 Structure-Function Studies of Human Interleukin-6

Recent work in our laboratory has shown that the four carboxy-terminal amino acids of human IL-6

$$- \mathrm{Leu}^{181} - \mathrm{Arg}^{182} - \mathrm{Gln}^{183} - \mathrm{Met}^{184}$$

are essential for its biological activity [20]. We therefore constructed at the cDNA level deletion mutants of human IL-6 lacking 1, 2, 3 or 4 amino acids from the carboxy-terminal end of the molecule. After in vitro transcription and cell-free translation in a reticulocyte system the biological activity of the deletion mutants was determined by two independent bioassays. In the mouse B9 cell proliferation assay as well as in the γ-fibrinogen-induction assay using the human hepatoma cells HepG2, we found that already the removal of the last amino acid methionine 184 resulted in an 80% loss of biological activity. Essentially no further reduction in biological activity was seen when methionine and glutamine were removed. The deletion of methionine, glutamine, and arginine 182 resulted in completely inactive IL-6. These results emphasize the importance of the carboxy-terminal end of IL-6 for its biological function.

Fig. 3. Reversed-phase HPLC of tryptic IL-6 fragments and identification of the O- and N-glycosylation sites of human IL-6. (A) Human IL-6 from NIH/3T3 cells (50 μg) digested with trypsin was separated on a reversed-phase HPLC column equilibrated with 0.1% trichloroacetic acid. The peptides were eluted by an acetonitrile gradient as indicated. The *numbers* above the peaks designate the tryptic fragments after their identification by sequence analysis. As schematically shown in Fig. 2, human IL-6 produced by transfected NIH/3T3 cells was desialylated and subsequently resialylated using either an O-glycan specific α-2,3-sialyltransferase (B) or an N-glycan-specific α-2,6-sialyltransferase (C) and CMP-[^{3}H]NeuAc. [^{3}H]NeuAc-labeled IL-6 digested with trypsin was separated by reversed-phase HPLC and radioactivity was determined in the various fractions

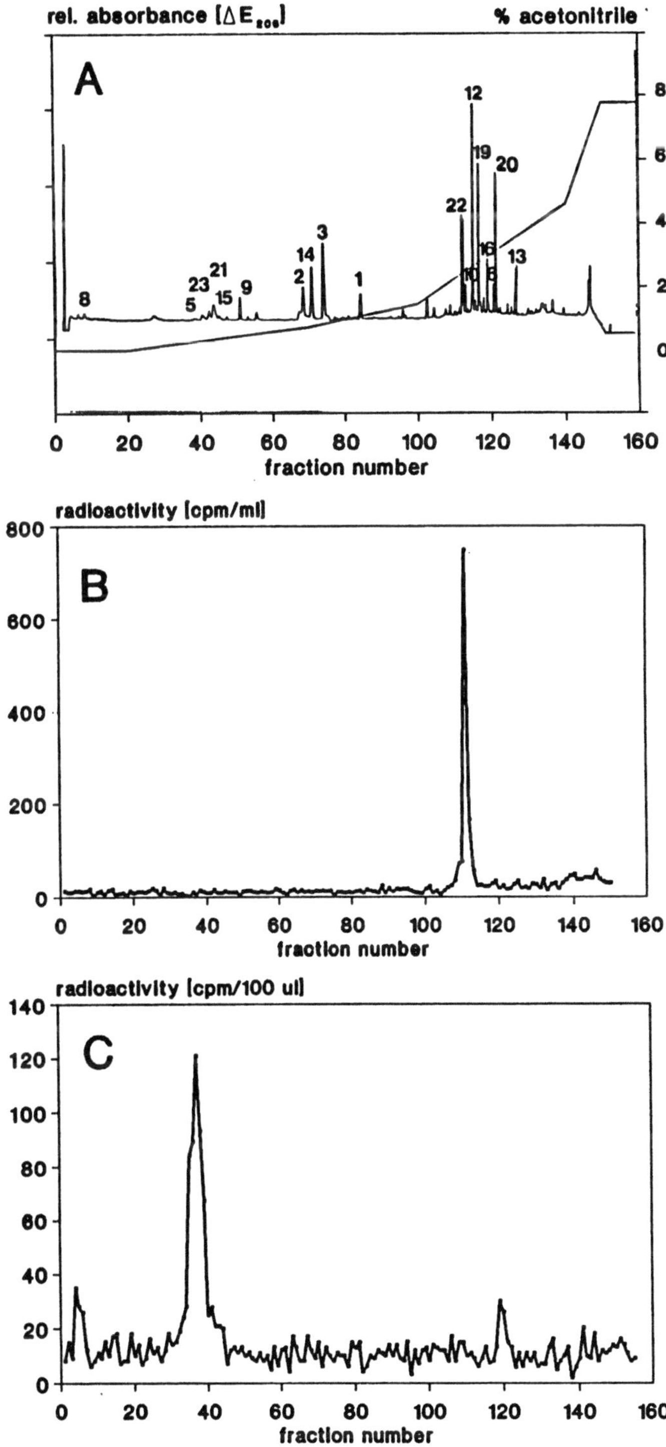

rel. absorbance [ΔE206]
% acetonitrile
A
fraction number
radioactivity [cpm/ml]
B
fraction number
radioactivity [cpm/100 ul]
C
fraction number

Table 1. Tryptic fragments of Interleukin-6

#	sequences	identified	pot. O-sites
1	MAPVPPGEDSK	*	o
2	DVAAPHR	*	
3	QPLTSSER	*	o
4	IDK		
5	QIR	*	
6	YILDGISAL	*	o
7	K		
8	ETCNK	*	o
9	SNMCESSK	*	o
10	EALAENNLNLPK	*	
11	MAEK		
12	DGCFQSGFNEETCLVK	*	o
13	IITGLLEFEVYLEYLQNR		o
14	FESSEEQAR	*	o
15	AVQMSTK	*	o
16	VLIQFLQK	*	
17	K		
18	AK		
19	NLDAITTPDPTTNASLLTK	*	o
20	LQAQNQWLQDMTTHLILR	*	o
21	SFK	*	o
22	EFLQSSLR	*	o
23	ALR	*	
24	QM		

When point mutations were introduced into the full length IL-6 molecule at positions 182 and 184, crucial for biological activity, the findings with the deletion mutants were essentially confirmed [21]. In addition, the study of these point mutations provided evidence for the importance of a positive charge (arg 182) and an α-helical structure of the C-terminus for biological activity of human IL-6 [21].

4.1 Pleiotropic Action of Interleukin-6

Besides its effect on hepatocytes, IL-6 acts on several other target cells. As shown in Fig. 5 (for reviews see [22–24], IL-6 functions as a terminal differentiation factor in B-cells. IL-6 is also involved in the activation and proliferation of cytotoxic T-cells.

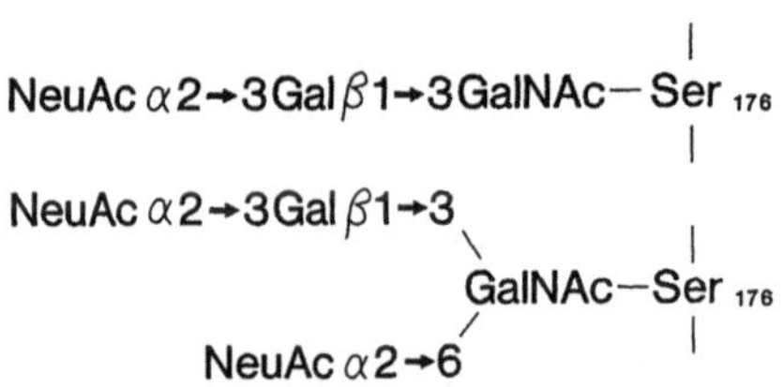

Fig. 4. Predicted structures of the O-linked carbohydrate side chains derived from a component analysis. Peptide #19 was subjected to methanolysis, reacetylation, and trimethylsilylation. Monosaccharide derivatives were analyzed by GLC on a 30-m DB1 capillary column with a Carlo Erba HRGC 5300 Mega series instrument

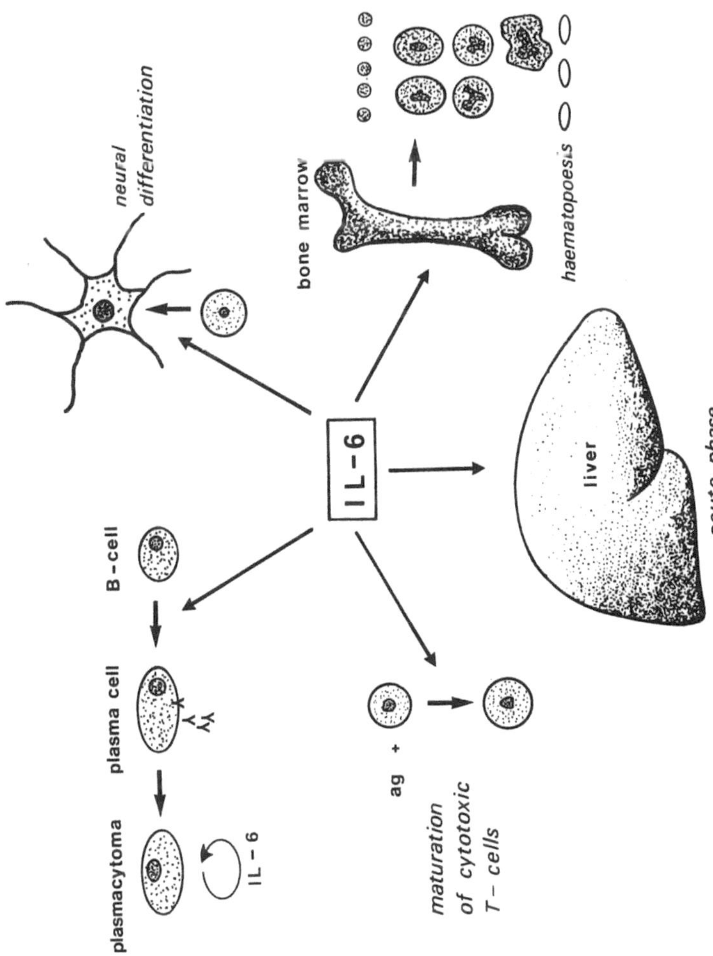

Fig. 5. Pleiotropic actions of IL-6

In cells of neuronal origin IL-6 has been found to change the morphology of axon-extending cells. In hematopoietic stem cells, IL-6 stimulates multipotent colony formation and in hepatocytes IL-6 has been described as the major inducer of acute-phase protein synthesis [8–13].

4.2 Plasma Half-Life and Target Cells of Interleukin-6

When iodinated or [^{35}S]methonine-labeled recombinant human IL-6 was intraveneously injected into a rat, the radioactivity as well as the biological activity of IL-6 disappeared very rapidly from the circulation. The kinetics of the clearance was biphasic; it consisted of a rapid initial disappearance corresponding to a half-life of about 3 min and of a second, slow one corresponding to a half-life of about 55 min [25]. Twenty min after intravenous injection, about 80% of the ^{125}I-rhIL-6 had disappeared from the circulation and was found in the liver. ^{125}I-rhIL-6 was exclusively localized on the surface of parenchymal cells 20 min after injection, suggesting the existence of an IL-6 receptor on hepatocytes [25]. One hour after i.v. injection the radioactive ^{125}I-rhIL-6 had disappeared from the liver and had accumulated in skin reaching 35% of the injected ^{125}-I-rhIL-6 5–8 h after injection [26].

5 Hepatic Interleukin-6 Receptor

Yamasaki et al. [27] were the first to succeed in the cDNA-cloning of an IL-6 receptor from human natural killer cells. In order to determine whether hepatocytes express the same receptor as lymphatic cells, we used the IL-6-receptor cDNA of natural killer cells as a probe for the screening of several human cDNA libraries of hepatic origin. From a cDNA library constructed with poly(A)RNA of dexamethasone-treated human hepatoma cells (HepG2), we isolated and sequenced several positive cDNA clones [28]. The cDNA sequence of the hepatic IL-6-receptor was found to be identical to the one expressed in leukocytes. A protein sequence was deduced from the nucleotide sequence [27]. The IL-6 receptor protein thus consists of a cytoplasmic domain of 82 amino acids, a transmembrane domain of 28 amino acids, and an extracellular domain of about 340 amino acids.

Using the IL-6 receptor cDNA as a probe, we studied the regulation of the hepatic IL-6 receptor in hepatoma (HepG2) cells. Treatment of these cells with dexamethasone, but not with IL-6, IL-1α, IL-1β, or TNF-α led to a time- and dose-dependent upregulation of IL-6-receptor mRNA levels [28, 29]. By using ^{125}I-rhIL-6 and a non-cleavable bivalent cross-linker, subsequent immunoprecipitation, SDS-PAGE, and autoradiography, this effect was also seen at the protein level, i.e., glucocorticoids stimulate the expression of the hepatic IL-6 receptor. Under conditions of IL-6 receptor upregulation by dexamethasone, γ-fibrinogen mRNA induction by IL-6 is stronger and occurs earlier than without dexamethasone, indicating that the expression of the IL-6 receptor might be a rate-limiting step in acute phase protein induction [29].

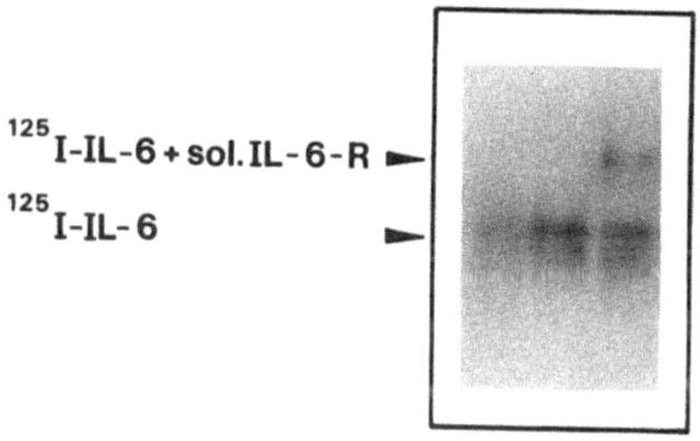

PBS	+	–	–
SN untransf. cells	–	+	–
SN transf. cells	–	–	+
^{125}I-IL-6	+	+	+

Fig. 6. Binding of ^{125}I-rhIL-6 to a soluble IL-6 receptor. A cDNA coding for a soluble IL-6 receptor was stably transfected into NIH/3T3 mouse fibroblasts. Supernatants from these cells, which synthesize and secrete a soluble IL-6-receptor, were assayed for IL-6 binding. Incubation with supernatants from transfected cells but not from control cells or PBS led to the formation of a slower migrating complex after separation on a native polyacrylamide gel, indicating the presence of a soluble receptor/ligand complex

When the hepatic IL-6 receptor cDNA was stably expressed in HepG2 cells, these cells became more sensitive to low concentrations of IL-6 with respect to the induction of γ-fibrinogen mRNA [28]. Interestingly, the same effect was seen when a genetically engineered soluble IL-6 receptor was added to HepG2 cells.

In most recent experiments (Schooltink et al., in preparation) we have expressed the soluble human IL-6 receptor in *E. coli* and also in mouse fibroblasts. The soluble receptor binds ^{125}I-rhIL-6 as shown by its different mobility in a native polyacrylamide gel (Fig. 6). It is also of interest that the human IL-6-receptor overexpressed in *E. coli* is nonglycosylated and is still capable of binding ^{125}I-rhIL-6.

5.1 Structure of the Hepatic IL-6 Receptor

In order to identify the IL-6 binding protein(s) in liver cells, ^{125}I-rhIL-6 was cross-linked to HepG2 cells with the homobifunctional cross-linker bis-(sulfosuccinimidyl) suberate. After SDS/PAGE under reducing conditions and autoradiography three bands with molecular masses of 100, 120, and 200 kDa (Fig. 7) were observed. An obvious interpretation of this finding would be that the IL-6 receptor is composed of monomers (100 kDa) and homodimers (200 kDa). In the model shown in Fig. 8, the 120-kDa polypeptide complex consists of an IL-6 dimer cross-linked to one 80-kDa IL-6 receptor polypeptide. Indeed, we were able to present evidence for such a structure in NIH/3T3 cells, which had been transfected with the human IL-6 receptor cDNA coding for the 80-kDa polypeptide [30]. A more complicated structure seems to exist in HepG2 cells. Cross-linking of ^{125}I-rhIL-6 also results in the appearance of three IL-6-containing complexes of apparent molecular masses of 100, 120, and 200 kDa. When the three ^{125}I-rhIL-6-containing protein bands were digested with

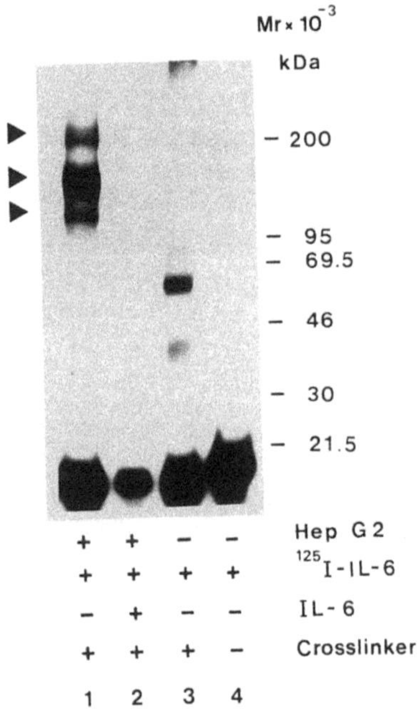

Fig. 7. Cross-linking of ^{125}I-rhIL-6 to HepG2 cells. 2×10^7 HepG2 cells were incubated with ^{125}I-rhIL-6 in the absence *(lane 1)* or presence *(lane 2)* of unlabeled rhIL-6 and cross-linked with bis-(sulfosuccinimidyl)-suberate. ^{125}-I-rhIL-6-containing complexes were immunoprecipitated with a specific polyclonal antiserum against rhIL-6 and subject to SDS/PAGE and autoradiography. Controls: ^{125}I-rhIL-6 with *(lane 3)* or without cross-linker *(lane 4)*

Staphylococcus aureus V8 protease, the cleavage patterns of the 100- and 120-kDa ^{125}I-rhIL-6-containing complexes differed (Fig. 9) and may therefore reflect the existence of nonidentical polypeptides. The digestion of the 200-kDa ^{125}I-rhIl-6-containing complex with high concentrations of V8 protease, however, resulted in a peptide pattern, which is similar to the digestion patterns of both the 100- and 120-kDa bands. Thus, the 200-kDa ^{125}I-rhIL-6-containing complex is likely to be composed of the 80- and 100-kDa polypeptides, forming a heterodimeric receptor complex. To determine wether both polypeptide components of the IL-6 receptor are able to bind IL-6 independently, we solubilized HepG2 cells by Triton X-100 before or after the addition of ^{125}I-rhIL-6 and cross-linking. Only when the ligand had been added prior

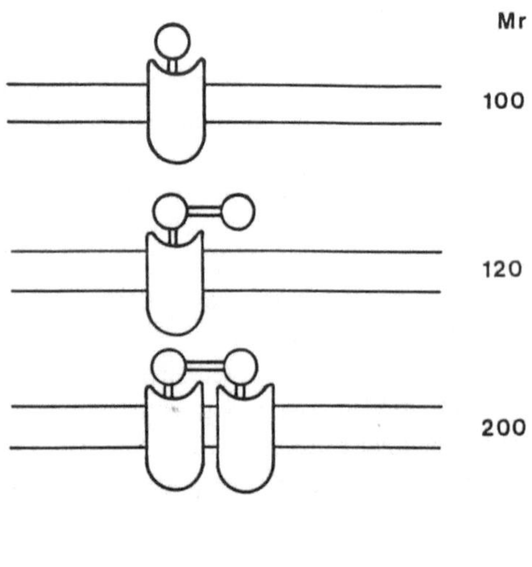

Fig. 8. Possible organization of the IL-6 receptor in the plasma membrane; o, IL-6; ᴗ, IL-6 receptor; =, cross-linker

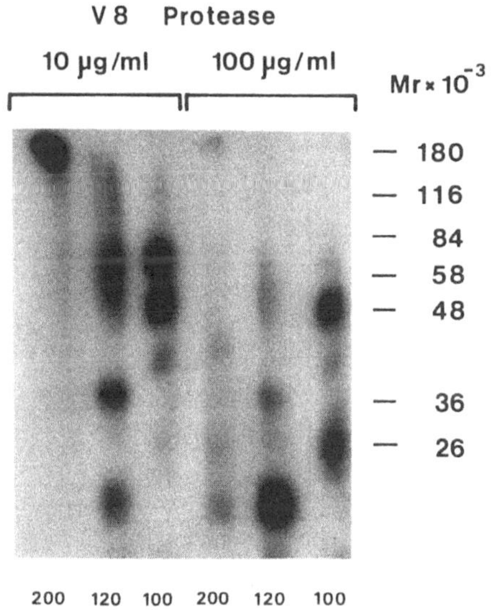

Fig. 9. *Staphylococcus aureus* V8 protease digestion of cross-linked ^{125}I-rhIL-6-containing complexes. The cross-linked ^{125}I-rhIL-6-containing protein complexes of molecular masses of 100, 120, and 200 kDa (see Fig. 7) were excised from the SDS-polyacrylamide gel. Gel slices were loaded on wells of a 10% SDS-polyacrylamide gel and overlayed with V8 protease (at concentrations of 10 or 100 mg/ml) in a mixture of 0.1 M Tris/HCl, pH 6.8, containing 20% glycerol, 0.1% SDS, and 0.1% bromophenol blue. Samples were advanced by electrophoresis into the stacking gel and left there for 30 min with the power turned off. Electrophoresis was then completed in the usual fashion and autoradiograms were prepared

to cross-linking and solubilization, all three ^{125}I-rhIL-6-containing complexes were detected. No such coprecipitation was observed with the anti-80-kDa serum, when cells were first solubilized and then exposed to the ligand and cross-linked. The 120-kDa band is found, however, when the supernatant of this immunoprecipitation was subjected to a second immunoprecipitation with anti-IL-6 serum.

6 Signal Transduction

We (unpubl. data) and others have shown that cAMP, cGMP, PKC, inositol phosphates and changes in intracellular calcium concentrations are not involved in the transduction of the IL-6 signal in rat and human hepatoma cells. As a first approach to elucidate the mechanism(s) of IL-6 signaling, we overexpressed the human IL-6 receptor in mouse fibroblasts (NIH/3T3) and in human hepatoma cells (HepG2). The stably transfected NIH/3T3 cells expressed high levels of IL-6 receptor mRNA and about 20000 functional IL-6 receptors per cell [30]. The binding of ^{125}I-rhIL-6 to these transfected fibroblasts can be visualized by autoradiography. When these transfected NIH/3T3 cells overexpressing the IL-6 receptor were incubated with rhIL-6, an inhibition in cell growth was observed. No such inhibition was seen when untrans-

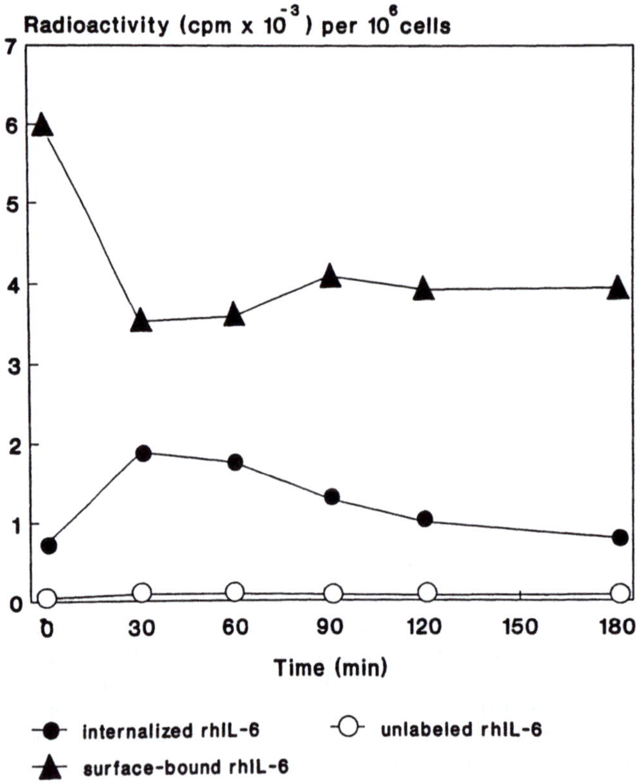

Fig. 10. Internalization of ^{125}I-rhIL-6 by transfected NIH/3T3 cells. Transfected NIH/3T3 cells were incubated with ^{125}I-rhIL-6 at 4 °C for 2 h and the temperature was then shifted to 37 °C. At the times indicated total cell-association ^{125}I-rhIL-6 (▲) and internalized ^{125}I-rhIL-6 (resistant to trypsin incubation) (●) were determined. As a control (○) we carried out the internalization experiment in the presence of a 100-fold excess of unlabeled rhIL-6

fected control cells, which do not express an IL-6 receptor, were incubated with rhIL-6. Similar effects of IL-6 on cell growth have been described for myeloblasts, human breast carcinoma cells [31], lymphoma cells, and HeLa cells (see review [23].

When the fate of ^{125}I-rhIL-6 after binding to the surface receptors of either transfected HepG2 or NIH/3T3 cells was studied, we found an internalization of the ligand. Transfected NIH/3T3 cells were loaded for 2 hours with ^{125}I-rhIL-6 at 4 °C. Then the temperature was shifted to 37 °C and the cells were incubated for various periods of time. It can be seen from Fig. 10 that the cell-associated radioactivity (filled triangles) declined, whereas the radioactivity resistant to trypsin treatment (internalized IL-6, filled circles) increased. It remains to be elucidated whether IL-6 internalization and signal transduction are functionally linked. An internalization after binding has been described for various cytokines [32, 33], but no connection to the biological response has yet been demonstrated.

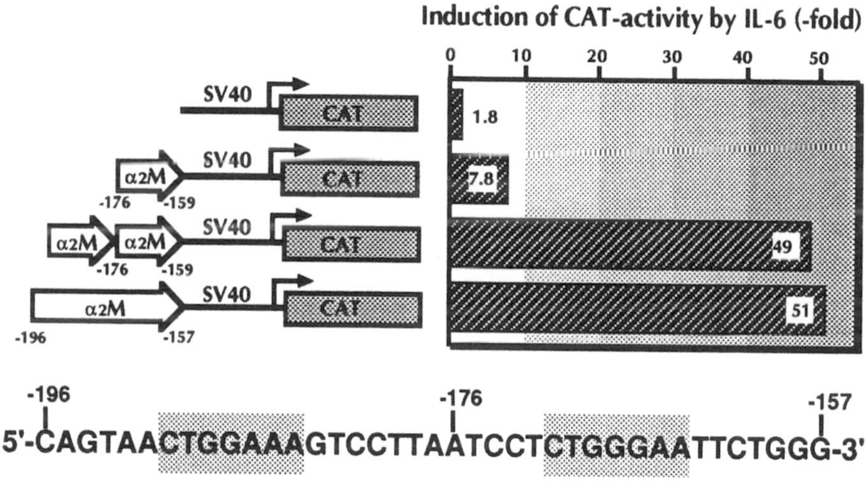

Fig. 11. Transient transfection of HepG2 cells with rat α_2-macroglobulin promoter constructs. Plasmids containing the indicated regions of the rat α_2-macroglobulin promoter fused to the SV40 promoter and the CAT reporter gene were transiently transfected into HepG2 cells by calcium phosphate precipitation. After treatment with IL-6 for 20 h, the cells were harvested and the CAT activity in the extracts was determined by measuring the rate of chloramphenicol acetylation

7 Acute Phase Protein Regulation by Interleukin-6

In order to understand the mechanism of acute phase protein expression by IL-6, we have concentrated our work on α_2-macroglobulin, the major acute phase protein in the rat. In earlier studies, we found that the rise in α_2-macroglobulin serum levels of IL-6-treated rats is preceded by an increase in liver α_2-macroglobulin mRNA concentrations [14, 34]. We and others have isolated the gene for α_2-macroglobulin by chromosome walking [35, 36, 37], characterizing its 5'-end [36], and constructed a series of plasmids in which various portions of the α_2-macroglobulin promoter sequences were fused to the bacterial reporter gene chloramphenicol acetyltransferase (CAT). These plasmids were transiently transfected into HepG2 cells and the promoter activity was determined by measuring the expressed CAT activity. We identified an IL-6-responsive element between base pairs – 215 and – 165 upstream of the transcriptional start site which contains a tandem of the heptanucleotide motif CTGG(G/A)AA found in the IL-6-responsive elements of other acute phase protein genes [38, 39, 40]. The isolated IL-6-responsive element, ligated to an SV40 promoter in front of the CAT gene, gave rise to a 51-fold stimulation of CAT activity by IL-6 (Fig. 11). When half of the element (– 159 to – 176) was similarly used, only a 7.8-fold stimulation of CAT activity was achieved in transfected HepG2 cells. An analogous construct containing two such motifs, however, resulted in a 49-fold stimulation of CAT activity after the action of IL-6. This finding indicates the importance of the presence of the two tandemly arranged heptanucleotide motifs for full inducibility of the rat α_2-macroglobulin promoter by IL-6. By gel retardation assays

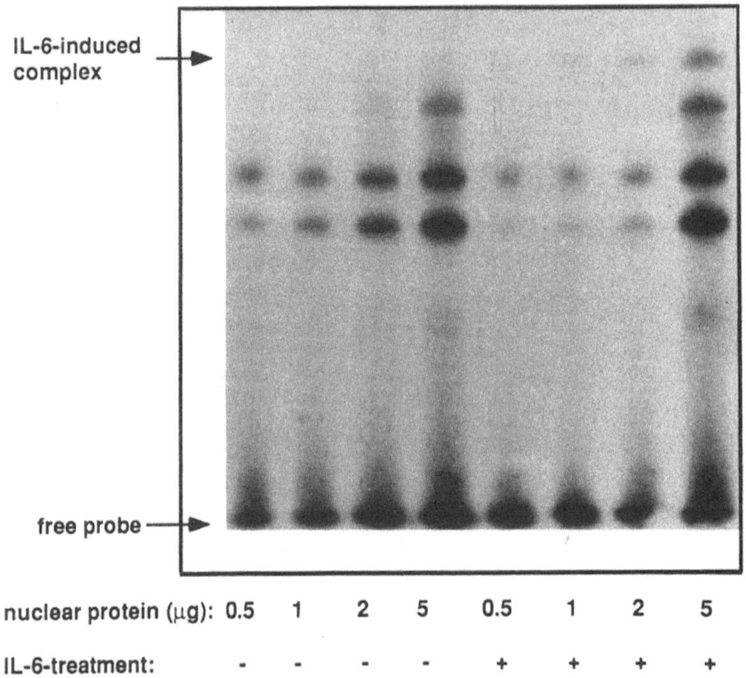

Fig. 12. Gel retardation assay using nuclear extracts from HepG2 cells. The indicated amounts of nuclear protein from HepG2 cells, either untreated or treated with IL-6 for 16 h, were incubated with a [^{32}P]-labeled synthetic oligonucleotide containing twice the sequence –176 to – 159 of the rat α_2-macroglobulin promoter. The protein-DNA complexes formed were separated by electrophoresis through a 4% polyacrylamide gel under nondenaturing conditions at 10 V/cm. The gel was dried and the bands were visualized by autoradiography

using nuclear extracts from HepG2 cells, treated either with or without IL-6, we could demonstrate the induction by IL-6 of a nuclear factor which binds specifically to the – 159 to – 176 element (Fig. 12). In contrast to the situation in rat, α_2-macroglobulin is not an acute phase protein in man, where serum levels of this protein are constitutively high. In collaboration with Y. Sakaki (Nagasaki, Japan) and P. Marynen (Leuven, Belgium), we isolated the 5'-end of the human α_2-macroglobulin gene and fused it to the CAT reporter gene. In HepG2 cells transfected with this construct, CAT activity – to our surprise – could be induced by IL-6 (not shown). By preparing deletion mutants of the promoter we localized an IL-6-responsive element in the region from – 248 to – 211. This element conferred IL-6 inducibility to an SV40 promoter (Fig. 13). Although there is no obvious sequence homology to the rat α_2-macroglobulin IL-6-responsive element, gel retardation assays with the human IL-6-responsive element resulted in a band pattern similar to that shown for the rat element in Fig. 12, indicating the binding of the same or a similar IL-6-induced factor to the two elements.

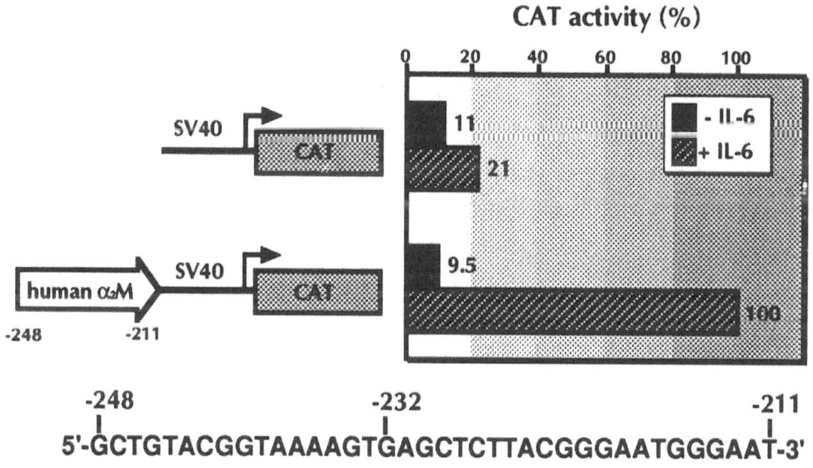

Fig. 13. Identification of an IL-6-responsive element in the human α_2-macroglobulin promoter. A plasmid containing the indicated region of the human α_2-macroglobulin promoter fused to the SV40 promoter and the CAT gene was used for transient transfection in HepG2 cells as described in Fig. 11

8 Summary

Interleukin-6 (IL-6) is a multifunctional cytokine involved in immune defense, hematopoesis, and acute phase protein regulation. As a secretory protein IL-6 is synthesized as a precursor with a signal peptide of 28 amino acids. IL-6 is O-glycosylated at threonine 138 and only about 10% of the cytokine carries N- (at asparagine 45) and O-linked carbohydrate side chains. Compared to nonglycosylated IL-6, the glycosylated cytokine exhibits a three to four times higher specific activity.

From structure-function studies it became evident that the last three C-terminal amino acid residues of IL-6 play an important role in biological activity, in particular the positive charge of arginine 182 and the α helical structure at the C-terminus.

IL-6 exerts its action on liver cells via a specific surface receptor. We have cloned the cDNA of the human hepatic IL-6-receptor and shown that glucocorticoids upregulate IL-6-receptor mRNA.

Hepatoma cells transfected with the human IL-6-receptor cDNA, expressing between 10000 to 20000 functional receptors, were found to be more sensitive to low IL-6 concentrations than untransfected cells, i.e., they synthesized larger amounts of acute phase protein mRNA.

Hepatoma cells (HepG2) as well as liver cells overexpressing the human hepatic IL-6 receptor internalize IL-6. Whether internalization and signal transduction are linked processes remains to be elucidated.

When the 5'-end of the major acute phase protein of the rat, α_2-macroglobulin, linked to the reporter gene CAT, was transfected into HepG2 cells, stimulation of CAT activity by IL-6 was observed. By using deletion mutants in the promoter region of the rat α_2-macroglobulin gene, we were able to identify an IL-6-responsive element consisting of two adjacent heptanucleotides (underlined):

– 196 – 157
5' CAGTAACTGGAAAGTCCTTAATCCTTCTGGGAATTCTGGC3'.

Future studies are necessary in order to link the stimulation of the IL-6 membrane receptor with acute phase protein gene activation in the nucleus.

Acknowledgments. This work was supported by grants from the Deutsche Forschungsgemeinschaft, Bonn, and the Fonds der Chemischen Industrie, Frankfurt. The authors would like to thank E. Gaeta and I. Hagemann for their help with the preparation of this manuscript.

References

1. Kushner I (1982) Ann NY Acad Sci 389:39–49
2. Koj A (1974) In: Allison AC (ed) Structure and function of plasma proteins, vol. 1. Plenum Press, London, pp 73–131
3. Heinrich PC, Castell JV, Andus T (1990) Biochem J 265:621–636
4. Van Leuven F (1982) Trends Biochem Sci 7:185–187
5. Barrett AJ, Starkey PM (1973) Biochem J 33:709–724
6. Flohe S, Heinrich PC, Schneider J, Wendel A, Flohe L (1991) Biochem. Pharmacol 41:1607–1614
7. Bauer J, Birmelin M, Northoff GH, Northemann W, Tran-Thi TA, Ueberberg H, Decker K, Heinrich PC (1984) FEBS Lett 177:89–94
8. Gauldie J, Richards C, Harnish D, Lansdorp P, Baumann H (1987) Proc Natl Acad Sci USA 84:7251–7255
9. Andus T, Geiger T, Hirano T, Northoff H, Ganter U, Bauer J, Kishimoto T, Heinrich PC (1987) FEBS Lett 221:18–22
10. Andus T, Geiger T, Hirano T, Kishimoto T, Tran-Thi TA, Decker K, Heinrich PC (1988) Eur J Biochem 173:287–293
11. Castell JV, Gomez-Lechon MJ, David M, Hirano T, Kishimoto T, Heinrich PC (1988) FEBS Lett 232:347–350
12. Castell JV, Gomez-Lechon MJ, David M, Andus T, Geiger T, Trullenque R, Fabra R, Gerok W, Heinrich PC (1988) FEBS Lett 242:237–239
13. Castell JV, Gomez-Lechon MJ, David M, Fabra R, Trullenque R, Heinrich PC (1990) Hepatology 12:1179–1186
14. Geiger T, Andus T, Klapproth J, Hirano T, Kishimoto T, Heinrich PC (1988) Eur J Immunol 18:717–721
15. Marinkovic S, Jahreis GP, Wong GG, Baumann H (1989) J Immunol 142:808–812
16. Hirano T, Yaksukawa K, Harada H, Taga T, Watanabe Y, Matsuda T, Kashiwamura S, Nakajima K, Koyama K, Iwamatsu A, Tsunasawa S, Sakiyama F, Matsui H, Takahara Y, Taniguchi T, Kishimoto T (1986) Nature (London) 324:73–76
17. May LT, Ghrayeb J, Santhanam U, Tatter SB, Sthoeger Z, Helfgott DC, Chiorazzi N, Grieninger G, Sehgal PB (1988) J Biol Chem 263:7760–7766
18. Gross V, Andus T, Castell J, Vom Berg D, Heinrich PC, Gerok W (1989) FEBS Lett 247:323–326
19. Schiel X, Rose-John S, Dufhues G, Schooltink H, Heinrich PC (1990) Eur J Immunol 20:883–887

20. Krüttgen A, Rose-John S, Dufhues G, Bender S, Lütticken C, Freyer P, Heinrich PC (1990) FEBS Lett 273:95–98
21. Lütticken C, Krüttgen A, Möller C, Heinrich PC, Rose-John S (1991) FEBS Lett 282:265–267
22. Van Snick J (1990) Annu Rev Immunol 8:253–278
23. Hirano T, Kishimoto T (1990) In: Sporn MB and Roberts AB (eds) Handbook of experimental pharmacology, vol. 95/1. Peptide growth factors and their receptors. Springer, Berlin Heidelberg New York, pp 633–665
24. Hirano T, Akira S, Taga T, Kishimoto T (1990) Immunol Today 11(12):443–449
25. Castell JV, Geiger T, Gross V, Andus T, Walter E, Hirano T, Kishimoto T, Heinrich PC (1988) Eur J Biochem 177:357–361
26. Castell JV, Klapproth J, Gross V, Walter E, Andus T, Snyers L, Content J, Heinrich PC (1990) Eur J Biochem 189:113–118
27. Yamasaki K, Taga T, Hirata Y, Yawata H, Kawanishi Y, Seed B, Taniguchi T, Hirano T, Kishimoto T (1988) Science 241:825–828
28. Schooltink H, Stoyan T, Lenz D, Schmitz H, Hirano T, Kishimoto T, Heinrich PC, Rose-John S (1991) Biochem J 277:659–664
29. Rose-John S, Schooltink H, Lenz D, Hipp E, Dufhues G, Schmitz H, Schiel X, Hirano T, Kishimoto T, Heinrich PC (1990) Eur J Biochem 190:79–83
30. Rose-John S, Hipp E, Lenz D, Legres L, Korr H, Hirano T, Kishimoto T, Heinrich PC (1991) J Biol Chem 266:3841–3846
31. Chen L, Mory Y, Zilberstein A, Revel M (1988) Proc Natl Acad Sci USA 85:8037–8041
32. Mizel B, Kilian PL, Lewis JV, Paganelli KA, Chizzonite RA (1987) J Immunol 138:2906–2912
33. Grenfell ST, Smithers N, Miller K, Solari R (1989) Biochem J 264:813–822
34. Northemann W, Heisig M, Kunz D, Heinrich PC (1985) J Biol Chem 260:6200–6205
35. Northemann W, Shiels BR, Braciak TA, Hanson RW, Heinrich PC, Fey GH (1988) Biochemistry 27:9191–9203
36. Kunz D, Zimmermann R, Heisig M, Heinrich PC (1989) Nucleic Acids Res 17:1121–1138
37. Ito T, Tanahashi H, Misumi Y, Sakaki Y (1989) Nucleic Acids Res 17:9425–9435
38. Oliviero S, Cortese R (1989) EMBO J 8:1145–1151
39. Prowse KR, Baumann H (1988) Mol Cell Biol 8:42–51
40. Huber P, Laurent M, Dalmon J (1990) J Biol Chem 265:5695–5701

Transcriptional Control of Liver Acute Phase Genes by Interleukin-6 and Leukemia Inhibitory Factor

G. HOCKE[1], G. BAFFET[1], M.-Z. CUI[1], T. BRECHNER[1], D. BARRY[1], A. GOEL[1], R. FLETCHER[1], C. ABNEY[1], M. HATTORI and H. FEY[1]

1 Introduction

The acute phase genes code for the acute phase proteins, a set of blood proteins with characteristic changes in their plasma concentrations occuring during acute and chronic inflammations. This is part of the overall acute phase response, the collective term for a complex set of systemic defense reactions of the host organism against tissue injury and infections. The acute phase proteins are mainly produced by liver hepatocytes and provide protective value for the organism in its defense against the original insult. Their functions and their regulation by mediators of inflammation have been adequately reviewed (Baumann 1989; Baumann and Gauldie 1990; Fey and Gauldie 1990; Heinrich et al. 1990; Toniatti et al. 1990). Briefly, macrophages, monocytes, and secondary cell types, including endothelial cells and keratinocytes produce the cytokines interleukin 1 (IL1), interleukin 6 (IL6), leukemia inhibitory factor (LIF), tumor necrosis factor (TNF), and interferons at peripheral sites of injury. These hormones participate both in local defense reactions and are disseminated with the bloodstream over the whole organism, triggering a systemic response. Liver hepatocytes have specific cell surface receptors for these hormones, and binding of the bloodborne mediators to the receptors induces the expression of the acute phase genes. The gene activation is primarily transcriptional, but posttranscriptional components may also occur. The intracellular signal cascades triggered by these cytokines are still incompletely understood. IL1 and IL6 receptors, but not the LIF receptor, have been purified and cloned. However, the postreceptor steps in their respective signal cascades are still not well described. The analysis of these signal cascades in different cell types, which participate in host defense reactions and respond to these mediators, is an important task for contemporary inflammation research. The inflammatory response can actually be considered a beautiful paradigm for a fundamentally interesting problem in gene regulation: how can one cell and ultimately one gene respond to several hormonal signals that arrive simultaneously? In the past few years, research in gene regulation has made good progress in understanding how a particular gene responds to a particular hormone, for example, steroid hormones, glucocorticoids, retinoic acids, and interferons. However, we are only just beginning to discover how one gene may respond to multiple hormones. During an inflammatory response, many of the cytokines described above are simultaneously present in the bloodstream at rapidly changing concentrations. Liver cells and other hormone-responsive cells must

[1] Department of Immunology, Research Institute of Scripps Clinic, La Jolla, CA 92037, USA

42. Colloquium Mosbach 1991
Molecular Aspects of Inflammation

Class	Genes		Inducers
1	human:	C-reactive protein (CRP); serum-amyloid A (SAA); complement C3; hemopexin; haptoglobin & others;	IL1; IL1+IL6; IL1+IL6+ glucocorticoids;
	rat:	SAA; C3; hemopexin, haptoglobin; α_1 acid glycoprotein	
2	rat:	α_1 antitrypsin; α_1 antichymotrypsin; thiostatin (=α_1 cysteine proteinase inhibitor); fibrinogen; α_2 macroglobulin (α_2M)	IL6; IL6+gluco-corticoids; LiF;

Fig. 1. Two classes of liver acute phase genes categorized by the hormones that are their main inducers

rapidly respond to the changes in the composition of this complex cocktail of inflammatory mediators to fine-tune and adapt the response to the needs of the situation.

In this report, we have chosen the rat α_2-macroglobulin (α_2M) gene as a prototype acute phase gene. It represents a class of acute phase genes (class 2; Baumann and Gauldie 1990; Fig. 1), which are regulated mainly by IL6 and LIF, by combinations of IL6 and glucocorticoids, and most potently by combinations of IL6, glucocorticoids, and LIF. We report a series of experiments that allowed us to identify an IL6-responsive cis-acting DNA control element, the IL6-response element (IL6-RE). We describe a nuclear protein that specifically binds at the IL6-RE, and is altered by the IL6 signal cascade in a characteristic manner. We propose to call this protein the IL6-RE binding protein (IL6 RE-BP) and postulate that it is a transcription control factor and mediates the IL6 signal to the transcriptional machinery assembling over the α_2M gene's promoter. We report the surprising new result that the LIF-response element (LIF-RE) is contained within the 18-bp sequence centered around the IL6-RE, and thus is possibly identical with the IL6-RE. Preliminary results indicate that the protein factors binding at this element and mediating the IL6 and LIF responses are probably identical.

These results suggest the exciting new concept that the IL6 and LIF signal cascades may converge at a common endpoint, although they originate at different IL6 and LIF specific cell surface receptors. Therefore, the protein factor binding at the IL6-RE may serve as a molecular integrator to add up the signals from both cascades. This is an attractive hypothesis, particularly relevant to inflammation research, because it may help us to understand more clearly how multiple hormonal signals arising in an inflammatory situation are ultimately integrated into one single readout at the gene level.

2 Results and Discussion

The acute phase genes have been divided into two classes according to the hormones that are their principal inducers (Baumann and Gauldie 1990; Fig. 1). Class 1 genes are mainly regulated by IL1, and by combinations of IL1 plus IL6 or IL1 plus IL6 plus glucocorticoids. Class 2 genes are mainly regulated by IL6, but also by LIF, by combinations of IL6 plus glucocorticoids, and most potently by combinations of IL6 plus LIF plus glucocorticoids.

To identify the IL6-RE, a series of plasmid constructs were produced that contained the α_2M gene's own promoter region plus variable extents of its promoter-upstream 5'-flanking region. The luciferase gene was inserted as a reporter gene in these constructs. The constructs were used to map the response element both by transient and stable transfection experiments in the rat hepatoma cell line FAO. This cell line had previously been shown to respond, "faithfully", i.e., in a similar manner as rat livers in a living animal, to these hormones with respect to the induction of the α_2M gene. The result of a stable transfection experiment is shown in Fig. 2A, the result of a series of transient expression experiments in Fig. 2B. A cell line was stably transfected with a construct carrying 1151 bp of the 5'-flanking sequences plus a neomycin (G418) resistance plasmid, and several stably transformed lines were derived by single cell cloning. The line FAO/HB3 was used as a representative, and the other lines showed similar behavior. When this line was treated with glucocorticoids alone (10^{-7} M dexamethasone, dex), a less than twofold activation of the promoter occurred. When it was treated with IL6 alone, a 10- to 20-fold activation took place, and simultaneous treatment with IL6 plus glucocorticoids resulted in a 30- to 40-fold activation (Hattori et al. 1990; Fig. 2A). Thus, IL6 alone was sufficient to induce this promoter. IL6 plus glucocorticoids acted synergistically rather than simply additively. We concluded that the 1150-bp flanking region contained response elements sufficient to generate an IL6 response and the synergistic response to both hormones.

A series of 5'-deletion constructs with progressive deletions of the 5' boundary from −1151 to −48 bp were produced and transiently transfected into FAO rat hepatoma cells. The cells were then either treated with dexamethasone (dex) alone or IL6 plus dex, and the specific inducibility by IL6 was determined, defined as the ratio of the induction obtained with IL6 plus dex over that obtained with dex alone. The result (Fig. 2B) showed that a sequence important for IL6 induction was located between −209 and −150 bp, approximately. Removal of this region resulted in a 60% loss of IL6-specific inducibility. IL6 clearly was the main inducer of the α_2M promoter. Increasing doses of dex alone from 10^{-10} to 10^{-6} M never gave rise to more than a two-fold induction in the stably transformed cell line FAO/HB3; however, when a constant dose of IL6 (100 units/ml) was combined with increasing concentrations of dex, a clear dose response was observed with the maximum synergistic effect reached at approximately $1-5 \times 10^{-7}$ M dex (Fig. 3). A series of internal deletion constructs were produced (Fig. 4A) and transient transfections into FAO cells confirmed that the area most critically required for IL6 action was located between −156 and −209 bp upstream of the transcription start site of the α_2M gene. Deletion of this region caused a 63% loss in hormonal inducibility (Fig. 4A). A series of linker scan mutants were produced, in which individual blocks of six to ten nucleotides over this

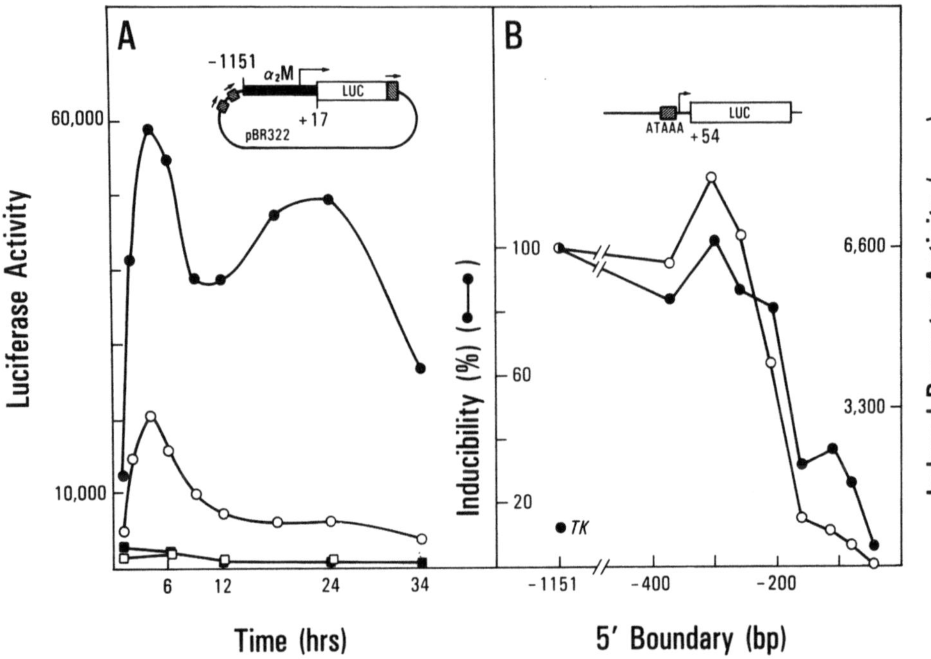

Fig. 2. Mapping of the IL6-RE of the rat α_2M gene by **A** stable and **B** transient transfection. FAO rat hepatoma cells were stably transfected with the construct shown in the *insert* and a neomycin (G418) resistance plasmid. Neomycin-resistant, clonally derived single cell lines were established. This experiment was performed with the line FAO/HB3 carrying multiple copies of this construct stably associated with the host cell genome (Hattori et al. 1990). Promoter activity after hormonal treatment was measured as luciferase activity generated from the transfected construct. -□- Treatment with buffer alone; -■- treatment with 10^{-7} M dex; -○- treatment with 100 U/ml IL6 alone; -●- treatment with 10^{-7} M dex plus 100 U/ml Il6. **B** The result of a 5' deletion series is shown. Each construct with a progressively shorter 5' boundary of the α_2M gene control region (*abscissa*) was transfected into FAO cells, and the cells were treated either with dex alone or with IL6 plus dex, as described (Hattori et al. 1990). The absolute promoter activities (normalized luciferase activity per standard amount of protein extract, -○-) and specific inducibilities (-●-) are plotted. Specific inducibility by IL6 is defined for each construct as the ratio of inductions obtained with IL6 + dex over that obtained with dex alone. Normalized to 100% for the longest construct

critical region were substituted with nonrelevant sequences (Fig. 4B). These constructs were again used in transient transfection and expression experiments in FAO cells, and the cells were subsequently treated with IL6 and corresponding controls. The mutant m205, replacing the hexanucleotide sequence –159 to –164 (CTGGGA), produced the single largest effect, a 57% loss in inducibility by IL6. Therefore, we called this element the hexanucleotide core of the IL6-RE. Other authors have also reached similar conclusions, based on a less extensive set of mutants that did not allow them to define the element with similar precision as the results presented here and elsewhere (Kunz et al. 1989, Hattori et al. 1990).

An initial series of experiments was performed attempting to demonstrate that this element was not only necessary, but also sufficient to confer IL6 responsiveness to

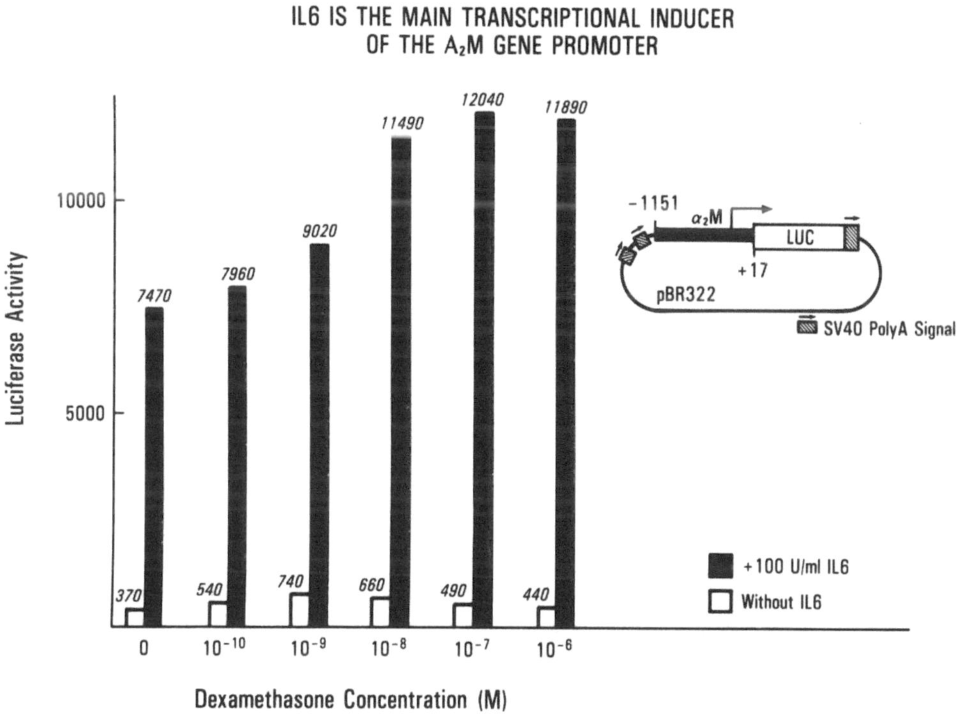

Fig. 3. Interleukin 6 is the main inducer of the α_2M gene promoter, and synergizes with glucocorticoids. The stably transfected cell line FAO/HB3, carrying the same construct as shown in Fig. 2A, was treated with increasing concentrations of dex either in the absence (*open bars*) or in the presence (*filled bars*) of 100 units/ml of IL6

other promoters, which were not normally IL6-responsive in their original state. For these experiments the minimal promoter of the herpes virus thymidine kinase (TK) gene was chosen, containing only its TATA-box region extending to –105 bp, but not its own enhancer region. Multiple copies of the α_2M element from –159 to –209 were sufficient to confer IL6 responsiveness to the TK promoter (Fig. 5). The element functioned in an orientation-independent manner, and thus had some of the defining properties of a hormone-regulated enhancer. These results were later confirmed using different promoters (Figs. 9, 10). The TK promoter was not the best choice for these experiments, because even this minimal promoter showed a 3.7-fold inducibility by IL6. Thus, the differential effects obtained for the hormone with constructs using this promoter plus enhancers over the promoter alone were not optimal. However, the experiments did establish that the element was sufficient to confer IL6 inducibility to heterologous promoters in cis.

DNAse I footprinting and methylation interference studies were performed with nuclear protein extracts from normal and acute phase rat livers (Fig. 6; Hattori et al. 1990). The results showed two windows of protection against DNAse I digestion (I and II), on both DNA strands. Window II was centered over the hexanucleotide core CTGGGA, and extended by about six to seven nucleotides beyond it on either side.

A

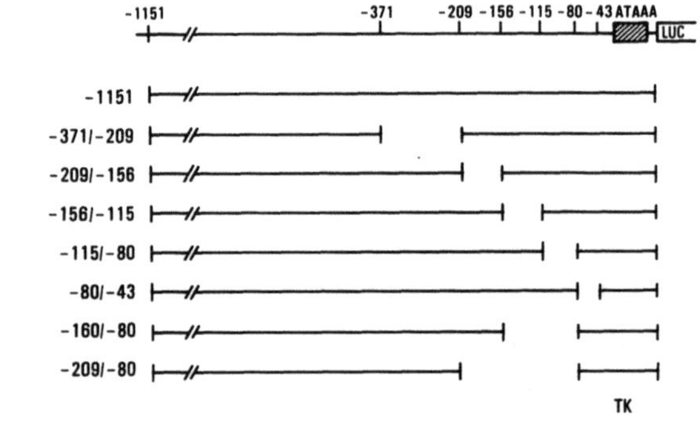

	Relative Luciferase Activity			
	IL6			
	-	+	Fold	(%)
-1151	100	1815±33	18.2	100
-371/-209	87±18	1383±190	15.9	87
-209/-156	75±20	511±64	6.8	37
-156/-115	64±15	970±242	15.2	84
-115/-80	44±3	871±72	19.8	109
-80/-43	20±1	322±21	16.1	88
-160/-80	80±16	1100±216	13.8	76
-209/-80	69±14	214±58	3.4	19
TK	465±12	1098±177	2.4	13

B

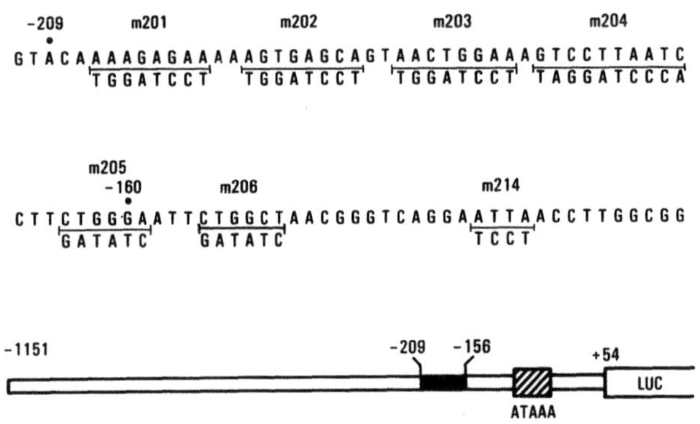

	-	+	Fold	(%)	CM (%)
WILD	100	2426±49	24.3	100	100
m201	74±7	1879±163	25.4	105	85
m202	70±12	1457±209	20.8	86	88
m203	78±5	1521±183	19.5	80	86
m204	66±15	1769±313	26.8	110	90
m205	56±8	582±78	10.4	43	40
m206	66±12	2305±123	34.9	144	116
m214	58±9	1171±164	20.2	83	94
TK	433±105	1431±234	3.3	14	-

Fig. 4. Mapping the IL6-RE by internal deletions and linker scan mutations. **A** Internal deletions were constructed as shown. The plasmids were transfected into FAO cells and either kept untreated or treated with IL6. Luciferase activities were then determined as a measure of the transcriptional activity of each construct. The relative increases are given in the column headed *Fold*. They are normalized to the wild-type construct defined as 100% in the column headed *%*. **B** A series of linker scan mutants were produced, in which the short sequence blocks shown under the wild-type sequence were substituted by oligonucleotide-directed, site-specific mutagenesis. *TK* Thymidine kinase minimal promoter as a control

Window I was centered over a related sequence located 20 bp upstream, CTGGAAA, which we called the IL6-RE core homology, because it was identical in 6/7 positions with the core sequence. Methylation interference studies showed that the G-residues in both the core and core homology regions were involved in protein DNA contacts in similar characteristic patterns (Fig. 6). These data suggested that possibly the same protein or proteins recognizing these sequences in very similar fashions were binding at the core and core-homology elements. The binding studies initially defined the core-homology experiment, whereas the functional data (transfection and mutagenesis) pointed to the core element as the functionally dominant element, since mutation of the core-homology sequence led only to a 20% loss in IL6 inducibility (Fig. 4B,

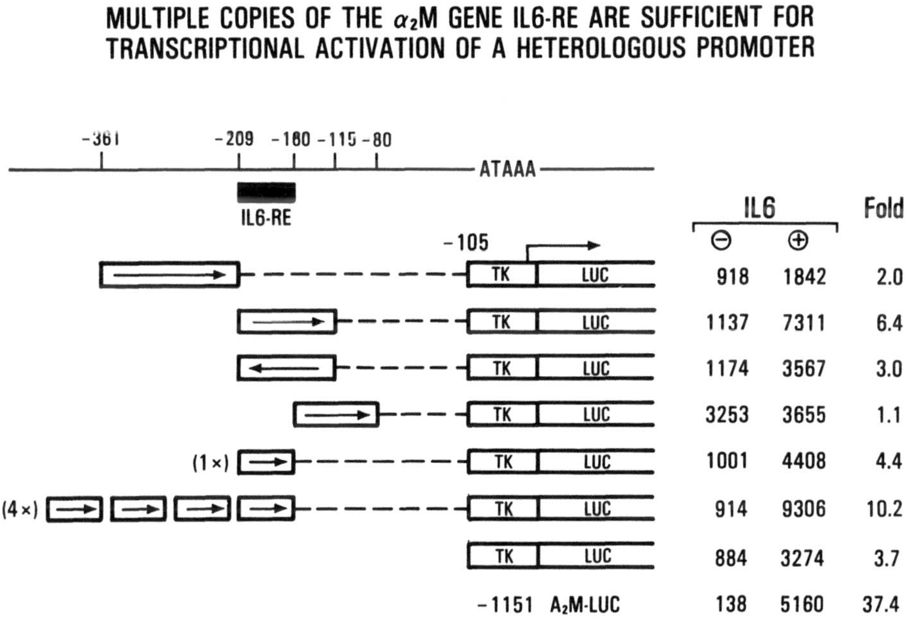

Fig. 5. Multiple copies of the IL6-RE region of the α_2M gene are sufficient to confer IL6 control to heterologous promoters. Restriction fragments as shown in *open boxes* with *arrows inside* were linked to the thymidine kinase (*TK*) minimal promoter and transfected into FAO hepatoma cells. The cells were then either kept untreated or treated with IL6 and the luciferase activities are given as well as the relative induction in the column headed *Fold*. The positive control was the plasmid shown in the *bottom row* that contained 1151 bp of the 5'-flanking region. This was the same plasmid as shown in Figs. 2A and 3, giving a 37.4-fold induction with IL6

mutant 203). However, we believe that in the real promoter both the core and the core-homology elements contribute functionally to the overall IL6 response, probably in a cooperative manner achieved through their close proximity. We have performed other studies (data not shown) clearly showing that the proteins binding at the core and core-homology elements are capable of cooperative interactions. The definition of these elements is schematically summarized in Fig. 7.

To improve the demonstration that the IL6-response region defined above (–159 to –209), containing both the core and core-homology elements was not only necessary, but also sufficient to confer IL6 responsiveness, a series of plasmid constructs were produced in which different minimal promoters were utilized (Fig. 8). The top row in Fig. 8 represents the same construct p4 x QTKLuc as the third row from the bottom in Fig. 5. This construct was based on the minimal, enhancerless TK promoter. The next series of constructs used the SV40 early minimal (enhancerless) promoter including sequences up to –105 bp 5' of its transcription start site. In construct p2 x T.SVLuc two tandem copies of an 18-bp oligonucleotide centered around the IL6-RE core and covering the region of footprint II (Fig. 6) were used. Construct p4 x T.SVLuc carried four tandem copies of the 18-bp core element. In construct p2 x mT.SVLuc two tandem copies of an 18-bp oligonucleotide with a mutation of the hexanucleotide core were used (same substitution mutation as in m205; Fig. 4B). A similar series of con-

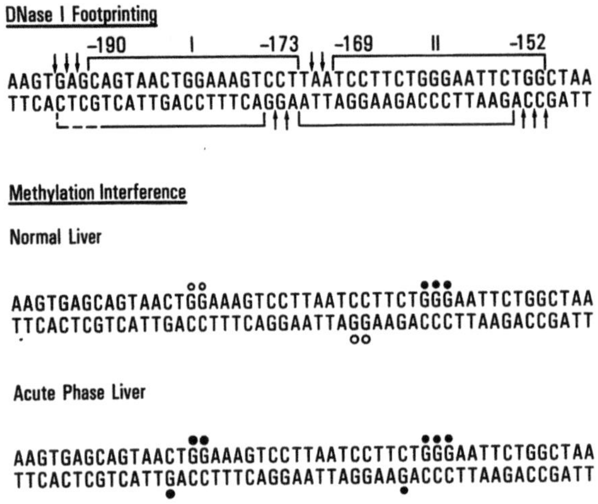

Fig. 6. Summary of DNAse I footprinting and methylation interference data. The windows *I* and *II* protected on each DNA strand against DNAse I by binding of nuclear proteins are shown by *brackets*. Window *II* was centered over the IL6-RE core sequence CTGGGAA and window *I* over the core-homology sequence CTGGAAA. These windows represented sequence-specific binding. When footprinting experiments were performed with a mutant DNA fragment, in which the core sequence was substituted (mutant m205 from Fig. 4B), then window *II* was lost and window *I* remained. Nucleotides in contact with protein as revealed by methylation interference are marked by *closed circles*. Weak contacts are shown by *open circles*. Original data were published by Hattori et al. (1990)

structs were produced using the minimal promoter of the α_2M gene itself, from –48 to + 53 bp. These constructs and the corresponding enhancerless negative controls (pα_2M Luc, pSV232 Δ5' Luc, pTKLuc) and positive controls (pSV2ALuc; Fig. 8) were used for transfection and transient expression studies in the human hepatoma cell lines Hep3B and HepG2. These lines were chosen because they also responded faithfully to the hormones, but presented the added advantage of being transfectable with greater efficiency.

The experiments with Hep3B cells (Fig. 9a) showed three important results. (1) The IL6-RE element was sufficient to confer IL6 responsiveness not only to the TK

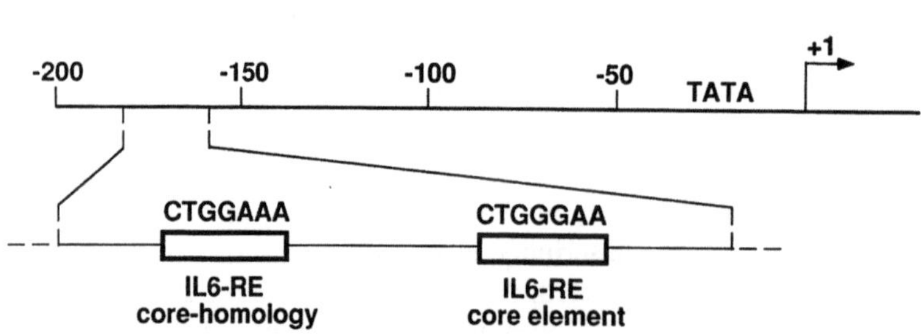

Fig. 7. The IL6 response region of the α_2M gene consists of two elements. Schematic summary of the mutagenesis and transfection results and the results from protein-DNA binding studies

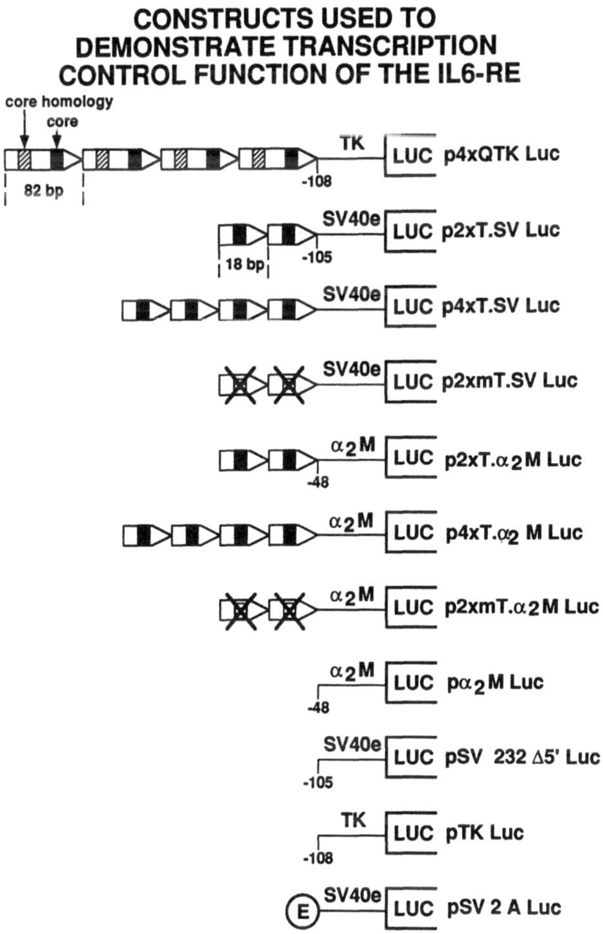

Fig. 8. Constructs produced to demonstrate that the IL6-RE core is sufficient to confer IL6 responsiveness to heterologous promoters. *TK* Minimal TK promoter to –108 bp. *SV40e* minimal SV40 early promoter to –105 bp; α_2M minimal α_2M promoter from –48 to +53 bp; *shaded box* core-homology element; *black box* core element; *crossed out box* mutated core element with the substitution m205 as shown in Fig. 4B. E SV40 early promoter's own enhancer. The construct pSV 2 A Luc contained the complete SV40 early promoter and enhancer region and was used as a positive control

promoter, but also to the SV40 early promoter and the α_2M genes own minimal promoter. It thus had an intrinsic, promoter-independent capability to confer IL6 responsiveness; (2) the same 18-bp element conferred not only the IL6 response, but also the synergistic response to IL6 plus glucocorticoids to heterologous promoters in cis; and (3) a mutation of the hexanucleotide core of the IL6-RE abolished not only the IL6 responsiveness, but also the synergistic response to IL6 plus glucocorticoids. Therefore, the action of glucocorticoids could either be mediated by another protein, possibly the glucocorticoid receptor (GR), interacting with this 18-bp element, or by a modification of the protein binding there, which is achieved as a result of glucocorticoid action.

Additional experiments were performed attempting to answer the question whether the GR interacted either directly with the 18-bp IL6-RE sequence or was part of a multiprotein complex assembled over this sequence, without binding directly to the DNA itself. So far, we have been unable to demonstrate either of these two modes of interactions for the GR, although we do know that the synergistic effect is mediated

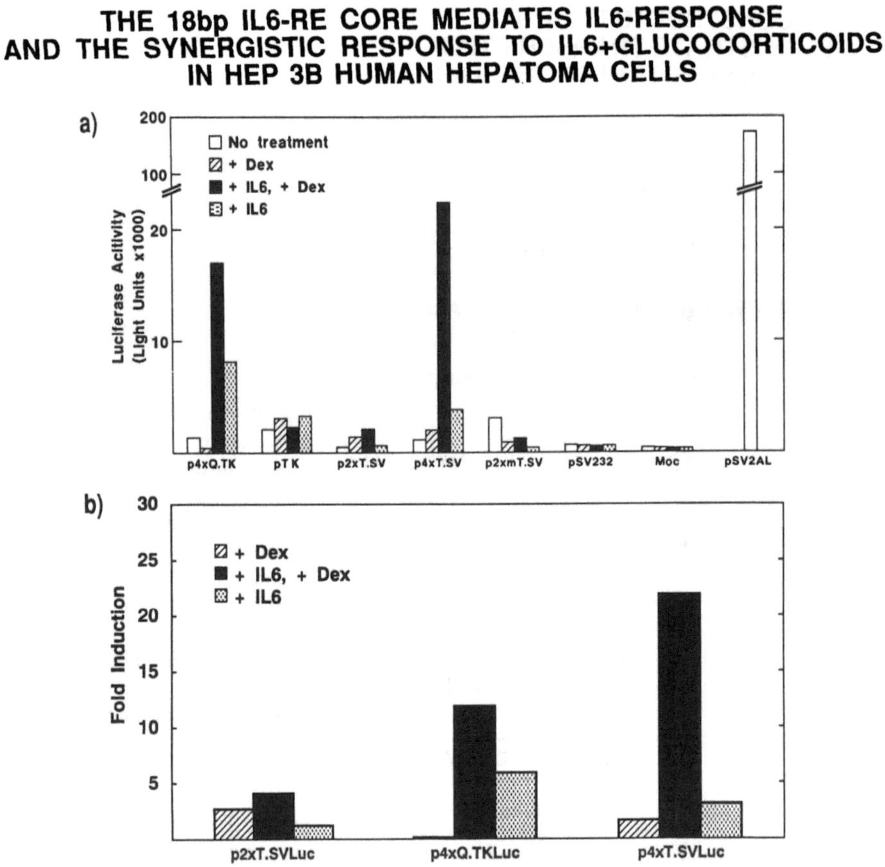

Fig. 9. The 18-bp IL6-RE core is sufficient to mediate the IL6 response to heterologous promoters and the synergistic response to IL6 plus glucocorticoids. The constructs shown in Fig. 8 were transfected into Hep3B human hepatoma cells, and the resulting luciferase activities are shown in **a.** The relative inductions ("fold" induction) are evaluated in **b**

by the GR, because it was prevented by the drug RU486, a GR-antagonist (G. Hocke et al., unpubl. data). Therefore, at present the most likely interpretation is that the GR does not act directly through the IL6-RE, but achieves the synergistic action with IL6 indirectly, by activation of an intermediate gene. The product of this gene then in turn most likely modifies the IL6 signal cascade to produce the synergistic effect.

The same constructs were also transfected into the HepG2 human hepatoma cell line and transient expression assays were performed (Fig. 10). This cell line also responded faithfully to IL6 and glucocorticoids, and showed even greater transfection efficiency than the Hep3B line used in Fig. 9. The construct p4 x T.SVLuc, carrying four tandem copies of the 18-bp element, comprising the IL6-RE core region and the SV 40 early minimal promoter (Fig. 8), gave rise to a 1.3-fold stimulation with dex alone, 9.7-fold with IL6 alone, and 11.5-fold with IL6 plus dex. These values are mean values over three independent experiments. Thus, in the HepG2 cell line it be-

Transient-expression; 18 hour exposure to 400 u/ml IL6 plus 10^{-7} M dex

Hep3B	p4xT SV LUC	Dex alone	1.7-fold
		IL6 alone	3.2-fold
		IL6 plus Dex	22-fold
HepG2	p4xT SV LUC	Dex alone	1.3-fold
		IL6 alone	9.7-fold
		IL6 plus Dex	11.5-fold
	p4xT α_2M LUC	Dex alone	1.6-fold
		IL6 alone	120-fold
		IL6 plus Dex	183-fold

Fig. 10. The 18-bp IL6-RE core also mediates the IL6 response and synergism with glucocorticoids in HepG2 cells. The constructs as explained in Fig. 8 were transfected into HepG2 human hepatoma cells, and a strong response (120-fold induction with IL6 alone, 183-fold with IL6 plus dex) was obtained

came more obvious than in Hep3B that the 18-bp element was capable of conferring IL6 inducibility to the heterologous SV40 promoter, while the synergistic effect with dex was less pronounced than in Hep3B cells. This result is probably explained by the known fact that HepG2 cells have lower levels of glucocorticoid receptor than Hep3B cells (Baumann et al. 1990). The most clear-cut results were obtained in HepG2 cells with the construct based on the α_2M gene's own minimal promoter, p4 x T.α_2MLuc. Here, a 1.6-fold induction was obtained with dex alone, a 130-fold induction with IL6 alone, and a 183-fold increase with IL6 plus dex (Fig. 10). These values are again mean values over three independent experiments. The result unambiguously established that the 18-bp IL6-RE core element was sufficient to confer a strong transcriptional activation by IL6 alone and also a synergistic activation by IL6 plus glucocorticoids. The interesting observation that this element functioned more strongly in combination with the α_2M gene's own minimal promoter than with the heterologous TK or SV40 early minimal promoters suggested that there may be another so far unrecognized element hidden in the α_2M minimal promoter that interacted particularly well with the IL6-RE 18-bp core element, and that may be absent from the TK and SV40 early minimal promoters. The identity of this "hidden" element is currently unknown and merits future investigation.

The IL6-response region in its natural configuration, comprising both the IL6-RE core and core-homology sites, served as a binding site for nuclear protein factors, as shown already by the DNAse I footprinting and methylation interference experiments (Fig. 6). This concept was supported by additional gel mobility shift experiments. (Fig. 11). An 82-bp DNA restriction fragment containing the response region (fragment Q, Fig. 8, top row) was labeled radioactively with ^{32}P, and incubated with nuclear protein extracts from normal and acute phase rat livers (Fig. 11, tracks 1, 2) and untreated or IL6-treated Hep3B human hepatoma cells (Fig. 11, tracks 3, 4). A weak complex I was found with normal rat liver and untreated Hep3B extracts, and a much stronger complex II of altered mobility with acute phase liver or IL6-treated Hep3B cell extracts. These complexes II obtained with rat and human nuclear proteins had identical mobility, suggesting binding of identical or very similar proteins.

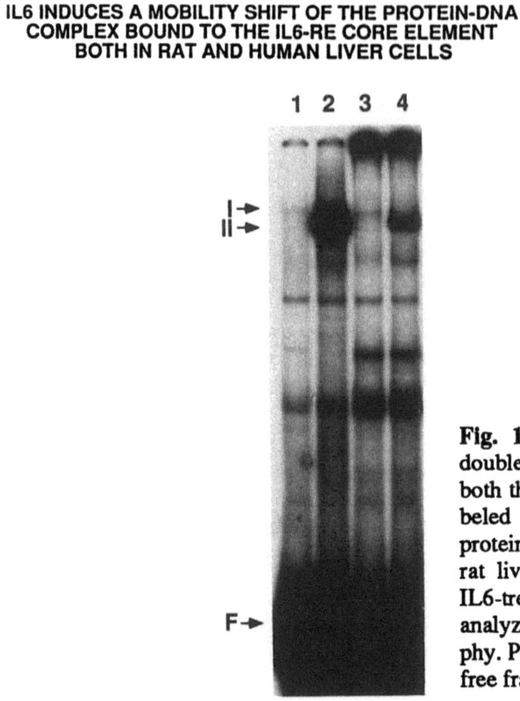

Fig. 11. Gel mobility shift experiments. A double-stranded DNA fragment comprising both the core and core-homology sites was labeled with ^{32}P and incubated with nuclear protein extracts from normal and acute phase rat livers (*tracks 1 and 2*) and untreated or IL6-treated Hep3B cells. The complexes were analyzed by electrophoresis and autoradiography. Positions of complexes *I* and *II* and of the free fragment (*F*) are marked

In Hep3B cells this complex II was induced to maximum intensity by IL6 after 18 h. This kinetics is consistent with the concept that complex II may participate in the IL6-mediated transcriptional activation of the α_2M gene promoter. We have no direct proof for this proposal, since the role of complex II has not been directly established in functional assays, for example, cell-free transcription systems depleted for complex II and then resubstituted with purified complex II. In absence of this type of direct evidence, we cite several circumstantial pieces of evidence in favor of this concept:

1. Complex II was assembled over a sequence shown by functional transfection and mutagenesis studies to be critical for transcriptional activation by IL6.
2. Complex II was inducible by the hormone with a kinetics compatible with the kinetics of transcriptional activation of the promoter by IL6 (Fig. 2).
3. Complex II was inducible in rat livers and rat and human hepatoma cells and had indistinguishable gel mobility (Fig. 11); and finally,
4. The complexes II obtained with rat and human nuclear proteins bound DNA with the same sequence specificity (gel mobility shift competition studies, data not shown).

Our current model, which attempts to visualize the key features of complexes I and II, is given in Fig. 12. The proteins capable of binding at the IL6-RE core and core-homology sites were present already in normal rat liver nuclei and untreated hepatoma cells, as evidenced by their capability to generate footprint windows I and II and the characteristic binding pattern in methylation interference experiments (Fig. 6). The reason why they only caused a weak gel shift complex I is unknown, and the protein

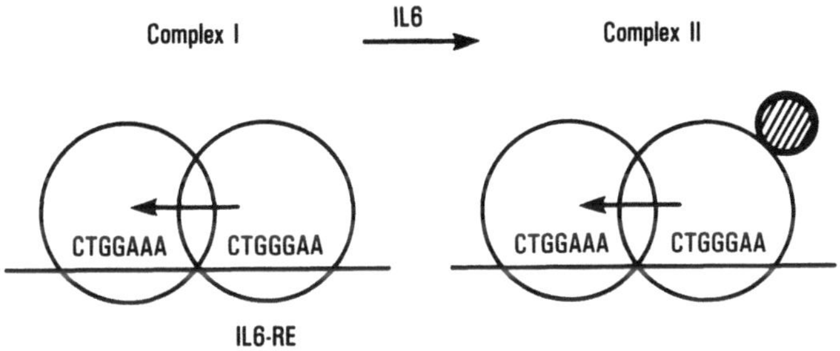

Fig. 12. Model of the composition of complexes *I* and *II*. We do not yet have direct evidence for the composition of complexes *I* and *II*. Circumstantial evidence suggests that probably the same protein represented by the *open circle* binds both at the core and core-homology sites. The core site has greater affinity and is likely to be the first site occupied. Its occupancy facilitates binding of a second molecule at the core homology site by cooperative effects, symbolized by an *arrow*. The modification leading from complex *I* to complex *II* is unknown, symbolized by a *shaded circle*

composition of gel shift complexes I and II has not been analyzed definitively. We have preliminary evidence (data not shown) suggesting that probably the same protein binds at the core and core-homology sites, and that binding between two protein molecules occupying these sites occurs in a cooperative manner symbolized by the arrow in Fig. 12. The core site had a higher affinity than the core-homology site (quantitative gel mobility shift competition data; data not shown) and is probably the first of the two sites to be occupied. Once it is occupied, the second molecule binds at the core-homology site with about 80-fold greater affinity due to cooperative interaction with the molecule already occupying the core site. The IL6 signal causes an unknown alteration to these proteins that leads to the characteristic increase in intensity and mobility of complex II as compared with complex I. We have yet no direct definitive evidence to show that complex II consisted of the same proteins as complex I, but we have circumstantial data suggesting that it most likely does. Strictly speaking, the possibility has not been ruled out that complex II consisted of a different set of polypeptides binding at the same sequence that replace the original set forming complex I after arrival of the IL6 signal. However, we propose that this is not the case. Rather, the presently available data indicate that the same proteins were modified by the IL6 signal. This modification could consist of a covalent alteration, such as addition or removal of a charged group, leading to the altered gel mobility, or in a noncovalent alteration, such as addition or removal of one or several components from complex I leading to complex II. Preliminary data (Ito et al. 1989) suggest that an alteration in the phosphorylation state of the nuclear factors binding at the IL6-RE may be involved. It will become very important in the future to purify and clone these nuclear factors, and then to determine the nature of this hormone-induced alteration.

This hormone-induced alteration requires ongoing protein synthesis. When protein synthesis was blocked by treatment of the cells with cycloheximide, complex II was not formed, although the proteins were present (Fig. 13). We have obtained identical data with cycloheximide treatment both in FAO rat hepatoma cells (Fig. 13) and Hep3B human hepatoma cells (data not shown). The only difference was that in un-

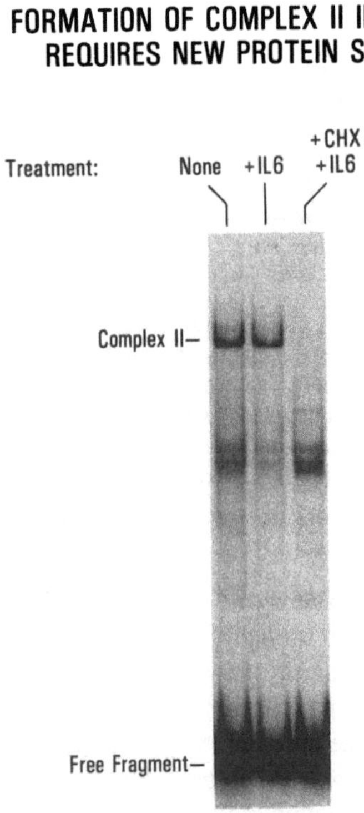

Fig. 13. Ongoing protein synthesis is required for the formation of complex *II*. Gel mobility shift experiment as in Fig. 11, but with nuclear protein extracts from FAO rat hepatoma cells. These cells already produce complex *II* in their untreated state, because they secrete endogenous IL6, although at low levels (Northemann et al. 1990). Treatment of these cells with cycloheximide (10 µg/ml) prevented formation of complex *II* (+ *CHX*). Similar results were obtained when Hep3B cells were treated with IL6 or with IL6 plus cycloheximide. In this case, complex II formation was also blocked (G. Hocke et al., unpubl. data)

treated Hep3B cells we detected complex I, whereas in untreated FAO cells, complex II was formed probably because FAO cells secreted low levels of endogenous IL6 (Northemann et al. 1990; Baffet et al. 1991) sufficient to lead to the constitutive formation of complex II. We have observed the same phenomenon in other rat hepatoma and human B-myeloma cell lines that constitutively secreted endogenous IL6 (Hocke et al., data not shown). Our current best interpretation of these data is that ongoing protein synthesis is required to form complex II, not in order to continuously resynthesize a labile DNA-binding protein, but rather to perform the modification function required for the formation of complex II that must be carried out by a labile protein requiring continuous de novo protein synthesis. The DNA-binding protein was stable during 6 h of treatment of the cells with cycloheximide and retained its capacity to bind to the IL6-RE sequence as evidenced in Southwestern blot experiments (Fig. 14). However, it lost its ability to form complex II under gel mobility shift conditions. Thus, the ability to specifically bind to the IL6-RE and to form complex II were not strictly the same: more was needed to form complex II than simple sequence-specific DNA binding.

Nuclear protein extracts from normal and acute phase rat livers and from untreated and IL6-treated Hep3B cells were separated by electrophoresis in SDS polyacrylamide gels and subsequently transferred to nitrocellulose filters (Western blot). The filters were then incubated with a radiolabeled 24-bp double-stranded oligonucleotide (TB1) representing the 18-bp IL6-RE core region. Subsequently, the filters were

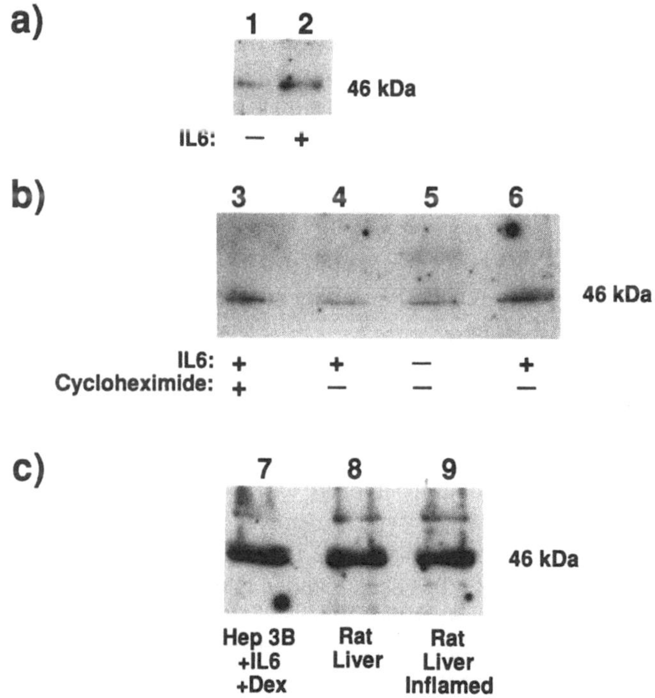

Fig. 14. The IL6-RE binding protein (IL6 RE-BP) has a molecular weight of 46 kDa. *Tracks 1* to *7* were obtained with nuclear protein extracts from Hep3B human hepatoma cells; *Track 8* with nuclear proteins from normal rat liver; *track 9* with nuclear proteins from a rat liver taken at the peak of an experimentally induced acute phase response. Cycloheximide treatment in *track 3* was for 6 h with 10 μg/ml

washed and exposed for autoradiography. By this technique (Southwestern blotting; Fig. 14) a protein of 46 kDa was revealed that specifically bound to the hexanucleotide core sequence. Sequence specificity of binding was demonstrated by competition-Southwestern blot with molar excesses of nonradioactive wild-type and mutant oligonucleotides (data not shown). The same 46-kDa protein was detected both in Hep3B human hepatoma cells and rat liver nuclear extracts. It was present both before and after IL6 treatment and both in normal and acute phase rat liver nuclear extracts (Fig. 14). Even after treatment with cycloheximide for 6 h the protein was still detected (Fig. 14B, tracks 3 and 4). Similar results were obtained by an independent technique (Ultraviolet cross-linking; data not shown) also demonstrating the existence of a 46 ± 3 kDa protein in both rat livers and human heptoma cells that survived cycloheximide treatment and was present in untreated and IL6-treated hepatic cells. These data formed the basis for the statement that the DNA-binding protein did not require ongoing protein synthesis for its own replacement. Rather, ongoing protein synthesis was required to provide the modifying activity that altered complex I to complex II. Note that the molecular weight of 46 kDa significantly differs from the molecular weight reported for the only other factor clàimed to mediate transcriptional regulation by IL6, the factor IL6-DBP/LAP (Akira et al. 1990; Descombes et al. 1990; Poli et al. 1990). The reported molecular weight for this factor was 32 kDa. We have performed extensive additional experiments using both gel shift and gel shift competition studies with binding site oligonucleotides for both factors and studies with antibodies against IL6-DBP/LAP and the related protein DBP1, kindly provided by U. Schibler and colleagues (data not shown). The result of those studies was that the 46-kDa IL6 RE-BP binding at the IL6-RE core of the rat α_2M gene and IL6-DBP/LAP were not identical. The two proteins did not cross-react with each others' respective DNA-binding sites and only IL6-DBP/LAP but not the 46-kDa protein reacted with a specific anti-DBP/LAP serum.

This distinction between the 46-kDa protein and IL6-DBP/LAP is important, because it further supports the notion that there are two types of IL6-responsive DNA control elements: type I elements and type II elements (Fig. 15). Type I elements have a poorly defined consensus sequence agTgNNGYAA, which is the binding site for IL6-DBP/LAP. These elements occur primarily in class 1 acute phase genes, which are regulated mainly by IL1 (Poli et al. 1989). Type II elements have the much better defined consensus sequence CTGGGA, as shown in the IL6-RE core of the rat α_2M gene (Hattori et al. 1990; Kunz et al. 1989; this chap.), and in the rat γ-fibrinogen and α_1-acid glycoprotein genes (Baumann et al. 1990; Won and Baumann 1990). They are binding sites for the 46-kDa protein shown in Fig. 14. Currently our best explanation for these findings is that IL6-DBP/LAP-NF IL6 is a protein that primarily mediates responsiveness to IL1. It probably has a degree of cross-reactivity for IL6, without really being primarily an IL6-response factor. This is reflected by the fact that type I IL6-RE's are found mainly in class 1 acute phase genes that respond mainly to IL1. The principal IL6-responsive nuclear factor probably is the 46-kDa protein shown here; we predict this, because it binds specifically at type II IL6-response ele-

Type I	Hpx A	−117 to −108	A G T G A T G T A A
	Hp A	−167 to −158	T G T G A A G C A A
	Hp C	−60 to −68	A T T T C G T A A
	CRPα	−56 to −47	A G T G G C G C A A
		Consensus	a g T g N N G Y A A
Type II	α_2M (Rat)		C T G G G A
	γFbg		
	α_1AGP		

Induction of Type I by IL6 not inhibited by Cycloheximide

Fig. 15. Two types of IL6 response elements. Type I elements have the poorly defined consensus sequence as shown. They are found mainly in class 1 acute phase genes (Fig. 1) that are primarily regulated by IL1 (Poli et al. 1989). The induction of *type I* elements by IL6 is not inhibited by cycloheximide. *Type II* elements have the better conserved consensus CTGGGA, as in the core element of the rat α_2M gene. They are found mainly in class II acute phase genes (Fig. 1) that are primarily regulated by IL6 ments that are found primarily in class 2 acute phase genes, which are mainly regulated by IL6. IL6-DBP/LAP is a member of the C/EBP family of factors. The structure of the 46-kDa IL6 RE-BP is unknown, because this factor has not yet been cloned and sequenced. It has not been excluded that this factor could be yet an additional member of the C/EBP family of factors. If this were the case, then it may also be capable of forming heterodimers with IL6-DBP/LAP. This could possibly provide a molecular explanation for the synergistic regulation of class 1 acute phase genes by IL6 and IL1.

The fact that two different classes of acute phase genes are regulated by IL6 through two different cis-elements and two different factors has led us to formulate the hypothesis of "signal divergence." This hypothesis states that the IL6 signal must activate two divergent signal pathways within the same hepatic cells: the one pathway activates class 1 acute phase genes via type I elements and IL6-DBP/LAP. The other pathway activates class 2 genes via type II elements and the 46-kDa protein.

Whereas "signal divergence" clearly exists (one hormone –two pathways), the converse also occurs: two hormones, one pathway. This concept is called signal convergence, and we encounter it again during the acute phase response of the rat α_2M gene. This gene and other class I acute phase genes also respond to LIF. LIF is a cytokine of different primary sequence than IL6 with no apparent sequence similarity: both are two entirely different peptide hormones, but of similar size, approximately 180 amino acids. Preliminary data indicate that the IL6 and LIF receptors on the hepatocyte surface are distinct. For example, our own experiments show that Hep3B cells respond to IL6 but not to LIF, probably because they lack the LIF receptor. Surprisingly, both the IL6 and LIF signal cascades converge at a common endpoint: the IL6-RE core of the rat α_2M gene. We have recently shown that the same construct p4 x T.α_2M Luc (Fig. 8) carrying four tandem copies of the 18-bp IL6-RE core sequence that generated a 145-fold induction by IL6 in HepG2 cells, also mediated a 39.9-fold induction by LIF in these cells. Thus, the IL6-response element cross-reacts with LIF. This provides a paradigm for the situation that we refer to as "signal convergence": two hormones, one common endpoint, the 18-bp IL6-RE. We have not yet formally proven that the IL6 and LIF cascades activate the same protein interacting with this sequence. The alternative has not been rigorously ruled out that each cascade terminates in a different factor, and that both bind at the same IL6-RE core. However, we believe that this is very unlikely, because we have additional data (not shown) demonstrating that treatment of hepatoma cells with LIF generates the same gel mobility shift complex II as treatment with IL6.

During our studies of acute phase genes we have encountered another illustrative example for the concept of signal convergence or "signal integration" in the case of the human complement C3 gene. This is a class 1 acute phase gene that is induced synergistically by IL1 and IL6. In collaboration with D. Wilson and G. Darlington, we have demonstrated the existence of a 58-bp region in the 5'-flanking sequence of the human C3 gene that mediates both activation by IL1 and by IL6 and the synergistic response to both of these hormones. Inspection of this sequence revealed the presence of a putative IL1-response element, a binding site for NF-IL6, the human equivalent of rat IL6-DBP/LAP. Partially overlapping this sequence a putative IL6-response element was found, CTGGGG, which matched the consensus IL6-RE in 5/6

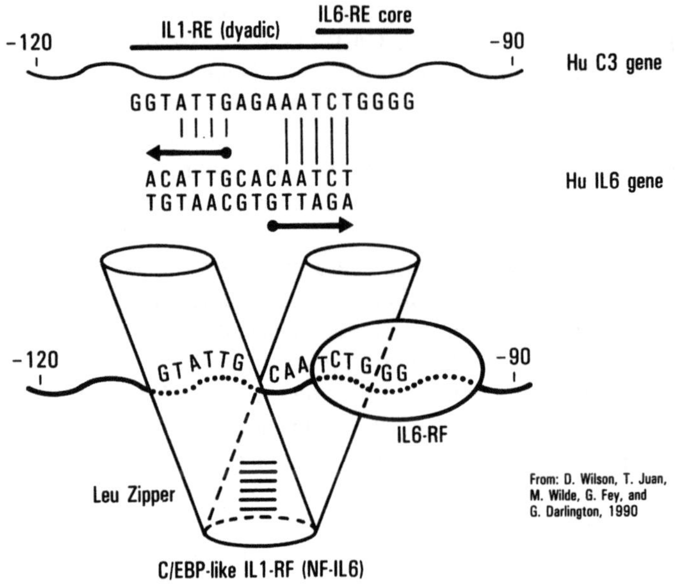

Fig. 16. Synergistic activation of the human C3 gene promoter by IL1 and IL6 through cooperative binding of IL1- and IL6-response factors at overlapping sites. This model represents the predictions made by Wilson et al. (1990) on the basis of mutagenesis and transfection results. A 58-bp region was shown to confer both responsiveness to IL1 and IL6 and a strong synergistic response to both of these hormones. This region contained putative binding sites for an NF-IL6-like factor, a characteristic dyadic sequence symbolized by *arrows* as found in the human IL6 gene, and an IL6-RE core sequence. Both sequences overlapped by the three nucleotides TCT. A mutation of these three nucleotides abolished the IL1 and IL6 responsiveness and the synergistic effect. Hence, the prediction that the synergism of the hormones may be due to cooperative binding of these factors at partially overlapping sites. *IL6-RF* = IL6-response factor; *IL1-RF* = IL1-response factor

positions including the critical G-residues for protein binding as shown in Fig. 6 (Fig. 16). A mutation in the three nucleotides that constitute the overlap (TCT) eliminated the IL1 response, the IL6 response, and the synergistic response. Our explanation for these findings is that this 58-bp region contains overlapping binding sites for IL1- and IL6-responsive factors that are arranged in such a close special arrangement that the two factors can physically contact each other (Fig. 16). We anticipate that the factors may actually bind in a cooperative manner, because this is a characteristic property both of members of the C/EBP family (Descombes et al. 1990) and of the 46-kDa IL6 RE-BP (unpubl. data). Therefore, the synergistic action of IL1 and IL6 on the C3 gene may be explained by cooperative binding of the two factors. This hypothesis remains to be proven (Wilson et al. 1990). If correct, this would be another very illustrative case of signal convergence or signal integration. It is very satisfactory that the mechanistic studies of transcriptional regulation of acute phase genes by multiple cytokines may eventually lead to a molecular explanation of signal convergence or signal integration by understanding the integrative properties of the transcription factors involved. This is an exciting development because it is a beginning of an understanding of a characteristic feature of inflammation: the simultaneous response to multiple mediators of inflammation, and because it reminds us of the excitement created in solid state physics, when semiconductors were first discovered as devices capable of signal integration. The analogy between semiconductors and biological integrators may provide a fertile theoretical framework guiding us in our future analysis of "molecular inflammation" and "molecular signal integration."

Acknowledgements. This work was supported by NIH grants AI22166 and AI23351 (to G. H. F.). G. H. was the recipient of a postdoctoral research fellowship from DFG (Deutsche Forschungsgemeinschaft). G. B. was the recipient of fellowships from INSERM, the French government research organization, and the Fondation Phillippe. T. B. was the recipient of a NATO postdoctoral research fellowship, sponsored by the DAAD (German Academic Exchange Service). We thank Dr. Wolfgang Northemann for experimental contributions and Dr. Ueli Schibler, University of Geneva, Switzerland for gifts of recombinant purified IL6-DBP/LAP, DBP 1, and the corresponding antisera. We thank Keith Dunn for expert preparation of the manuscript. This is publication number 6819-IMM from the Department of Immunology. Research Institute of Scripps Clinic.

References

Akira S, Isshiki H, Sugita T, Tanabe O, Kinoshita S, Nishio Y, Nakajima T, Hirano T & Kishimoto T (1990) A nuclear factor for IL6 expression (NF-IL6) is a member of the C/EBP family. EMBO J 9:1897–1906

Baffet G, Braciak TA, Fletcher RG, Gauldie J, Fey GH & Northemann W (1991) Autocrine activity of interleukin 6 secreted by hepatocarcinoma cell lines. Mol Biol Med 8:141–156

Baumann H (1989) Hepatic acute phase reaction in vivo and in vitro. In Vitro Cell Dev Biol 25:115–126

Baumann H & Gauldie J (1990) Regulation of hepatic acute phase plasma protein genes by hepatocyte stimulating factors and other mediators of inflammation. Mol Biol Med 7:147–160

Baumann H, Jahreis GP & Morella KK (1990) Interaction of cytokine- and glucocorticoid-response elements of acute phase plasma protein genes. J Biol Chem 265:22275–22281

Descombes P, Chojkier M, Lichtsteiner S, Falvey E & Schibler U (1990) LAP, a novel member of the C/EBP gene family, encodes a liver-enriched transcriptional activator protein. Genes Dev 4:1541–1551

Fey G & Gauldie J (1990) The acute phase response of the liver in inflammation. In: Popper H & Schaffner F (eds) pp 89–116 Progress in liver disease 9th edn WB Saunders, Philadelphia

Hattori M, Abraham LJ, Northemann W & Fey GH (1990) Acute phase reaction induces a specific complex between hepatic nuclear proteins and the interleukin 6 response element of the rat α_2-macroglobulin gene. Proc Natl Acad Sci USA 87:2364–2368

Heinrich PC, Castell JV & Andus T (1990) Interleukin 6 and the acute phase response. Biochem J 265:621–636

Ito T, Tanahashi H, Misumi Y & Sakaki Y (1989) Nuclear factors interacting with an interleukin 6 responsive element of the rat α_2-macroglobulin gene. Nucl Acids Res 17:9425–9535

Kunz DR, Zimmermann R, Heisig M & Heinrich PC (1989) Identification of the promoter sequences involved in the interleukin 6 dependent expression of the rat α_2-macroglobulin gene. Nucl Acids Res 17:1121–1138

Northemann W, Hattori M, Baffet G, Braciak TA, Fletcher RG, Abraham LJ, Gauldie J, Baumann M & Fey GH (1990) Production of interleukin 6 by hepatoma cells. Mol Biol Med 7:273–286

Poli V, Silengo L, Altruda F & Cortese R (1989) The analysis of the human hemopexin promoter defines a new class of liver specific genes. Nucl Acids Res 17:9351–9365

Poli V, Mancini FP & Cortese R (1990) IL6-DBP, a nuclear protein involved in interleukin 6 signal transduction, defines a new family of leucine zipper proteins related to C/EBP. Cell 63:643–653

Toniatti C, Arcone R, Majello B, Ganter U, Arpaia G & Ciliberto G (1990) Regulation of the human C-reactive protein gene, a major marker of inflammation and cancer. Mol Biol Med 7:199–212

Wilson DR, Juan TS-C, Wilde MD, Fey GH & Darlington GJ (1990) A 58 base pair region of the human C3 gene confers synergistic inducibility by interleukin-1 and interleukin-6. Mol Cell Biol 10:6181–6191

Won KA & Baumann H (1990) The cytokine response element of the rat α_1-acid glycoprotein gene is a complex of several interacting regulatory sequences. Mol Cell Biol 10:3965–3978

Disturbance of the Hemostasis and Fibrinolysis Balance by Tumor Necrosis Factor

F. BACHMANN[1], and R. MEDCALF[1]

1 Introduction

The maintenance of fluidity in the human blood circulation, on the one hand and the provision of a blood coagulation system which is activated when lesions of the vascular wall occur, on the other hand, represent a formidable achievement of nature. The balance between local coagulation and systemic fluidity is established through a complicated cascade system of clotting proteases, coagulation cofactors, inhibitors of blood coagulation, and proteins attached to the cell membranes of endothelial cells, macrophages, and platelets which may act as receptors for coagulation factors or as inhibitors of the coagulation cascade. A second system intervenes during and after the formation of fibrin, the fibrinolytic system. It serves to proteolytically degrade fibrin and to reestablish patency in a thrombosed blood vessel. The key cell for these processes is the endothelial cell (Table 1).

Several of the constituents of blood coagulation and the fibrinolytic system belong to the most highly regulated proteins in the human mechanism. The expression of the four constituents listed in Table 1 is influenced by endotoxin, several cytokines, hormones, and growth factors, and particularly by IL-1 and tumor necrosis factor (TNF) (Nawroth and Stern 1986, 1987). TNF exists in an α-form (cachectin) and a β-form (lymphotoxin) (Beutler and Cerami 1987). It is a polypeptide of 157 amino acids,and

Table 1. Hemostasis factors produced by the endothelial cell

General action	Component	Specific action
Procoagulant	Tissue factor	Forms a complex with Factor VII facilitating the activation of F. VII to F. VIIa
Anticoagulant	Thrombomodulin	Inactivates thrombin; the complex thrombin-thrombomodulin activates protein C
Fibrinolytic	t-PA	Activates plasminogen
Antifibrinolytic	PAI-1	Inactivates t-PA and two-chain u-PA

[1] Division of Hematology, Department of Medicine, University Medical Centre, 1011 Lausanne, Switzerland

42. Colloquium Mosbach 1991
Molecular Aspects of Inflammation

contains one disulfide bond (Aggarwal et al. 1985). The biologically active form is a trimer. TNF binds to two distinct receptors of 55 (Loetscher et al. 1990) and 75 kD (Hohmann et al. 1989) which do not cross-react immunologically (Brockhaus et al. 1990). The main sources of TNF are monocytes/macrophages. TNFα gene transcription is highly inducible by virus and lipopolysaccharide (LPS) in macrophages and certain other cell types. The promoter region of the TNFα gene contains three NF-$_{K}$B-binding sites which are implicated in the constitutive, and in maximal virus- and LPS-induced transcription. However, deletion of these sequences, which bind the transcription factor NF-$_{K}$B, still allows some stimulation of TNF expression by viruses and LPS (Goldfield et al. 1990).

2 Regulation of the Coagulation Balance by TNF

In the physiological state, tissue factor is hardly expressed on the surface of endothelial cells but thrombomodulin is. Thus, the endothelial cell presents an anticoagulant surface to the circulating blood. Tissue factor (TF), a 47-kD membrane-bound glycoprotein, is a receptor for blood coagulation factor VII (F. VII). Conversion of TF-bound F. VII to its activated form F. VIIa takes place autocatalytically in the presence of Ca^{2+}, and this reaction is enhanced by traces of F. Xa and/or F. IXa (Yemerson 1991). The presence of TF-bound F. VIIa on endothelial cells triggers the intrinsic coagulation cascade via activation of F. IX and the extrinsic coagulation system via activation of F. X.

Thrombomodulin (TM) is a 100-kD endothelial cell membrane glycoprotein which binds circulating thrombin. Bound thrombin undergoes a conformational change and is no longer able to activate F. V and F. VIII or to clot fibrinogen (Preissner et al. 1990). On the other hand, the efficiency of TM-bound thrombin to convert protein C to activated protein C is increased 20 000-fold compared with that occurring when protein C and thrombin are mixed in a test tube (Comp 1991; Esmon 1987). Activated protein C cleaves F. Va and F. VIIIa and has thus potent anticoagulant properties.

After the injection of endotoxin into man, circulating TNF levels greatly increase (Michie et al. 1988; von Deventer et al. 1990). Endotoxin, as well as TNF induces the expression of TF activity in peripheral blood monocytes (Conkling et al. 1988) and on the surface of endothelial cells (Bevilacqua et al. 1986; Conway et al. 1989; Moore et al. 1987; Nawroth and Stern 1987; Scarpati et al. 1989). Simultaneously, TM expression is downregulated (Moore et al. 1987; Nawroth and Stern 1987; Conway and Rosenberg 1988). Endothelial cells undergo other changes. Polynuclear neutrophils attach much more readily to human umbilical vein endothelial cells which have been incubated with TNF (or IL-1) and transendothelial passage of neutrophils is greatly enhanced (Moser et al. 1989).

The time course of expression enhancement of TF by cultured endothelial cells is similar after stimulation with phorbol ester (PMA) and TNF. Maximal ten-fold increase in mRNA expression of two messages (due to alternative or incomplete splicing) occurs after 4 h (Conway et al. 1989). Nuclear runoff assays reveal only a three- to four fold increase in transcription (Scarpati and Sadler 1989). This suggests mRNA stabilization in addition to transcriptional regulation. Treatment of endothelial cells with cycloheximide, which suppresses protein synthesis, shows that protein synthesis is not necessary for TF expression. TF mRNA is rapidly degraded. It contains A + U-rich sequences within the 3' untranslated region as TM and some other cytokines such as GM-CSF, and the β-chain of PDGF. The instability of the TF mRA conferred by the AUUA sequences appears to be reduced upon treatment with cycloheximide (Conway et al. 1989; Crossman et al. 1990), possibly due to the disappearance of a labile repressor protein which inhibits the transcription of the TF gene in unstimulated endothelial cells.

TM activity on cultured endothelial cells is greatly diminished by TNF (Nawroth and Stern 1986; Conway and Rosenberg 1988; Moore et al. 1989). Conway and Rosenberg (1988) have demonstrated that TNF suppresses transcription of the TM gene within 40 to 60 min and that the decrease in TM mRNA caused by TNF in bovine aortic and human umbilical vein endothelial cells is of the same magnitude as the decrease in functional TM activity. TNF appears, in addition, to induce internalization and subsequent degradation of the TM molecule (Moore et al. 1989).

3 Regulation of the Fibrinolysis Balance by TNF

The principal fibrinolytic enzyme is plasmin, a trypsin-like serine protease. Its zymogen plasminogen is converted to its active form plasmin by two different plasminogen activators (PA), tissue-type PA (t-PA) and urinary-type PA (u-PA). The activities of the two PA's are highly regulated by two distinct plasminogen activator inhibitors (PAI), PAI-1 and PAI-2. t-PA, a serine protease of 68 kD (Bachmann and Kruithof 1984), and PAI-1, a 50-kD glycoprotein serpin, are constitutively expressed by the endothelial cell and released into the circulating blood. The biological half-lives of free t-PA and free PAI-1 are of the order of 3 to 5 min. In the blood, free t-PA and PAI-1 form enzyme-inhibitor complexes (second-order rate constant $10^7\ M^{-1}s^{-1}$) and more than 80% of circulating t-PA exists in the form of a complex with PAI-1 (Kruithof 1988).

Schleef et al. (1988) have shown that IL-1 as well as TNF diminish the expression of t-PA antigen levels in the supernatant of cultured bovine aortic endothelial cells. Simultaneously, both cytokines stimulated the production of PAI-1 activity, antigen, and mRNA.

We have studied the effect of TNF on the fibrinolytic system using the fibrosarcoma cell line HT-1080, which constitutively expresses t-PA, u-PA, PAI-1, and PAI-2 (Medcalf et al. 1988a). The 24-h conditioned medium of HT 1080 cells contains typically 25 to 40 U/ml (u-PA international units equivalency) of fibrinolytic activity as measured by the classical fibrin plate assay. TNF, at a concentration of 2.5 ng/ml suppressed over 99% of this activity. This was mainly due to a 2.5-fold increase in PAI-1 antigen, the principal plasminogen activator inhibitor of HT-1080 cells and an even larger increase in PAI-2 antigen. Simultaneously, t-PA antigen levels declined (Table 2). Secreted and cell-associated u-PA antigen levels, the predominant PA in this cell line, were unaffected by TNF.

Table 2. Influence of different concentrations of TNF on secreted and cell-associated t-PA, PAI-1, and PAI-2 antigen levels in HT-1080 cells

TNF added (ng/ml)	Fibrinolytic activity in conditioned medium (U/ml)	antigen level ng/10^6 cells					
		t-PA		PAI-1		PAI-2	
		Secreted	Cell-associated	Secreted	Cell-associated	Secreted	Cell-associated
0	30	5	< 0.1[a]	400	60	18	175
2.5	0.2	2.5	< 0.1	800	100	34	900
25	< 0.1	< 0.1	< 0.1	1000	140	104	1500
50	< 0.1	< 0.1	< 0.1	1000	90	100	1300
250	< 0.1	< 0.1	< 0.1	1040	100	124	1900

[a] Detection limit 0.1 U/ml.

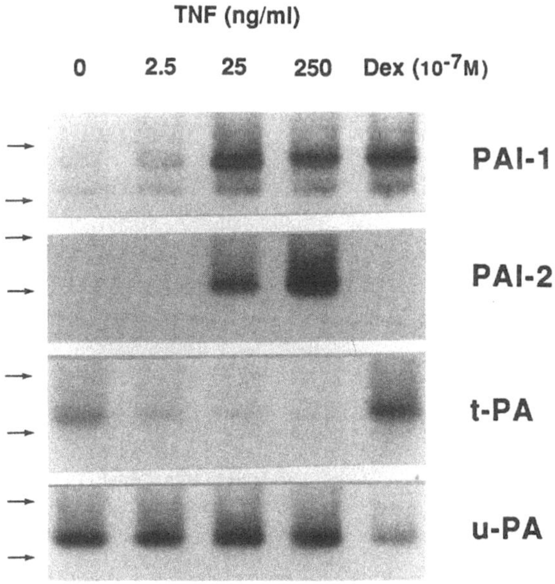

Fig. 1. Dose-dependent modulation of PAI-1, PAI-2, and t-PA mRNA by TNF. mRNA prepared from HT-1080 cells treated for 24 h with 0, 2.5, 25 or 250 ng/ml human TNF was assessed by Northern blot analysis. Dexamethasone (Dex, 10^{-7}M) was added to cells as a control for induction of PAI-1 and t-PA mRNA and suppression of u-PA mRNA. Relative changes in PAI-1, PAI-2, t-PA, and u-PA were determined by hybridization to random primer-labeled PAI-1, PAI-2, t-PA, and u-PA cDNA probes of similar specific activity. Successive hybridizations of each probe to the same membrane are illustrated. *Arrows* on the *left* represent the position of the 28 S and 18 S ribosomal RNA markers (Medcalf et al. 1988a).

Northern blot analysis revealed a dose-dependent increase in the 3.4- and 2.4-kb PAI-1 mRNA and in the PAI-2 mRNA and a concomitant downregulation of t-PA mRNA. u-PA mRNA levels were unaffected (Fig. 1). When HT-1080 cells were incubated with 50 ng/ml of TNF and mRNA extracted from cells after various incubation periods, increased PAI-1 and PAI-2 mRNA levels were detected after 8, respectively 4 h. Densitometric analysis of the Northern blots demonstrated a 16- and 100-fold increase in PAI-1 and PAI-2 mRNA after 24 h (Fig. 2), respectively.

To evaluate whether these increases in steady-state mRNA levels were due to an increased gene transcription rate, we determined relative changes in PAI-1, PAI-2, and t-PA gene template activity in isolated nuclei by means of the "run-on" transcription assay. Increases of TNF-induced gene transcription were recognizable after 2 h for PAI-2 and after 8 h for PAI-1. The maximal increase in gene template activity was 4-fold for PAI-1 and 20-fold for PAI-2 (Fig. 3).

These experiments clearly show that, in the HT-1080 cell line, TNF greatly induces the expression of both PAI's and downregulates t-PA. The net effect of this modulation is a greatly reduced or absent fibrinolytic activity. Taken together, the effects of TNF on blood coagulation and the fibrinolytic system are a unique example of a coordinate action of TNF toward the formation and stabilization of fibrin thrombi.

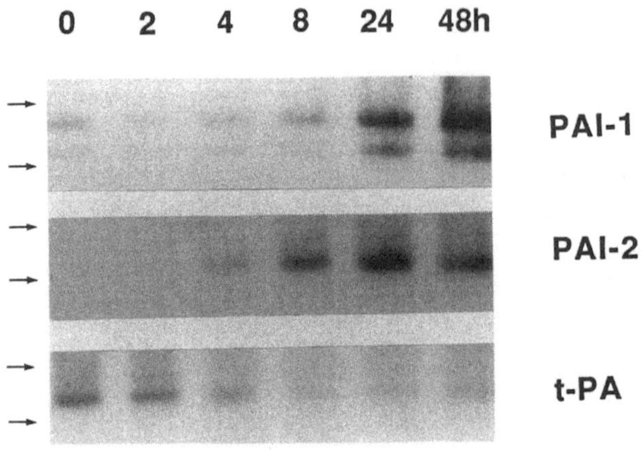

Fig. 2. Time course of PAI-1 and PAI-2 mRNA accumulation and suppression of t-PA mRNA by TNF in HT-1080 cells. mRNA was prepared from nontreated (0 h) cells or from cells treated with 50 ng/ml TNF for 2, 4, 8, 24, and 48 h. Northern blot analysis of PAI-1, PAI-2, and t-PA mRNA was performed as described in the legend to Fig. 1 (Medcalf et al. 1988a)

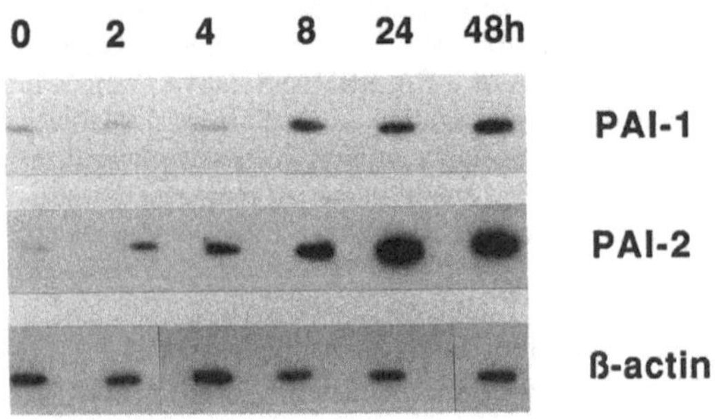

Fig. 3. Time course of TNF-induced modulation of PAI-1 and PAI-2 gene transcription. HT-1080 cell nuclei were prepared from nontreated (0 h) cells and from cells treated for 2, 4, 8, 24, or 48 h with 50 ng/ml TNF. Elongation and labeling of initiated RNA transcripts in isolated nuclei and hybridization to immobilized cDNA were performed following a modification of the originally described "run-on" transcription assay (Medcalf et al. 1988b). Exposure times were chosen to visually optimize the signals present (Medcalf et al. 1988a)

4 In Vivo Effects of Endotoxin and TNF in Human Volunteers

Several groups of investigators have studied the effect of the intravenous injection of *Escherichia coli* endotoxin or of TNF into human volunteers. Suffredini et al. (1989) used a dose of 4 ng/kg body weight and measured t-PA activity and antigen, PAI activity, plasmin α_2-antiplasmin complexes (PAP), neutrophil elastase, and von Willebrand factor in serial blood samples obtained from 19 healthy volunteers over 24 h.

Table 3. Changes in coagulation, fibrinolysis, complement, and leukocyte components after the injection of endotoxin or TNF into man

		Time course (min) of increase in component after			
System	*Component*	Endotoxin[a, b]		TNF[c]	
		Early rise	Peak levels	Early rise	Peak levels
Instrinsic coagulation	F. XII-C_1' inh.	Neg[b]		Neg	
	Kallikrein-C_1 inh.	Neg[b]		Neg	
	F. IX activ. peptide			Neg	
Common pathway of coagulation	F. X. activ. peptide			30	90
	F 1 + 2	120[b]	240[b]	30	240
	TAT complexes	120[b]	180[b]		
Profibrinolytic	t-PA activity	30[b]	120[b]		
	t-PA antigen	60[a]–90[b]	150[b]–180[a]		
Antifibrinolytic	PAI activity	120[a]–150[b]	300[a, b]		
	PAI antigen	150[b]	240[b]		
	PAP complexes	60[a]–90[b]	120[b]–180[a]		
Complement	C3a-des-arg	Neg[b]			
Leukocytes	α_1-Antitrypsin-elastase	120[a]	300[a]		
Platelet adhesion	Von Willebrand factor	60[a]–120[b]	180[a]–240[a]		

[a] Suffredini et al. (1989), 4 ng/kg endotoxin.
[b] Van Deventer et al. (1990), 2 ng/kg endotoxin.
[c] Van der Poll et al. (1990), 50 μgTNF/m^2.

Van Deventer et al. (1990) injected 2 ng/kg of *E. coli* endotoxin into six healthy subjects and measured circulating TNF and IL-6, t-PA and PAI-1 antigen levels, PAP complexes, von Willebrand factor, and complement factor C_3-des-arg over a period of 6 h after endotoxin injection. To determine whether activation of the blood coagulation system had taken place, they also measured early activation products of prothrombin (fragment 1 + 2) and thrombin/antithrombin complexes (TAT). These authors found that circulating TNF levels began to rise 30 to 45 min after the injection of a bolus of endotoxin and that peak levels were obtained after 60 to 90 min. The increases in circulating IL-6 followed that of TNF with a 15-min delay. The nadir of the drop of leukocytes was reached after 60 min. Body temperatures rose gradually and reached a peak of 38 °C after 2 h.

Van der Poll et al. (1990) injected 50 μg TNF/m^2 of body surface area and investigated particularly whether the intrinsic or rather the extrinsic pathway of coagulation was triggered by TNF.

Table 3 summarizes the results of these three studies. The earliest events are the generation of increased levels of the coagulation factor X activation peptide and of prothrombin fragment 1 + 2 in the absence of increases in activated F. XII (as determined by F.XII_a-C_1' inhibitor complexes), in kallikrein (as determined by kallikrein-C_1' inhibitor complexes), and in F. IX activation peptide. This suggests that endotoxin and TNF do not trigger the coagulation pathway via the contact phase but via the tissue factor pathway. This activation of the coagulation mechanism is accompanied by the formation of TAT complexes, and proves that thrombin has been generated in the circulation.

As far as the fibrinolytic system is concerned, the earliest events are increases in circulating t-PA activity and antigen. This is accompanied by the appearance of PAP complexes, an index of plasmin formation. PAI activity and PAI-1 antigen levels increase only in a later phase and are more pronounced than the transient and smaller t-PA increase.

The decrease in the neutrophil count is apparently associated with the release of leukocyte elastase as demonstrated by the appearance of circulation α_1-antitrypsin/elastase complexes. Complement does not appear to be activated as judged by the absence of increases in the C3a-des-arg fragment. Platelet counts also drop during endotoxinemia. It is conceivable that the approximately four-fold increase in von Willebrand factor may play a role in this respect.

5 Conclusion

Endotoxin and TNF cause profound changes in the coagulation and fibrinolytic system in cultured endothelial cells and after injection into human volunteers. It remains to be seen whether the reported alterations in coagulation factors and PA's and PAI's faithfully reflect the events which take place during clinical septicemia.

The salient features of the TNF effect are an increased expression of TF and a downregulation of TM on endothelial cells, leading to a procoagulant state and facilitating the formation of fibrin. Although TNF induces an early transient increase in t-PA activity, the net effect of TNF on the fibrinolytic system is a marked and long-lasting increase in PAI-1 expression which prevents the fibrinolytic degradation of fibrin. These effects of TNF well explain the often observed complication of disseminated intravascular coagulation with ensuing organ failure in gram-negative septicemia.

References

Aggarwal BB, Kohr WJ, Hass PE, Moffat B, Spencer SA, Henzel WJ, Bringman TS, Nedwin GE, Goeddel DV & Harkins RN (1985) Human tumor necrosis factors. Production, purification, and characterization. J Biol Chem 260:2345–2354

Bachmann F & Kruithof EKO (1984) Tissue plasminogen activator: chemical and physiological aspects. Semin Thromb Hemostasis 10:6–17

Beutler B & Cerami A (1987) Cachectin: more than a tumor necrosis factor. N Engl J Med 316:379–385

Bevilacqua MP, Pober JS, Majeau GR, Fiers W, Cotran RS & Gimbrone MA Jr (1986) Recombinant tumor necrosis factor induces procoagulant activity in cultured human vascular endothelium: characterization and comparison with the actions of interleukin 1. Proc Natl Acad Sci USA 83:4533–4537

Brockhaus M, Schoenfeld HJ, Schlaeger EJ, Hunziker W, Lesslauer W & Loetscher H (1990) Identification of two kinds of tumor necrosis factor receptors on human cell lines by monoclonal antibodies. Proc Natl Acad Sci USA 87:3127–3131

Comp PC (1991) Control of coagulation reactions. In: Hematology. Williams WJ, Beutler E, Erslev AJ & Lichtman MA (eds) McGraw-Hill, New York, pp 1304–1312

Conkling PR, Greenberg CS & Weinberg JB (1988) Tumor necrosis factor induces tissue factor-like activity in human leukemia cell line U937 and peripheral blood monocytes. Blood 72:128–133

Conway EM, Bach R, Rosenberg RD & Konigsberg WH (1989) Tumor necrosis factor enhances expression of tissue factor mRNA in endothelial cells. Thromb Res 53:231–241

Conway EM & Rosenberg RD (1988) Tumor necrosis factor suppresses transcription of the thrombomodulin gene in endothelial cells. Mol Cell Biol 8:5588–5592

Crossman DC, Carr DP, Tuddenham EGD, Pearson JD & McVey JH (1990) The regulation of tissue factor mRNA in human endothelial cells in response to endotoxin or phorbol ester. J Biol Chem 265:9782–9787

Esmon NL (1987) Thrombomodulin. Semin Thromb Hemostasis 13:454–463

Goldfield AE, Doyle C & Maniatis T (1990) Human tumor necrosis factor α gene regulation by virus and lipopolysaccharide. Proc Natl Acad Sci USA 87:9769–9773

Hohmann HP, Remy R, Brockhaus M & van Loon APGM (1989) Two different cell types have different major receptors for human tumor necrosis factor (TNFα). J Biol Chem 264:14927–14934

Kruithof EKO (1988) Inhibitors of plasminogen activators. In: Kluft C (ed) Tissue-type plasminogen activator (t-PA). Physiological and clinical Aspects, vol I CRC Press, Boca Raton, FL, pp 189–210

Loetscher H, Pan Y-ChE, Lahm H-W, Gentz R, Brockhaus M, Tabuchi H & Lesslauer W (1990) Molecular cloning and expression of the human 55 kd tumor necrosis factor receptor. Cell 61:351–359

Medcalf RL, Kruithof EKO & Schleuning W-D (1988a) Plasminogen activator inhibitor 1 and 2 are tumor necrosis factor/cachectin-responsive genes. J Exp Med 168:751–759

Medcalf RL, Van den Berg E & Schleuning W-D (1988b) Glucocorticoid-modulated gene expression of tissue- and urinary-type plasminogen activator and plasminogen activator inhibitor 1 and 2. J Cell Biol 106:971–978

Michie HR, Manogue KR, Spriggs DR, Revhaug A, O'Dwyer S, Dinarello CA, Cerami A, Wolff SM & Wilmore DW (1988) Detection of circulating tumor necrosis factor after endotoxin administration. N Engl J Med 318:1481–1486

Moore KL, Andreoli SP, Esmon NL, Esmon CT & Bang NU (1987) Endotoxin enhances tissue factor and suppresses thrombomodulin expression of human vascular endothelium in vitro. J Clin Invest 79:124–130

Moore KL, Esmon CT & Esmon NL (1989) Tumor necrosis factor leads to the internalization and degradation of thrombomodulin from the surface of bovine aortic endothelial cells in culture. Blood 73:159–165

Moser R, Schleiffenbaum B, Groscurth P & Fehr J (1989) Interleukin 1 and tumor necrosis factor stimulate human vascular endothelial cells to promote transendothelial neutrophil passage. J Clin Invest 83:444–455

Nawroth PP & Stern DM (1986) Modulation of endothelial cell hemostatic properties by tumor necrosis factor. J Exp Med 163:740–745

Nawroth PP & Stern DM (1987) Endothelial cell procoagulant properties and the host response. Semin Thromb Hemostasis 13:319–397

Nemerson Y (1991) Sequence of coagulation reactions. In: Hematology. Williams WJ, Beutler E, Erslev AJ & Lichtman MA (eds) McGraw-Hill, New York, pp 1295–1304

Preissner KT, Koyama T, Müller D, Tschopp J & Müller-Berghaus G (1990) Domain structure of the endothelial cell receptor thrombomodulin as deduced from modulation of its anticoagulant functions. Evidence for a glycosaminoglycan-dependent secondary binding site for thrombin. J Biol Chem 265:4915–4922

Scarpati EM & Sadler JE (1989 and 1990) Regulation of endothelial cell coagulant properties. Modulation of tissue factor, plasminogen activator inhibitors, and thrombomodulin by phorbol 12-myristate 13-acetate and tumor necrosis factor. J Biol Chem 264:20705–20713 and 265:14696 (correction)

Schleef R, Bevilacqua R, Sawdey MP, Gimbrone MA & Loskutoff DJ (1988) Cytokine activation of vascular endothelium. Effects on tissue-type plasminogen activator and type 1 plasminogen activator inhibitor. J Biol Chem 263:5797–5803

Suffredini AF, Harpel PC & Parrillo JE (1989) Promotion and subsequent inhibition of plasminogen activation after administration of intravenous endotoxin to normal subjects. N Engl J Med 320:1165–1172

van der Poll T, Büller HR, ten Cate H, Wortel CH, Bauer KA, von Deventer SJH, Hack CE, Sauerwein HP, Rosenberg RD & ten Cate JW (1990) Activation of coagulation after administration of tumor necrosis factor to normal subjects. N Engl J Med 322:1622–1627

van Deventer SJH, Büller HR, ten Cate JW, Aarden LA, Hack CE & Sturk A (1990) Experimental endotoxemia in humans: analysis of cytokine release and coagulation, fibrinolytic, and complement pathways. Blood 76:2520–2526

Cellular Injury by Oxidants

C. G. COCHRANE[1]

Abstract

Oxidants, generated by stimulated leukocytes, induce a variety of distinct biochemical changes in target cells. HOC1, produced by the action of peroxidase on H_2O_2 in the presence of chloride ion, acts at low molar concentration (10–20 μM) to damage proteins on cell membranes and destroy their function. H_2O_2 rapidly permeates cells and causes inhibition of ATP synthesis via both glycolytic and oxidative phosphorylation (mitochondrial) pathways. In the glycolytic pathway, damage is limited to the step involving glyceraldehyde 3-PO_4 dehydrogenase (GAPDH). This results from both an attack of H_2O_2 on GAPDH and, indirectly, by a reduction in concentration of the GAPDH cofactor, nicotinamide adenine dinucleotide (NAD). This latter effect was found to result from activation of the enzyme, poly-ADP ribose polymerase, an enzyme involved in DNA repair. Indeed, DNA damage in target cells was found at low concentrations of H_2O_2 (20–80 μM) in many cell types. Both strand breaks and base hydroxylation were observed, and resulted from the generation of hydroxyl radical from the H_2O_2 in the presence of transition metal. The consequence of the damage of DNA was found to be either cell injury and death or mutations of the base sequence and amino acid residues. These latter effects led to malignant transformations in cultured cells in both tissue cultures of the cells and in vivo in athymic mice. Exposure of a protooncogene, K-ras 4B, also led to the development of a malignant transformation by virtue of mutations in codon positions 12 and 61. Thus, oxidant effects on target cells can lead to damage to multiple functional pathways inside the cells, on the one hand, and malignant transformation via DNA damage, on the other.

1 Introduction

In studies of the inflammatory injury of tissues, oxidants, generated by stimulated leukocytes, appear to participate significantly in the attack on cells and supporting tissues. Several experimental means of determining the release of oxidants in vivo are now available including detection of H_2O_2 complexed to catalase or myeloperoxidase through inactivation of the complexed enzymes with 3 amino, 1,2,4 triazole; conversion of methionines (met) either as free ^{35}S met or met in molecules such as alpha$_1$-

[1] Department of Immunology, IMM12, Research Institute of Scripps Clinic, 10666 North Torrey Pines Road, La Jolla, CA 92037, USA

42. Colloquium Mosbach 1991
Molecular Aspects of Inflammation

PI, to met •SO; depletion of reduced glutathione levels in cells in the inflammatory site; formation of the oxidized adduct of the spin trap, DMPO; detection of oxidized DNA bases in the urine; detection of ethane in exhaled air; and the conversion of intracellular xanthine dehydrogenase to xanthine oxidase. This list is not complete. Despite these assays, the biochemical tracking of oxidant release in even semi-quantitative terms is imprecise and requires considerably more development.

Nevertheless, since it has become clear that oxidants are indeed released during inflammatory injury of tissues, several years ago we initiated studies to determine the means by which oxidants damage cells and tissues. Several forms of oxidants were employed: those generated from stimulated granulocytes (PMN's); or chemically, from purine and xanthine oxidase; reagent H_2O and H_2O_2 plus 10 μM Fe^{2+} in the presence of ascorbate. By those means, O_2^-, H_2O_2, and HOC1 were formed, and when amino acids such as glycine were added to HOC1, chloramines were rapidly generated. Our attention was focused on H_2O_2, HOC1, and •OH as a result of studies indicating that injury of cells was not inhibited by superoxide dismutase (but was by catalase), and the formation of chloramines protected cells against the effects of HOC1.

2 Injury to Cells by HOC1

HOC1 was found to damage and induce target cell lysis at low concentrations. Ten to 20 μM HOC1 caused oxidation of plasma membrane sulfhydryls and disturbance of various protein functions in the external membrane (Aruoma et al. 1989), such as inactivation of glucose and amino acid transporters, and loss of K^+ pumping capacity. It is likely that these effects are caused by SH oxidation in proteins, since similar effects are induced by para-chloromercuriphenylsulfonate, a sulfhydryl reagent that acts only at the cell surface. Higher doses (> 50 μM) of HOC1 induced generalized -SH oxidation and protein carbonyl formation. These latter events were associated with cell lysis. Thus, damage to the functions of the cell membrane precede, and occur at low concentrations of HOC1, while at higher concentrations, cells undergo lysis and generalized oxidative damage.

3 Oxidant Effects on Cellular Energy Systems

Early observations from our laboratory indicated that ATP levels were depleted in P388D1 and cultured endothelial cells after exposure to oxidants. Initial reduction occurred at 2–3 min, followed by a dose-dependent decrease reaching levels below 10% of control values (Fig. 1; Spragg et al. 1985).

The loss of ATP could result from either accelerated consumption or diminished formation. To examine the rate of consumption in P388D1 cells exposed to oxidants, the rate of loss of ATP was followed in cells exposed to H_2O_2 in which both glycolytic and mitochondrial synthesis were blocked (2 deoxyglucose + atractyloside). A

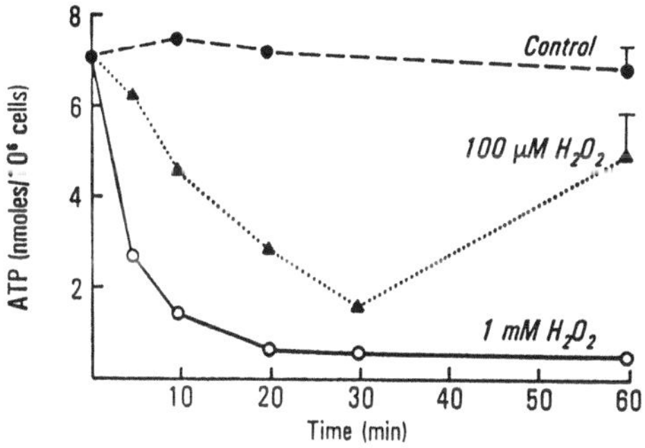

Fig. 1. Diminution of ATP in P388D$_1$ cells exposed to 100 μM and 1 mM H_2O_2 for the time intervals shown

calculated loss of 2.3 nmol ATP $min^{-1}10^{-6}$ cells compared favorably with the rate in cells not exposed to H_2O_2 and to the rate of synthesis in control cells of 2.1 nmol $min^{-1}10^{-6}$ cells (Hyslop et al. 1988). Thus, increase in catabolism was an unlikely explanation for the drop in ATP levels in these cells. However, this may vary between cell types.

4 Effect of Oxidants on the Glycolytic Synthesis of ATP

By contrast, both glycolytic and mitochondrial synthesis were profoundly affected by exposure to oxidants (H_2O_2, O_2^-, and HOCl) (Hyslop et al. 1988). With H_2O_2 in P388D1 cells, the glycolytic pathway was inhibited principally at the level of glyceraldehyde-3-phosphate dehydrogenase (GAPDH) (Jarrick et al. 1982). Of the 11 enzymes of the glycolytic pathway analyzed, only GAPDH showed a significant decrease in Vmax. Km of each enzyme was affected by less than 10%. Vmax of GAPDH in cells exposed to 0.5 mM H_2O fell from 78.1 ± 8.7 to 0.8 ± 0.1 nmol $min^{-1}10^{-6}$ cells and nearly total inhibition of activity at 1 mM H_2O_2. This was accompanied by an elevation of the components of the glycolytic pathway immediately preceding the GAPDH step, glyceraldehyde-3-phosphate, and dihydroxyacetone phosphate. The combined levels were increased approximately six fold in the 10 min after exposure to H_2O_2. Lactate production decreased in proportion (Hyslop et al. 1988).

The inhibitory effect on the GAPDH step resulted from three independent factors: a direct attack of H_2O_2 on GAPDH, a decrease in its essential cofactor, nicotinamide adenine dinucleotide (NAD), and a small shift in pH in the cell responding to H_2O_2. As to the first, not only a decrease in Vmax of GAPDH was measured in oxidant-exposed cells with an IC_{50} of approximately 100 μM H_2O_2, but isolated rabbit muscle GAPDH was also inactivated by H_2O_2 with an IC_{50} of 10 μM H_2O_2. Further studies have indicated that loss of two SH groups is associated with loss of activity of GAPDH and that S-S bonds are formed. The SH groups involved were found to be cysteines 149 and 153 which are known to be spatially opposed in one turn of an alpha helix. Disulfide formation between these cysteines inhibits activity (Hyslop et al., unpubl. observ.), possibly by preventing binding of NAD.

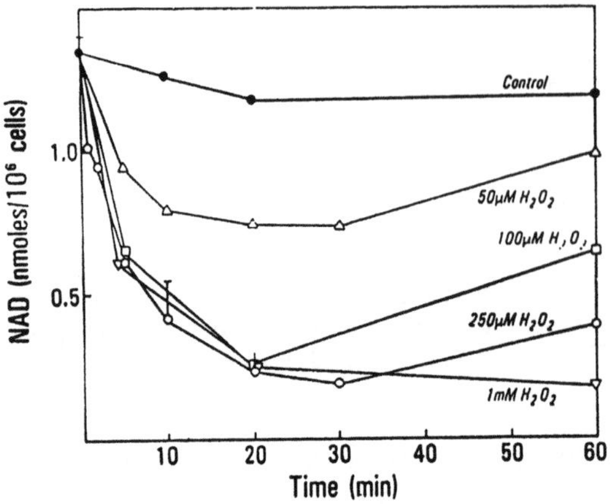

Fig. 2. Diminution in levels of nicotinamide adenine dinucleotide (NAD) in $P388D_1$ cells exposed to various initial concentrations of H_2O_2

The second major influence on GAPDH activity was found to be a fall in levels of the essential cofactor NAD (Fig. 2; Schraufstatter et al. 1986a; Hyslop et al. 1988). NAD levels fall relative to NADH in cells that contain poly-ADP ribose polymerase (polymerase). This latter enzyme becomes activated during H_2O_2 exposure and consumes NAD, its substrate. The two products of NAD during hydrolysis by the polymerase are nicotinamide (NA) and ADP-ribose which becomes polymerized and binds to proteins in the vicinity of its formation. NA formation in P388D1 cells paralleled stoichiometrically the loss of NAD. Ribosylation of proteins simultaneously occurred. These proteins were indistinguishable from nucleohistones by HPLC analysis.

The effects of the direct inactivation of GAPDH by H_2O_2, diminishing levels of cofactor NAD relative to NADH, and slight, but significant lowering of pH would appear to combine to produce the inhibition of the glycolytic pathway at this step.

A relationship between the fall in NAD levels and activation of poly-ADP ribose polymerase was observed employing inhibitors of the polymerase. 3-Aminobenzamide, nicotinamide, and theophylline each inhibited the polymerase as measured by NAD formation and protein ribosylation and prevented the loss of NAD (Schraufstatter et al. 1986b), with P388D1 cells, inactivation of the polymerase with 3-aminobenzamide was associated with not only maintenance of levels of NAD, but ATP as well. These cells also did not develop a flux of extracellular Ca^{2+} and viability was maintained.

It should be noted that treatment of cells with 3-aminobenzamide did not affect the half-life of H_2O_2 or certain polymerase-independent effects of H_2O_2 such as translocation of intracellular Ca^{2+}. Thus, in P388D1 cells, NAD cofactor of the GAPDH steps of the glycolytic pathway appeared under the conditions of these experiments to be a critical modulating factor in oxidant damage to the energy-producing system and viability of the cells.

5 Effect of Oxidants on Oxidative Phosphorylation

The second ATP-generating system, i.e., oxidative phosphorylation by the mitochondria, was also affected by exposure to oxidants, specifically, H_2O_2. Of interest, at H_2O_2 concentrations of 100–300 μM, an increase in net mitochondrial synthesis of ATP occurred. Nevertheless, over the same concentration range of H_2O_2 and above, ATP synthetase (ADP phosphorylation) activity consistently declines (Hyslop et al. 1988).

Thus, both glycolytic and oxidative phosphorylation pathways of ATP formation are affected by exposure of cells to oxidants. The balance of effect on one vs the other pathway will likely vary with cell type. Those cells heavily dependent upon mitochondrial oxidative phosphorylation will not be as subject to ATP loss caused by inhibition of GAPDH as those more dependent upon glycolysis. The reverse would be true for cells more dependent upon glycolysis.

6 Oxidant Damage of Cellular DNA

Since poly (ADP-ribose) polymerase is activated under conditions of DNA strand breakage, we reasoned that DNA damage might have occurred in cells exposed to oxidants. This has proved to be the case in all cell types assayed to date, and the DNA damage occurs within seconds of exposing the cells to oxidants (Schraufstatter et al. 1986a).

A major unanswered question was that of the mechanisms by which oxidants damage DNA. Information available from the literature suggested that O_2^- (Birnboim and Kanabus-Kominska 1987), as well as H_2O_2 and hydroxyl radical ($^\bullet OH$), could be responsible, although little information about HOC1, singlet O_2 ($^1O_2$), or N-chloramines was available. We first determined that reagent H_2O_2 could induce damage to Cellular DNA, as measured by an alkaline unwinding assay. Human peripheral lymphocytes and P388D1 cells were susceptible to micromolar concentrations of H_2O_2 (Fig. 3; Schraufstatter et al. 1986a). Damage to cellular DNA developed within 20 s after exposure of the cells to H_2O_2 and reached a maximal effect in minutes. With catabolic removal of the H_2O_2 from the medium, repair of the DNA occurred over the succeeding hours in most, but not all, cell types. Exposure of several cell types (P388D1 cells, human peripheral lymphocytes, GM1380 fibroblasts, and rabbit alveolar macrophages) to H_2O_2 revealed varied susceptibility of the DNA to damage by the H_2O_2. If H_2O_2 were the major oxidant species penetrating the cells to damage the DNA, then specific inhibition of catalase in the cytoplasm of the various cell types should produce cells of equal susceptibility. This proved to be the case. When these various cell types were first exposed to 3-amino-1,2,4-triazole to inhibit catalase (as well as myeloperoxidase in the neutrophils) and were then treated with various concentrations of H_2O_2, the cells showed greater susceptibility to DNA damage, and the degree of susceptibility was the same for all cell types. In addition, in the absence of aminotriazole, the susceptibility of the cellular DNA to various concentrations of

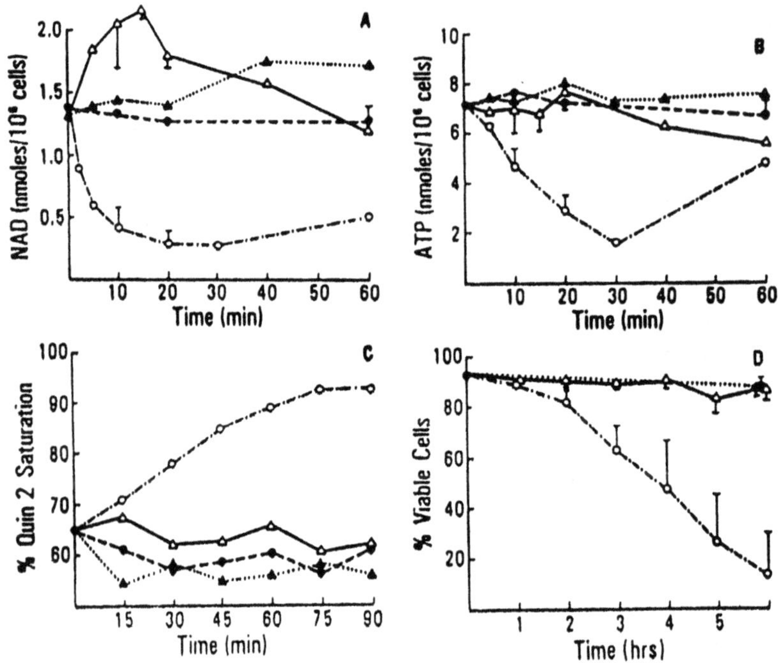

Fig. 3. The capacity of 2.5 mM 3-aminobenzamide (3 ABA) to inhibit biochemical alterations in 2 x 10^6 target P388D$_1$ cells induced by H_2O_2. **A** NAD levels ●---● untreated; O-·-O exposed to 250 μM H_2O_2; ▲···▲ treated with 2.5 mM 3 ABA; ∇—∇ treated with 2.5 mM 3 ABA followed by 250 μM H_2O_2. **B** ATP levels. *Symbols* same as in **A**. **C** Intracellular Ca^{2+} levels as indicated by Quin-2 fluorescence. *Symbols* same as in **A**. **D** Viability. *Symbols* same as in **A**

H_2O_2 was inversely proportional to the activity of catalase in the cells (Schraufstatter et al. 1988).

Several oxidants generated by human monocytes were then assessed for their capacity to damage DNA in peripheral human lymphocytes. The ratio of monocytes to lymphocytes was 1:3. A strong correlation existed between the concentration of H_2O_2 in the extracellular medium and the degree of DNA strand breaks. Addition of catalase to the medium completely prevented the strand breaks in each system, but addition of superoxide dismutase (SOD) did not. These data suggest that H_2O_2 in the medium to which the target cells were exposed was responsible for initiating the DNA damage (Schraufstatter et al. 1988).

HOC1, N-chloramine, H_2O_2, and $^1O_2$ were then compared for their capacity to damage isolated PM2 DNA. Isolated DNA was chosen to avoid intracellular variation in oxidant activity, such as from inactivation by cellular inhibitors and natural scavengers. While 100 μM H_2O_2 plus 30 μM iron induced an average of 55% strand breaks, 100 μM HOC1 induced 5%, N-chloramine (100 μM HOC1 plus 500 μM taurine) induced 6%, and a combination of 100 μM H_2O_2 and 100 μM HOC1 (to generate $^1O_2$) induced 20%. Similarly, addition of 10 μmM HOC1 to P388D1 cells failed to induce DNA damage, as measured by the alkaline unwinding assay.

These results support the conclusion that H_2O_2 is the dominant oxidant leading to DNA strand breaks. The data on cells in which intracellular catalase was inhibited by aminotriazole suggest that H_2O_2 penetrates the cell, thereby gaining access to DNA. The H_2O_2 could conceivably induce damage directly or indirectly. Therefore, experiments were undertaken to examine this aspect.

7 Participation of the Hydroxyl Radical in DNA Damage

Our attention turned to •OH as the free radical most likely generated by the presence of H_2O_2 in the vicinity of nuclear DNA. H_2O_2 alone, at concentrations as high as 100 µM, failed to induce significant numbers of strand breaks in isolated PM2 DNA. However, abundant strand breaks were observed in the presence of Fe^{2+} (10 to 30 µM), especially when either ascorbate or O_2^- was added to reduce Fe^{3+} to Fe^{2+}. These results, in combination with previous studies (e.g., Floyd 1981; de Mello Filho and Meneghini 1985; Frenkel et al. 1986), strongly implicated the generation of •OH from H_2O_2 as a mechanism. In whole P388D1 cells, DNA damage, measured as strand breaks, was prevented by the chelation of iron using phenanthroline or desferrithiocin.

The apparent association of •OH formation by the Fenton reaction and DNA damage was to this point based on indirect evidence. To determine more accurately if this association exists, we conducted experiments to measure •OH levels at a time when DNA strand breaks occurred. We used electron paramagnetic resonance (EPR) spectroscopy with the spin trap dimethylpyroline-N-oxide (DMPO) to detect the presence

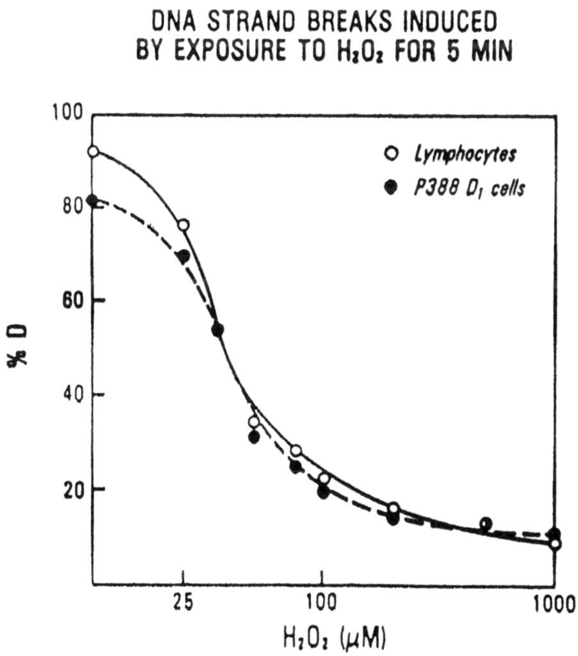

Fig. 4. DNA single strand breaks as determined by alkaline unwinding technique. $P388D_1$ cells (●) or human peripheral lymphocytes (○) (2×10^6/ml MGB) were incubated with different doses H_2O_2 for 5 min at 37 °C, H_2O_2 was removed by centrifugation, the cell pellet was resuspended in 450 µl ice-cold meso-inositol buffer, and 200 µl samples were lysed in alkaline medium. Ethidium bromide fluorescence was determined (excitation, 520 nm; emission, 590 nm) after a 45-min incubation at 15 °C. *D* represents percent double-stranded DNA at the end of this incubation period as described. Mean of three experiments determined in duplicate

of •OH. With H_2O_2, Fe^{2+}, and PM2 DNA, a four-line spectrum appeared that was characteristic of the DMPO adduct with •OH (Jackson et al. 1987). A similar spectrum was observed utilizing cigarette smoke (a rich source of H_2O_2) and asbestos (a rich source of iron). A correlation was observed between the amount of DMPO-OH observed by EPR and the percentage of DNA with strand breaks (Fig. 4). Since the DMPO-OH spectrum can be produced by mechanisms not involving •OH, the secondary •OH traps ethanol and dimethylsulfoxide (DMSO) were added to the mixtures described above. The resulting EPR spectra revealed the six-line DMPO-alpha-hydroxyethyl signal and the six-line DMPO-methyl signal, respectively. These patterns develop only when DMPO-OH is generated from •OH, confirming that •OH had been formed. Further confirmation was obtained by the observation that catalase (but not SOD), •OH scavengers, and iron chelators inhibited both •OH detection and DNA strand breaks. Complete inhibition of DNA strand breaks was observed with the addition of 3.5 μg catalase, 100 mM DMSO, 100 mM mannitol, 100 mM sodium benzoate, 4 mM phenanthroline, and 4 mM desferrithiocin (Jackson et al. 1987).

8 DNA Base Hydroxylation by Leukocytic Oxidants

The question was then raised whether damage by oxygen-free radicals was limited to single strand breaks in DNA, or if modifications in bases took place as well. To examine this, calf thymus DNA was exposed to neutrophils stimulated with phorbol myristate acetate in the presence or absence of Fe. The DNA was then analyzed by gas chromatography-mass spectrometry following hydrolysis and derivatization. Hydroxylation was observed in all four bases and the addition of Fe increased the yield such that 7 out of every 1000 bases were modified (Jackson et al. 1989). Similar observations were made independently by Aruoma et al. (1989).

9 DNA Damage by Products of Lipid Peroxidation

Damage to DNA by the H_2O_2/•OH reaction may indeed play a major role in cellular injury and in altered function in the inflammatory response. In addition, products of lipid peroxidation induced by the interaction of •OH with polyunsaturated fatty acids in the membranes of target cells may also contribute to the DNA damage. Ochi and Cerutti (1987) emphasized the relationship of lipid hydroperoxides to the clastogenic agents released by stimulated monocytes. These investigators have described the capacity of such factors, at concentrations of 10 to 20 μM, to induce DNA strand breaks in target cells. Our preliminary data support their conclusions. Unsaturated fatty acids (oleic, linoleic, linolenic, and arachidonic acids, 164 μM) induced DNA damage in GM1380 fibroblasts, and the amount of damage (leading to DNA strand breaks) corresponded to the degree of fatty acid unsaturation. The mechanism of DNA damage by this means remains uncertain, but the effect of the unsaturated fatty acids is clear.

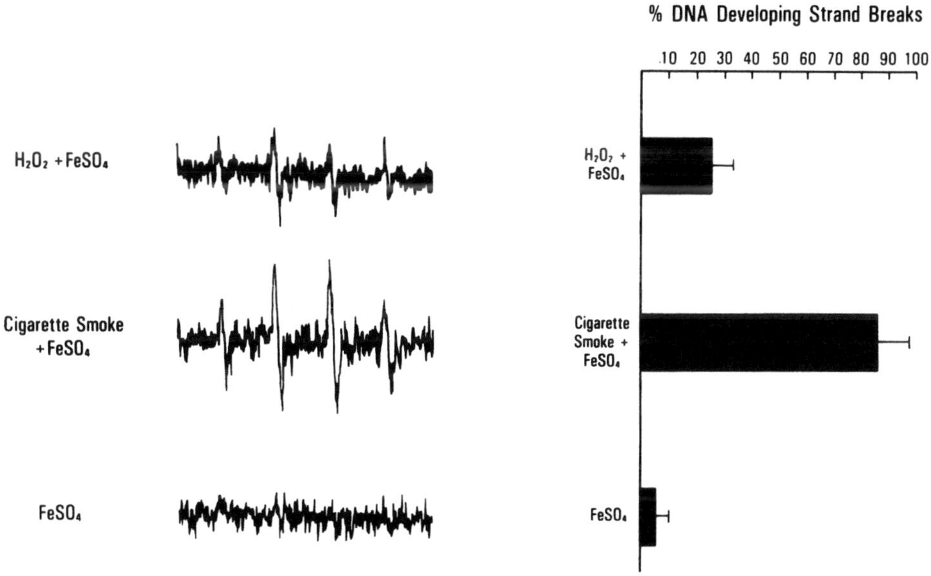

Fig. 5. Electron paramagnetic resonance spectral patterns developing from mixtures of 0.8 μg PM2 DNA, 0.05 mg crocidolite asbestos, 10 μl cigarette smoke in PBS, 65 μM H_2O_2, and/or $FeSO_4$ in a final volume of 100 μl of PBS, pH 7.4. 100 mM DMPO was added to form adducts with $^{\bullet}OH$

The importance of this mechanism in the inflammatory response remains to be determined.

10 Conclusions

The studies reviewed and the data described here support the hypothesis that, of various oxidants generated by leukocytes during the inflammatory response, H_2O_2 is the most likely oxidant to penetrate cells and reach the various cellular targets, including cell membranes, DNA, glyceraldehyde-3-phosphate dehydrogenase, Ca^{2+} reservoirs, and mitochondria. A tabulation of intracellular targets shown in those studies to be affected by oxidants, in particular H_2O_2 or H_2O_2-$^{\bullet}OH$ is shown in Fig. 5. Considering oxidant damage to DNA, concentrations of H_2O_2 as low as 25 μM cause damage to DNA in target cells. Such concentrations are physiologically relevant since millimolar concentrations of H_2O_2 can be generated in the immediate vicinity of the target cells by stimulated neutrophils. Studies suggest that H_2O_2 acts indirectly through $^{\bullet}OH$ to damage DNA.

10.1 Consequences of DNA Damage by Oxidants: Malignant Transformation

In this laboratory, in studies conducted by J. Jackson, two approaches to the development of the malignant potentiality of DNA have been taken after exposure of DNA to oxidants. In the first, cell lines bearing a low potential of malignancy were exposed to oxidants and examined over the course of time in culture and in vivo for the development of malignancies. DNA from any malignant cells would then be extracted and examined for mutations and deletions that may be associated with the malignant state. And second, a protooncogene DNA, Kirsten-*ras* of the 4th exon (K-ras 4B) and DNA of related *ras* protooncogenes were exposed to oxidants (identical to those in the first approach) and transfected into cells of low malignant potentiality to determine if malignant transformation develops. DNA of the transfected K-*ras* 4B gene in malignant cells was then examined for specific mutations, and the 21-kD gene product of K-*ras* 4B examined for changes in physical properties.

10.1.1 The Development of Malignancies in Cell Lines Exposed to Oxidants

In these studies, C3H T 10-1/2 and NIH 3T3 cells were exposed to several oxidants: O_2^- and H_2O_2, generated by purine and xanthine oxidase, oxidants generated from PMN leukocytes during stimulation with phorbol myristate acetate (PMA), and cigarette smoke dissolved in buffered salt solution. The oxidant-exposed and control NIH 3T3 and C3H T 10/1/2 cells were then grown in culture for 11–14 days and examined for the development of Type III malignant foci and C3H T 10-1/2 cells were injected into athymic mice and followed for 11 weeks to determine the presence of malignant transformations. The results are shown in Table 1.

10.1.2 Malignant Transformation of K-ras 4B Protooncogene After Exposure to Oxidation

Of the approximately 40 cellular oncogenes identified, ras oncogenes are the most frequent in human carcinomas. In addition, in 20–40% of colon, pancreatic, and lung epithelial malignancies, K-*ras* oncogene has been found using the polymerase chain reaction to amplify the oncogene from biopsied tumors. Other forms of *ras*, i.e., H and N ras, are not found or are found only rarely. The *ras* protooncogenes, probably functioning in signal transduction in normal cellular proliferation, encode a protein of 21 kD (188 or 189 residues) that binds guanine nucleotides, GDP, and GTP interconvertably. The protein is synthesized in the cytoplasm and becomes bound to the plasma membrane when active by isoprenylation of a cysteine (cys 186) lying four residues from the C-terminus.

Table 1. Oxidant-induced transformation of C3H 10 T-1/2 cells

Exposure	Tissue culture transformed foci	Tumors in athymic mice
None	2/85	0/26
PMA	4/98	2/24
PMA + PMN	21/113	7/43

Normal *ras* genes acquire malignant potentiality following mutations of amino acid residues at positions 12, 13, and 61. Naturally occurring cancers of human beings have demonstrated such mutations. The gene product of K-*ras* after mutations demonstrates a change in molecular size and exists predominantly in the active GTP-binding form, characteristic of the active form of G-proteins in general.

In this laboratory, J. Jackson exposed K-*ras* 4B DNA (1.0 µg/plate) to several sources of oxidants: xanthine oxidase and purine, phorbol myristate acetate (PMA)-stimulated human PMN's in the presence of 10 µM Fe^{2+} and cigarette smoke in buffered saline in the presence of Fe^{2+}. Controls included each reagent alone. After incubation 1 h at 37 °C the cells were heated to 65 °C to inactivate enzymes, DNA extracted, introduced into pZIP-Neo SV(X) 1 retrovirus-transforming vector, and transfected into NIH 3T3 cells by the calcium phosphate precipitation technique. Focus-forming transfection assays were monitored 14–21 days.

The results demonstrated that malignant foci developed in approximately 25% of plates containing cells treated with K-*ras* 4B exposed to xanthine oxidase and purine, 32% of plates with cells treated with K-*ras* 4B exposed to PMA stimulated human PMN's, and 32% of plates containing cells treated with K-*ras* 4B exposed to cigarette smoke. All controls were negative except those exposed to 10 µM Fe^{2+} alone, which reveal 4% transformation. Cells in positive foci from cells treated with xanthine oxidase and purine or PMA-stimulated human PMN's were amplified and their genomic DNA extracted and transfected into NIH-3T3 cells. Those yielded positive foci. DNA of the K-*ras* 4B of those cells, when amplified by the polymerase chain reaction, demonstrated mutations at codons 12 (10–70% of plates) and 61 (30–35% of plates). Of interest, in four transformed K-ras 4B DNA's, mutations were limited to codons 12 and 61. In addition, treatment of Harvey *ras* DNA identically to that of Kirsten-ras, demonstrated malignant transformation in NIH 3T3 cells in only 1/120 plates. This heavy preference of malignant transformation of K-*ras* 4B treated with oxidants parallels the incidence of transformed K- vs H-ras in naturally occurring cancers of the bowel, pancreas, and lung in human beings.

10.2 Summary of the Effects of Oxidants on DNA

Thus, exposure of DNA to oxidants induces base hydroxylation and strand breaks, the result, primarily of ·OH formed from H_2O_2 in the presence of transition metal. This damage to DNA can lead to a fall in ATP levels and death of the cell, or can lead to malignant transformation in surviving cells. In a single example of an oncogene, K-*ras* 4B, oxidants induce damage to the DNA that, upon transfection into NIH 3T3 cells, induces mutations at codons 12 and 61, i.e., the same mutations that are associated with malignant transformation in naturally occurring human cancers of the colon, pancreas, and lung.

Acknowledgment. This work was supported by United States Public Health Service Grants HL-23584 and GM-37696 and Office of Naval Research.

References

Aruoma OL, Halliwell B & Dizdaroglu M (1989) Iron ion-dependent modification of bases in DNA by the superoxide radical generating system hypoxanthine/xanthine oxidase. J Biol Chem 264:13024–13028

Birnboim HC & Kanabus-Kominska M (1987) The production of DNA strand breaks in human leukocytes by superoxide anion may involve a metabolic process. Proc Natl Acad Sci USA 82:6820–6824

de Mello Filho AC & Meneghini R (1985) Protection of mammalian cells by o-phenanthroline from lethal and DNA-damaging effects produced by active oxygen species. Biochem Biophys Acta 847:82–99

Floyd RA (1981) DNA-ferrous iron catalyzed hydroxyl free radical formation from hydrogen peroxide. Biochem Biophys Res Commun 99:1209–1215

Frenkel K, Chrzan K, Troll W, Teebor GW & Steinberg JJ (1986) Radiation-like modification of bases in DNA exposed to tumor promoter-activated polymorphonuclear leukocytes. Cancer Res 46:5533–5540

Hyslop PA, Hinshaw DB, Halsey WA Jr, Schraufstatter IU, Jackson JJ, Spragg R, Sauerheber R & Cochrane CG (1988) Mechanisms of oxidant mediated cell injury: the glycolytic and mitochondrial pathways of ADP phosphorylation are major intracellular targets inactivated by hydrogen peroxide. J Biol Chem 263:1665–1675

Jackson J, Schraufstatter IU, Hyslop PA, Vosbeck K, Sauerheber R, Weitzman SA & Cochrane CG (1987) Role of oxidants in DNA damage: hydroxyl radical mediates the synergistic DNA damaging effects of asbestos and cigarette smoke. J Clin Invest 80:1090–1095

Jackson JH, Gajewski E, Fuciarelli AE, Schraufstatter IU, Hyslop PA, Cochrane CG & Dizdaroglu M (1989) Damage to the bases in DNA induced by stimulated neutrophils. J Clin Invest 84:1644–1649

Jarrick BA, Nathan CF, Griffith OW & Cohn ZA (1982) Glutathione depletion sensitizes tumor cells to oxidative cytolysis. J Biol Chem 257 (3):1231–1237

Ochi T & Cerutti PA (1987) Clastogenic action of hydroperoxy-5, 8, 11, 13-icosatetranoic acids on the mouse embryo fibroblasts C3H/10 T1/2. Proc Natl Acad Sci USA 84:990–994

Schraufstatter IU, Hinshaw DB, Hyslop PA, Spragg RG & Cochrane CG (1986a) Oxidant injury of cells: DNA strand-breaks activate polyadenosine diphosphate-ribose polymerase and lead to depletion of nicotinamide adenine dinucleotide. J Clin Invest 77:1312–1320

Schraufstatter IU, Hyslop PA, Hinshaw DB, Spragg RG, Sklar LA & Cochrane CG (1986b) Hydrogen peroxide-induced injury of cells and its prevention by inhibitors of poly (ADP-ribose) polymerase. Proc Natl Acad Sci USA 83:4908–4912

Schraufstatter IU, Hyslop PA, Jackson JH & Cochrane CG (1988) Oxidant-induced DNA damage of target cells. J Clin Invest 82:1040–1050

Schraufstatter IU, Browne K, Harris A, Hyslop PA, Jackson JH, Quehenberger O & Cochrane CG (1990) Mechanisms of hypochlorite (HOC1) injury of target cells. J Clin Invest 85:554–562

Spragg RG, Hinshaw DB, Hyslop PA, Schraufstatter IU & Cochrane CG (1985) Alterations in adenosine triphosphate and energy charge in cultured endothelial and P388D1 cells following oxidant injury. J Clin Invest 76:1471–1476

Endothelium-Derived Relaxing Factor: Nitric Oxide

R. BUSSE[1] and A. MÜLSCH[1]

1 Introduction

Several mammalian cells express a newly discovered signal transduction pathway, which uses nitric oxide (NO) synthetized from L-arginine to modulate the function of target cells (for a compilation of recent findings, see Moncada and Higgs 1990). NO released from endothelial cells, the "endothelium-derived relaxant factor" (EDRF), activates the hemoproteins-soluble guanylyl cyclase in vascular smooth muscle cells and in platelets (Fig. 1). The resulting increase in cyclic GMP inhibits smooth muscle contraction and platelet activation by a reduction in intracellular free Ca^{2+} concentration (Rapoport and Murad 1983; Ignarro and Kadowitz 1985; Busse et al. 1987). In a similar manner, by stimulating the soluble guanylyl cyclase and the subsequent elevation of cyclic GMP, NO acts in certain regions of the central nervous systems of vertebrates (Knowles et al. 1989). Recent data indicate that a calcium-dependent NO synthesis is triggered in the granule cells of the cerebellum following activation of N-methyl-D-aspartate (NMDA) receptors by the major excitatory neurotransmitter, glutamate (Garthwaite et al. 1988, 1989; Bredt and Snyder 1989). It has been shown that NO released from presynaptic sites of neuronal cells can alter the synaptic efficacy and plasticity within a given neuroanatomical area (Gally et al. 1990) and is essential for the long term synaptic depression in the cerebellum (Shibuki and Okada 1991), a phenomenon implicated in the neuronal mechanism of cerebellar motor learning.

Cytotoxically activated macrophages constitute another important cell type in which NO is clearly identified as a paracrine effector molecule (Hibbs et al. 1987; Iyengar et al. 1987). NO is formed in these cells after induction of a Ca^{2+}-independent NO synthase by endotoxins from gram-negative-bacteria and/or by cytokines (Marletta et al. 1988; Hauschildt et al. 1990) and seems to account partially for the cytotoxic and microbiostatic activity of macrophages directed against tumor and other target cells. There are several experimental hints that the cytotoxic effect of NO is mediated by a direct inhibition of several key enzymes such as the Krebs cycle enzyme, aconitase (Drapier and Hibbs 1986), and the proximal two oxidoreductases of the mitochondrial electron transport chain (complexes I and II) (Drapier and Hibbs 1988). These enzymes contain iron-sulfur prosthetic groups that are essential for their catalytic function. NO with its strong affinity for iron inhibits mitochrondrial respir-

[1] Institut für Angewandte Physiologie, Universität Freiburg, Hermann-Herder-Str. 7, W-7800 Freiburg, FRG

42. Colloquium Mosbach 1991
Molecular Aspects of Inflammation

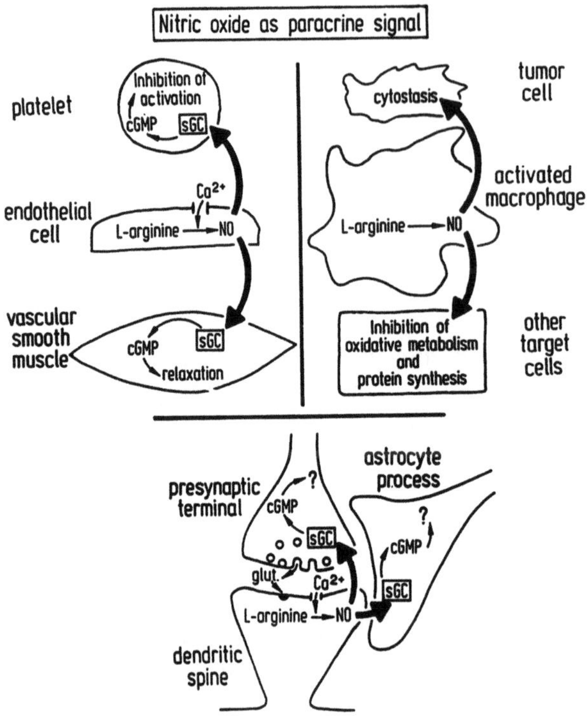

Fig. 1. Paracrine functions of NO in different cell systems. For further explanation, see text. *sGC* soluble guanylyl cyclase

ation and aconitase activity by binding to the iron-sulfur clusters of the enzymes: formation of nitrosyl-iron-sulfur complexes has been detected by electron paramagnetic resonance spectroscopy in cytokine-activated macrophages (Pellat et al. 1990). In addition to these examples, NO formation has now been found in a variety of other cells – several of which are listed in Table 1.

The finding of a rather widespread distribution of NO synthase activity associated with an activation of soluble guanylyl cyclase in the NO-generating cell themselves has strengthened the concept that NO is not only an intercellular messenger, but may also act as an autocrine signal (Martin et al. 1988; Schröder and Schrör 1989; Busse and Mülsch 1990b; Radomski et al. 1990a). However, the cellular consequences of the increase in soluble guanylyl cyclase activity in some of the different cell types have not yet been unequivocally identified and remain to be elucidated.

There appear to exist at least two forms of NO synthase (Table 1), a constitutive, Ca^{2+}-calmodulin-dependent enzyme (Bredt and Snyder 1990; Busse and Mülsch 1990a; Förstermann et al. 1991; Schmidt et al. 1991) and an inducible, Ca^{2+}-calmodulin-independent enzyme (Marletta et al. 1988; Billiar et al. 1990; Busse and Mülsch 1990b). All isoenzymes studied so far require as cofactors NADPH and 6R-tetrahydrobiopterin (Kwon et al. 1989; Tayeh and Marletta 1989; Mayer et al. 1990; Mülsch and Busse 1991). While in macrophages and brain tissue the NO synthase has been characterized exclusively as a cytosolic enzyme, recent studies indicate that in endothelial cells a considerable part of NO synthase activity is associated with the par-

Table 1. Nitric oxide-producing cells

I. Constitutive NO synthase	
Endothelial cells	Palmer et al. (1987)
Human neutrophils	Wright et al. (1989)
Rat adrenal gland	Palacios et al. (1989)
Rat forebrain	Knowles et al. (1989)
N1E-115 neuroblastoma cells	Förstermann et al. (1989)
Rat cerebellum	Garthwaite et al. (1988)
Kidney epithelial cells	Schröder and Schrör (1989)
Human platelets	Radomski et al. (1990b)
NANC nerves	Boeckxstaens et al. (1991)
II. Inducible NO synthase	
Murine endothelial cells	Kilbourn and Belloni (1990)
Rodent macrophages	Iyengar et al. (1987)
EMT-6 adenocarcinoma cells	Amber et al. (1988)
Rat kupffer cells	Billiar et al. (1989)
Rat hepatocytes	Curran et al. (1989)
Rabbit smooth muscle cells	Busse and Mülsch (1990b)
Rat mesangial cells	Pfeilschifter and Vosbeck (1991)

ticulate fraction (Boje and Fung 1990; Förstermann et al. 1991). The soluble NO synthase in the brain appears to exist as a homodimer of two presumably identical subunits with a molecular mass of about 150 kDa (Bredt and Snyder 1990; Mayer et al. 1990; Schmidt et al. 1991). Preliminary data indicate that the particulate enzyme in endothelial cells might be slightly different with a molecular mass of about 135 kDa (Förstermann et al. 1991).

A number of L-arginine analogs have been identified and characterized as specific inhibitors of these NO synthases (Marletta et al. 1988; Palmer et al. 1988; Schmidt et al. 1988; Mülsch and Busse 1990) (Fig. 2). In endothelial cells, N^G-nitro-L-arginine is a more potent inhibitor than N^G-methyl-L-arginine and imino-ethyl-L-ornithine. Since these compounds are highly stereospecific inhibitors – the D-enantiomers are completely ineffective on the NO synthase – they are excellent tools for investigating

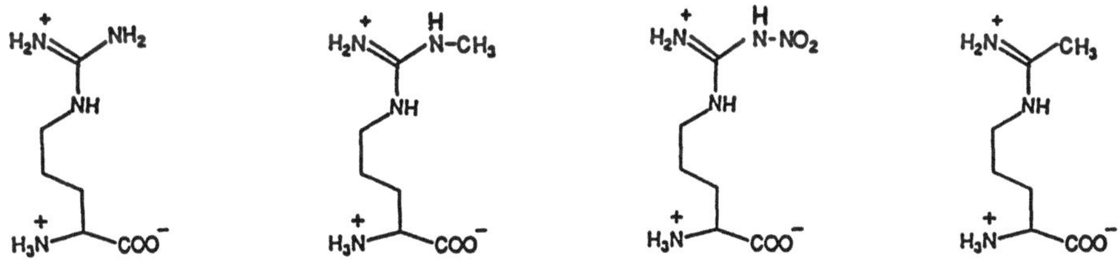

L-arginine

N^G-monomethyl-L-arginine L-NMMA

N^G-nitro-L-arginine L-NNA

N-iminoethyl-L-ornithine L-NIO

Fig. 2. Structural formulae of L-arginine and NO synthase-inhibiting L-arginine analogs the L-arginine-NO pathway in cells as well as in intact organisms (Rees et al. 1989).

In this review we will focus on the mechanisms by which NO synthesis is controlled in endothelial cells as well as on the induction by cytokines of an NO synthase in vascular smooth muscle cells. The relative importance of the different sources of NO from the vascular wall under physiological and pathophysiological conditions will be discussed.

2 Results and Discussion

2.1 Activation of NO Synthesis in Endothelial Cells

The decisive role of the endothelium in the control of vascular tone first became apparent in 1980 when Furchgott and Zawadzki (1980) reported that the endothelium is obligatory for the acetylcholine-induced relaxation in isolated precontracted rabbit aortic preparations. Removal of the endothelium by rubbing the intimal surface completely abolished this endothelium-dependent relaxation. Subsequent studies revealed that a variety of other endogenous vasoactive substances exert their dilator effect by the release of EDRF, among them the platelet-derived compounds ADP, ATP, and serotonin, the neurotransmitter substance P as well as bradykinin and histamine (Busse et al. 1985). Besides these compounds which induce endothelium-dependent dilation by activating receptor-dependent pathways in the endothelium, physical stimuli such as the fluid shear stress generated by the streaming blood at the luminal surface of the endothelium and the pulsatile stretching of the vascular wall trigger the release of EDRF (Pohl et al. 1986a, b). As will be discussed later, these physical stimuli may have more physiological significance in the control of organ blood flow than the receptor-dependent ones.

Despite the endeavors of many groups to identify the chemical structure of EDRF, its identity remained elusive for more than 6 years following Furchgott's key discovery. The hypothesis that EDRF is nitric oxide was first made by Furchgott at an international symposium in 1986 (Furchgott 1988) and independently forwarded by Ignarro at the same conference (Ignarro et al. 1988). One year later, Palmer and coworkers showed that cultured endothelial cells release NO in response to bradykinin in amounts sufficient to account for vascular smooth muscle relaxation induced by EDRF (Palmer et al. 1987). Although abundant pharmacological and biochemical evidence supports the concept that the oxidized nitrogen species formed by L-arginine oxygenation in endothelial cells is identical with free NO, some results do not appear consistent with this conclusion. Thus, it has been proposed that EDRF is identical with hydroxylamine (Thomas and Ramwell 1986; Schmidt et al. 1990), with a nitrosothiol (Myers et al. 1990), or with a mixture of these, called NO_x (Schmidt et al. 1990). Recent electron paramagnetic resonance (EPR) spectroscopic studies, performed in our laboratory in collaboration with A. Vanin and P. Mordvintcev, revealed that NO is the immediate product of endothelial L-arginine oxygenation (Mülsch et al. 1991). The method is based on the high affinity of NO for the ferrous di-

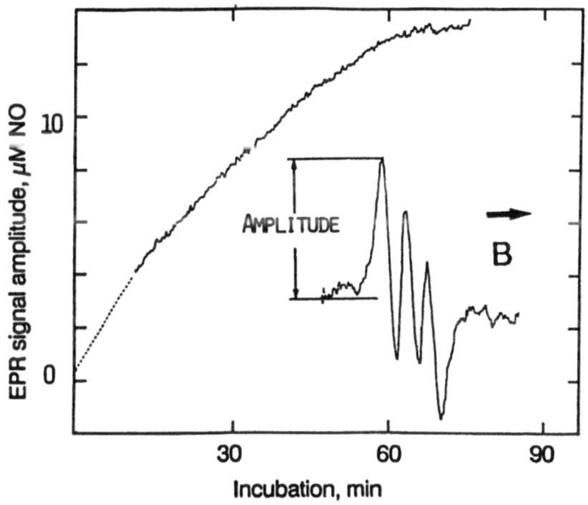

Fig. 3. Time course of NO formation by endothelial cytosol as detected by EPR spectroscopy. Diethyldithiocarbamate (DETC)-loaded heat-killed yeast cells were incubated at 37 °C with cytosol (0.26 mg/ml) from freshly isolated porcine aortic endothelial cells, L-arginine, and NO synthase cofactors (see text) in a flat quartz cuvette placed inside the cavity of the EPR spectrometer. After adjustment of instrument settings (indicated by *broken lines*) the magnetic field (*B*) was fixed to the first hyperfine line of the isotropic triplet EPR signal of the paramagnetic NO-Fe^{2+}-DETC complex (*inset*) and the increase in the signal amplitude was recorded with time. After 90 min, the complete EPR spectrum was recorded (*inset*) and the total amplitude was measured as indicated. The amplitude was calibrated in terms of NO concentration (μM) by recording the EPR signal from paramagnetic dinitrosyl ferrous thiosulfate standards

ethyldithiocarbamate complex, which therefore is an efficient and selective trap for NO. The paramagnetic mononitrosyl complex formed by incubation of endothelial cytosol with these traps in the presence of L-arginine and all cofactors required for NO synthase exhibits the characteristic EPR spectrum shown in Fig. 3, at $g_{\perp} = 2.04$ and $g_{\parallel} = 2.02$. The time course of NO accumulation could be followed at 37 °C, if the incubations were performed in a flat quartz cuvette inside the cavity of the EPR spectrometer, with the magnetic field fixed to one hyperfine line (insert, Fig. 3). Under these conditions, endothelial cytosol (0.25 mg protein/ml) generated about 10 μM paramagnetic nitrosyl complex within 1 h, after which time NO formation ceased.

As already mentioned, NO formation in endothelial cells is constitutive and is transiently stimulated by receptor-dependent agonists which increase intracellular Ca^{2+} concentration by an inositol-1, 4, 5-trisphosphate (IP_3)-mediated Ca^{2+} release from intracellular stores and an Ca^{2+}-influx through voltage-independent Ca^{2+} channels (Lückhoff and Busse 1986; Busse et al. 1988; Lückhoff et al. 1988). Figure 4 shows representative recordings of the time course of the intracellular free Ca^{2+} concentration as measured by the fluorescent Ca^{2+} indicator Indo-1 as well as NO release as measured by activation of a purified soluble guanylyl cyclase in cultured porcine aortic endothelial cells. As can be seen from these recordings, both responses exhibit a very similar time course, suggesting a tight coupling between the Ca^{2+} increase and NO formation. Removal of extracellular Ca^{2+} abolishes both the sustained phase of

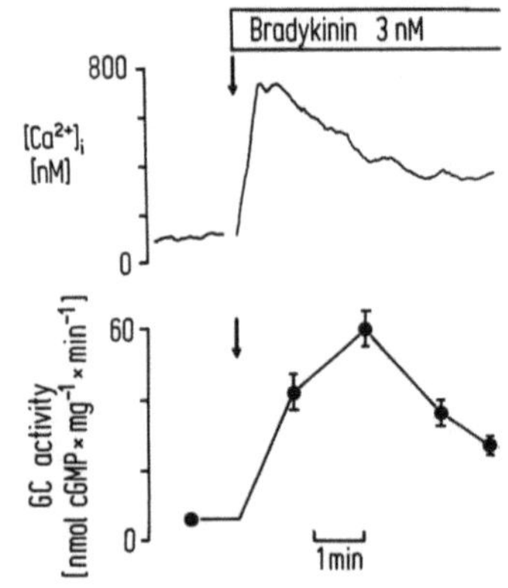

Fig. 4. Effects of bradykinin (Bk) on the intracellular free calcium concentration ($[Ca^{2+}]_i$) in, and NO release from cultured porcine endothelial cells grown on dishes. NO release was measured in the supernatant of the cells by the use of purified soluble guanylyl cyclase as an NO detector (GC activity)

the receptor-dependent Ca^{2+} increase and the autacoid release. This can be demonstrated even better with receptor-independent stimuli of NO formation, such as thimerosal, which induces a slowly developing, but sustained increase in intracellular free Ca^{2+}, associated with a long-lasting release of NO (Busse et al. 1989). As shown in Fig. 5, the thimerosal-induced NO formation instantly ceases, if Ca^{2+} is removed from the perfusion medium, but recovers completely after restoration of the extracellular Ca^{2+} concentration.

To investigate the molecular basis of agonist-induced NO synthesis, we measured NO synthase activity (by activation of purified soluble guanylyl cyclase) in the cytosol obtained from isolated porcine aortic endothelial cells. The NO synthase used L-arginine as a substrate (EC_{50} 10 μM) and required NADPH as a cofactor (EC_{50} 3 μM). In EGTA-buffered cytosol, the activity was increased about 2–3-fold by Ca^{2+} (EC_{50} 0.3 μM free Ca^{2+}). The calmodulin-binding peptides melittin and mastoparan inhibited the Ca^{2+}-dependent activation of NO synthase, indicating the participation

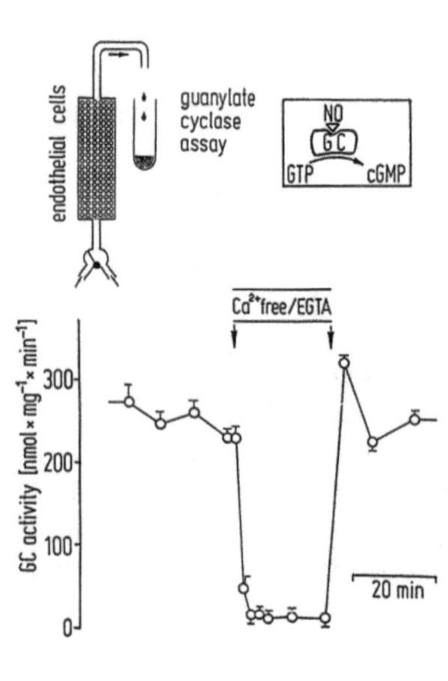

Fig. 5. Effect of removal of extracellular calcium on the thimerosal-induced release of NO from cultured endothelial cells. Above: Experimental setup used for the measurement of NO release from cultured endothelial cells. Endothelial cells were grown on microcarrier beads and continuously superfused with a saline solution. NO content was determined in the effluent by a guanylyl cyclase assay

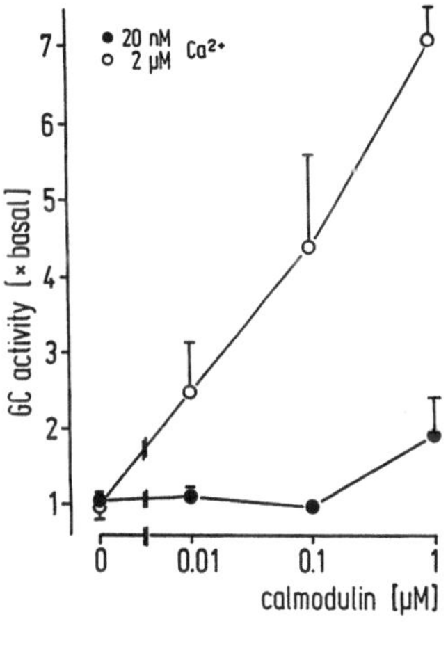

Fig. 6. Dependence on free Ca^{2+} and calmodulin of the activity of partially purified endothelial NO synthase. Endogenous calmodulin was removed from endothelial cytosol by anion exchange chromatography (0 to 0.5 M NaCl gradient, Mono Q sepharose, Pharmacia, Freiburg). NO synthase activity (measured by activation of a purified soluble guanylyl cyclase) in the eluate (0.1 mg/ml protein) was only detectable, if L-arginine (0.3 mM), NADPH (0.1 mM), Ca^{2+} (2 μM free concentration), and calmodulin (from porcine brain, EC_{50} 100 nM) were added during activity determinations. With low free Ca^{2+} (20 nM; 0.1 mM EGTA present) calmodulin induced only a weak activation at 1 μM). (Busse and Mülsch 1990a)

of an endogenous calmodulin-like activity in this process. In fact, if the endogenous calmodulin was separated from the NO synthase by anion exchange chromatography on Mono Q sepharose, Ca^{2+}-dependent activation of NO synthase was lost, but could be reconstituted by addition of calmodulin (Fig. 6). According to these findings, we conclude that the intracellular free Ca^{2+} acts as a second messenger to link agonist-induced stimulation of endothelial cells with increased NO formation. The Ca^{2+}-calmodulin complex functions as a trigger to enhance NO synthase activity upon agonist-induced increases in the intracellular free Ca^{2+} concentration. Other constitutive NO synthases, as in neuronal cells, seem to be activated by the same mechanism.

We (Mülsch and Busse 1991) and others (Mayer et al. 1990) recently found that tetrahydrobiopterin, a cofactor of inducible NO synthases, is also required by the constitutive enzyme in endothelial cells and cerebellum. As can be seen in Fig. 7, 6R-tetrahydrobiopterin potently enhanced the Ca^{2+}-dependent and the Ca^{2+}-independent NO synthase activity in the cytosol from porcine aortic endothelial cells (EC_{50} = 10 nM). The stereoisomeric 6S-tetrahydrobiopterin and the 6R, S-dihydrobiopterin were 100-fold less potent. The hypothesis has been advanced that tetrahydrobiopterin participates in the hydroxylation of the guanidino nitrogen of L-arginine in a manner

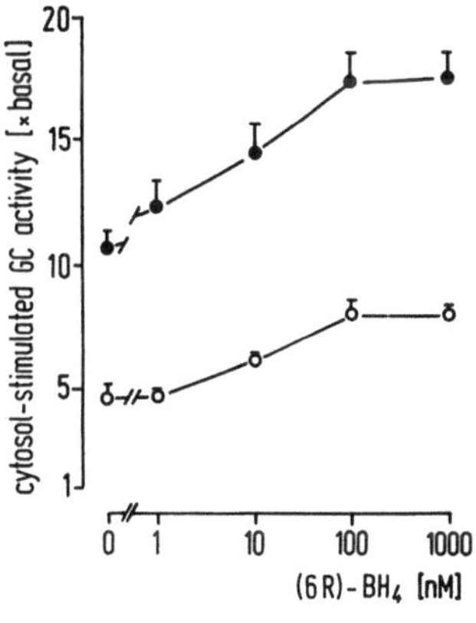

Fig. 7. NO synthase in endothelial cytosol requires 6R-tetrahydrobiopterin as a cofactor. The activity of NO synthase (as measured by activation of a purified soluble guanylyl cyclase) in cytosol (0.1 mg/ml protein) from freshly isolated porcine aortic endothelial cells in the presence of L-arginine (0.3 mM), NADPH (0.1 mM), and endogenous calmodulin was concentration-dependently enhanced by 6R-tetrahydrobiopterin (6R-BH_4) in the absence (○) and presence (●) of Ca^{2+} (2 μM)

similar to the hydroxylation of aromatic amino acids (Kwon et al. 1989). This mechanism involves an enzymatic or nonenzymatic, thiol-dependent recycling of tetrahydrobiopterin, since it is oxidized during catalysis. Unexpectedly, the purified NO synthase from cerebellum and endothelial cells was activated by tetrahydrobiopterin in the absence of thiols (A. Mülsch and R. Busse, unpubl. results), leaving the possibility that tetrahydrobiopterin is recycled by NO synthase itself.

2.2 Agonist-Independent NO Synthesis in Endothelial Cells

As mentioned above, the main determinants for a continuous NO release in vivo are hemodynamic stimuli, particularly the fluid shear stress at the luminal surface of the endothelium generated by the streaming blood ("viscous drag"). It can be derived from the Navier-Stokes equation that in a vessel segment, under the conditions of a steady laminar flow, shear stress is a linear function of the viscosity of the fluid and of the flow rate and is inversely related to the third power of the vessel radius. If shear stress is indeed the adequate stimulus responsible for this type of endothelium-dependent dilation, then not only changes of flow but also changes of fluid viscosity or of vascular radius should result in vasodilation and/or release of NO. A clear correlation of dilator responses and the magnitude of flow or viscosity changes has been demonstrated in various types of vessel (Smiesko et al. 1985; Pohl et al. 1986b;

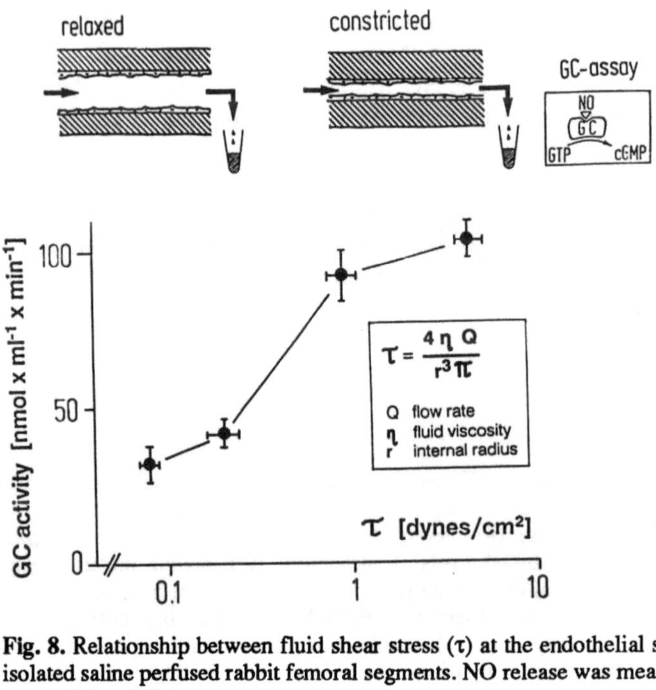

Fig. 8. Relationship between fluid shear stress (τ) at the endothelial surface and NO release in isolated saline perfused rabbit femoral segments. NO release was measured by guanylyl cyclase assay (GC activity)

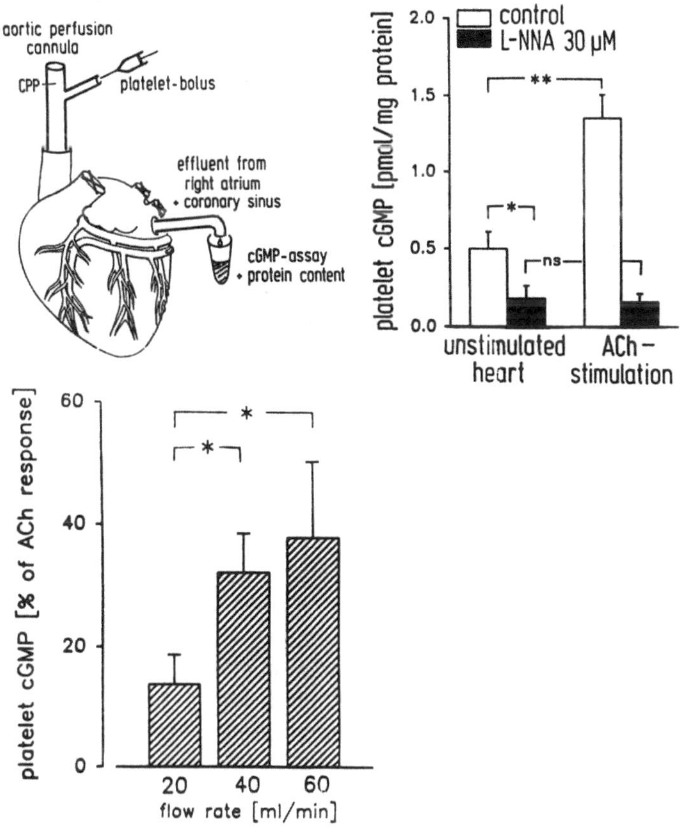

Fig. 9. *Upper panel* Basal release of NO from the intact rabbit coronary vascular bed. *Left* Experimental setup. The increase in cyclic GMP content of washed platelets passing through the coronary bed of saline perfused rabbit heart was used as index of NO release. *Right* Platelet cyclic GMP content after passage through the coronary bed under control conditions and during acetylcholine infusion, in the absence and in the presence of the NO synthase inhibitor N^G-nitro-L-arginine (L-NNA). *Lower panel* Shear stress-induced release of NO in the coronary vascular bed. Platelet cyclic GMP content, expressed as percent of the platelet cyclic GMP content during acetylcholine infusion (1 µM), was measured at different flow rates, and therefore at different levels of shear stress

Melkumyants et al. 1987; Tesfamarian and Cohen 1988). Likewise, reduction of vascular diameter at constant flow should be associated with an enhanced release of EDRF. This is demonstrated in Fig. 8. The shear stress acting on the endothelial surface of rabbit femoral artery segments was gradually increased by constricting the segments with either endothelin-1 or norepinephrine under constant flow. NO release into the perfusate as assessed by a guanylyl cyclase assay correlated positively with the increase in shear stress.

Mechanoreception is a widespread sensory modality of living organisms. In general, local changes in the conductance of the cell membrane resulting in changes of the membrane potential have been described in several types of mechanoreceptor cells as primary responses to the deforming forces and displacements in the process of sensory transduction (Thurm 1983). This process of sensory transduction has to be understood in endothelial cells, which as nonexcitable cells are not primarily designed as mechanoreceptors. One might speculate that the viscous drag is sensed by structures located at the outer leaflet of the cell glycocalix that are coupled to the cytoskeleton or to regulating proteins at the level of membrane channels. This view is supported by the finding that treatment of endothelium-intact segments with neuraminidase significantly decreased the shear stress-dependent NO release. This enzyme cleaves the glycosidically bound sialic acid, which represents the terminal residues of the cell-coating glycoproteins and glycolipids. Consistent with the concept of an involvement of ion channels in the endothelial response to shear, shear stress-activated K^+ currents have been observed in cultured endothelial cells grown in capillary flow tubes. These currents, which lead to a hyperpolarization of the cell membrane, varied in magnitude and duration as a function of shear stress and represented a fast and early event after initiation of shear stress increase (Olesen et al. 1988). Since hyperpolarization facilitates the transmembrane influx of Ca^{2+} into endothelial cells (Lückhoff and Busse 1990), this mechanism may contribute to the maintenance of intracellular Ca^{2+} elevation and consequently NO formation in response to shear stress. Shear stress-dependent NO release is not confined to large conduit arteries but can also be observed in an intact vascular bed (Fig. 9). The functional significance of this continuous NO formation in the control of nutritive blood flow at increased tissue oxygen demands as a counteracting principle to neuro- and myogenic contraction mechanisms has been documented in recent studies (Pohl and Busse 1988; Pohl et al. 1990).

2.3 Induction of NO Synthesis in Vascular Smooth Muscle Cells

In endotoxemia part of the lethal complications arise from the severe hypotension generally associated with endotoxic shock. Also, the doses of cytokines applicable to patients during tumor therapy is limited by their hypotensive action. Since bacterial toxins and cytokines induce a Ca^{2+}-calmodulin-independent NO synthase in macrophages and other cells, it has been suggested that under these conditions an NO synthase induced in the vascular wall could be responsible for the systemic hypotensive effects. Indeed, we and others demonstrated a significant decrease in the response to contractile agonists of endotoxin- and cytokine-pretreated vascular segments (Fig. 10; Beasley 1990; Beasley et al. 1990; Busse and Mülsch 1990b; Fleming et al. 1990). This effect was observed after at least 4 h of preincubation with cytokines and was completely inhibited by the presence of cycloheximide during that time, indicating the requirement for protein synthesis. The cytokine-induced attenuation of contractile responses was acutely inhibited by L-NNA in the perfusion medium, suggesting the participation of the L-arginine-NO pathway. However, removal of the endothelium

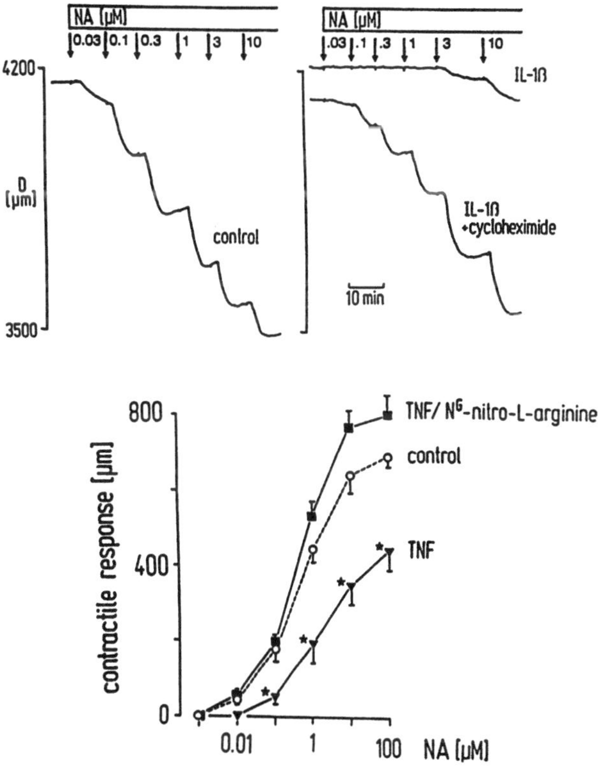

Fig. 10. Cytokines attenuate the contraction to noradrenaline (*NA*) of endothelium-denuded rabbit aortic segments. Endothelium-denuded segments from rabbit thoracic aorta were incubated for 20 h (37 °C) in M 199 medium (10% calf serum); *left* in the absence (control) and *right* presence of interleukin-1β (IL-1β; 200 U/ml) or IL-1β and cycloheximide (20 μg/ml); and *below* in the absence (○) and presence (▼) of tumor necrosis factor α/interferon γ (TNF/IF; 3000/500 units/ml). Cumulative dose-response-curves to noradrenaline (*NA*) were obtained by perfusion (40 ml/h) of the segments with oxygenated Tyrode's solution in an organ bath (37 °C). The outer diameters of the segments were recorded photoelectrically. *Upper right* The attenuation by IL-1β of the contractile response to NA is restored by coincubation with cycloheximide. Original tracings. *Below* The attenuation by TNF/IF of the contractile response to NA is abolished by N^G-nitro-L-arginine (■; 30 μM). *Significantly different from control ($p < 0.05$). (Busse and Mülsch 1990b)

had no effect. Though NO synthase can be induced in endothelial cells by cytokines and endotoxins (Kilbourn and Belloni 1990; Radomski et al. 1990b), the extent of induction is obviously insufficient to account for the cytokine-induced inhibition of contraction. In contrast, we detected NO synthase activity in the cytosol from native and cultured smooth muscle cells from rabbit aorta after prolonged incubation with cytokines (Fig. 11). This NO synthase required tetrahydrobiopterin and NADPH as cofactors, but not Ca^{2+} and calmodulin, which indicates its relationship with the macrophage-type enzyme. The induction of this NO synthase was completely pre-

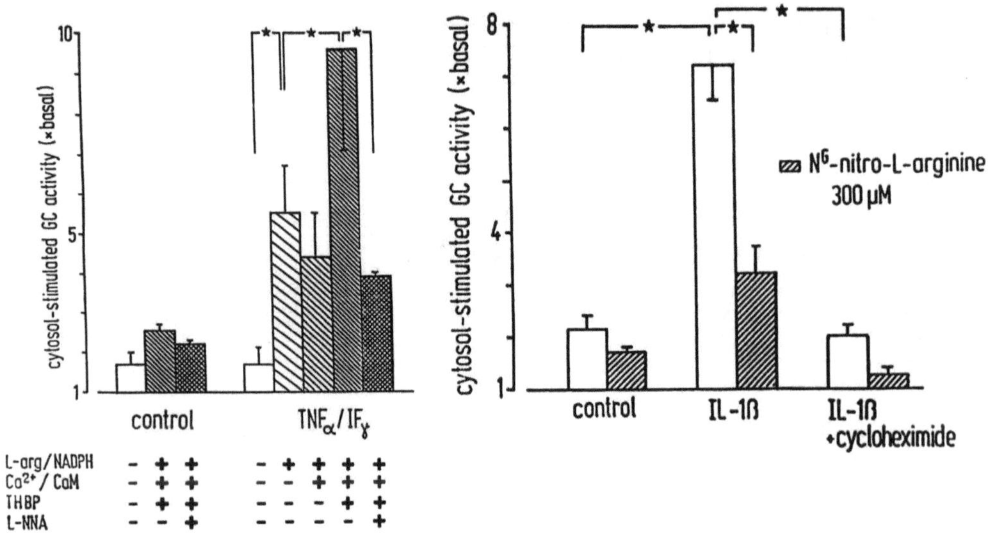

Fig. 11. Cytokines induce a macrophage-type NO synthase in cultured vascular smooth muscle cells from rabbit aorta. Cultured smooth muscle cells were incubated with cytokines as described in Fig. 10. NO synthase activity in the cytosolic fraction (100 000 x g supernatant; 0.13 mg/ml protein) from homogenized cells was detected by activation of purified guanylyl cyclase. *Left* L-arginine (0.3 mM) and NADPH (0.1 mM)-dependent NO formation was detected in TNF/IF-treated, but not in untreated cells. NO formation was enhanced by 6R-tetrahydrobiopterin (THBP; 1 µM), but not by Ca^{2+}/calmodulin(2 µM/1 µM), and was inhibited by N^G-nitro-L-arginine (L-NNA; 0.3 mM). * $p < 0.05$. *Right* Induction by IL-1β of cytosolic NO synthase is prevented by cotreatment of smooth muscle cells with cycloheximide. *$p < 0.05$. (After Busse and Mülsch 1990b)

vented by cycloheximide, and its activity was inhibited by L-NNA. Furthermore, dexamethasone and TGF_β interfered with the induction of NO synthase by cytokines in vascular smooth muscle cells, probably at the level of gene transcription. Similar data on the inhibition of the induction of an NO synthase by these two compounds have been reported in mesangial cells (Pfeilschifter and Vosbeck 1991) and in endothelial cells (Radomski et al. 1990b). We conclude that cytokine-induced NO formation by increasing the intracellular cyclic GMP content leads to a pronounced attenuation in vascular tone and, therefore, may be causally related to the severe hypotension observed in endotoxic shock and in tumor therapy with cytokines.

Figure 12 summarizes in a simple scheme the mechanisms leading to NO synthesis in the vascular wall. Under physiological conditions the vascular tone in vivo depends decisively on the functional integrity of the endothelium associated with an appropriate production of basal and receptor-stimulated NO release. The basal release of NO is caused by physical stimuli such as fluid shear stress and is a major factor in maintaining adequate vascular tone and thereby, of nutritive blood flow to active tissues. Agonist-induced NO release can be regarded as an NO synthesis reserve to manage a transient and local requirement for higher NO concentrations, for instance, in counteracting formation of microthrombi. Under certain pathophysiological condi-

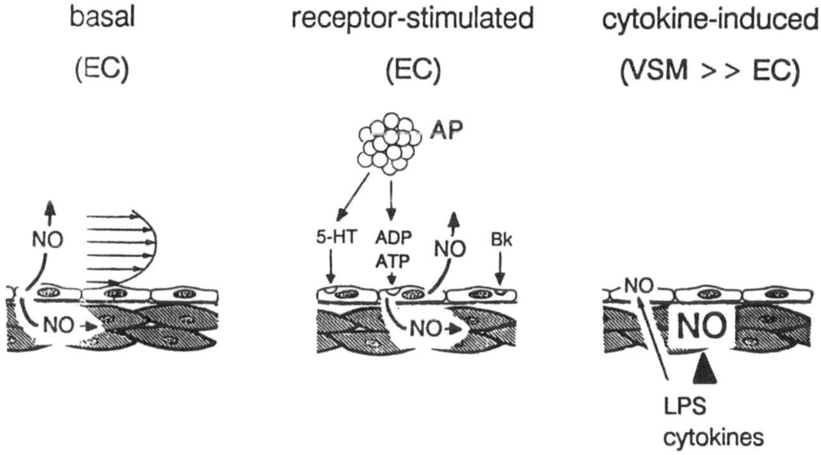

Fig. 12. Schematic diagram of the principal NO-stimulating mechanisms in the vascular wall. For further explanation, see text. *EC* Endothelial cells; *VSM* vascular smooth muscle cells; *AP* aggregating platelets; *5-HT* serotonin; *Bk* bradykinin; *LPS* lipopolysaccharide (endotoxin)

tions associated with increased levels of cytokines, an NO synthase of the macrophage type can be induced in vascular smooth muscle cells. Once fully activated, NO is synthesized by this Ca^{2+}-calmodulin-independent enzyme at a very high rate. This high NO output pathway in the vascular smooth muscle overruns the influence of endothelial NO formation on vascular tone. Though in the first instance this mechanism may be desirable for defense against an invading pathogen, it becomes fatal in the end by excessive vasodilatation and severe systemic hypotension.

References

Amber IJ, Hibbs JB, Taintor RR & Vavrin Z (1988) Cytokines induce an L-arginine-dependent effector system in nonmacrophage cells. J Leukocyte Biol 44:58–65

Beasley D (1990) Interleukin-1 and endotoxin activate soluble guanylate cyclase in vascular smooth muscle. Am J Physiol 259:R38–R44

Beasley D, Cohen RA & Levinsky NG (1990) Endotoxin inhibits contraction of vascular smooth muscle in vitro. Am J Physiol 258:H1187–H1192

Billiar TR, Curran RD, Stuehr DJ, West MA, Bentz BG & Simmons RL (1989) An L-arginine-dependent mechanism mediates Kupffer cell inhibition of protein synthesis in vitro. J Exp Med 169:1467–1472

Billiar TR, Curran D, Stuehr DJ, Stadler J, Simmons RL & Murray SA (1990) Inducible cytosolic enzyme activity for the production of nitrogen oxides from L-arginine in hepatocytes. Biochem Biophys Res Commun 168:1034–1040

Boeckxstaens GE, Pelckmans PA, Bogers JJ, Bult H, Deman JG, Oosterbosch L, Herman AG & Vanmaercke YM (1991) Release of nitric oxide upon stimulation of nonadrenergic noncholinergic nerves in the rat gastric fundus. J Pharmacol Exp Ther 256:441–447

Boje KM & Fung HL (1990) Endothelial nitric oxide generating enzyme (S) in the bovine aorta-subcellular location and metabolic characterization. J Pharmacol Exp Ther 253:20–26

Bredt DS & Snyder SH (1989) Nitric oxide mediates glutamate-linked enhancement of cGMP levels in the cerebellum. Proc Natl Acad Sci USA 86:9030–9033

Bredt DS & Snyder SH (1990) Isolation of nitric oxide synthetase, a calmodulin-requiring enzyme. Proc Natl Acad Sci USA 87:682–685

Busse R & Mülsch A (1990a) Calcium-dependent nitric oxide synthesis in endothelial cytosol is mediated by calmodulin. FEBS Lett 265:133–136

Busse R & Mülsch A (1990b) Induction of nitric oxide synthase by cytokines in vascular smooth muscle cells. FEBS Lett 275:87–90

Busse R, Trogisch G & Bassenge E (1985) The role of endothelium in the control of vascular tone. Basic Res Cardiol 80:475–490

Busse R, Lückhoff A & Bassenge E (1987) Endothelium-derived relaxant factor inhibits platelet activation. Naunyn-Schmiedebergs Arch Pharmacol 336:566–571

Busse R, Fichtner H, Lückhoff A & Kohlhardt M (1988) Hyperpolarization and increased free calcium in acetylcholine-stimulated endothelial cells. Am J Physiol 255:H965–H969

Busse R, Lückhoff A & Pohl U (1989) Generation and transmission of endothelium-dependent vasodilator signals. In: Catravas JD, Gillis CN & Ryan US (eds) Vascular endothelium: receptors and transduction mechanisms. Plenum Press, New York, pp 225–236

Curran RD, Billiar TR, Stuehr DJ, Hofmann K & Simmons RL (1989) Hepatocytes produce nitrogen oxides from L-arginine in response to inflammatory products from Kupffer cells. J Exp Med 336:1769–1774

Drapier JC & Hibbs JB (1986) Murine cytotoxic activated macrophages inhibit aconitase in tumor cells. Inhibition involves the iron-sulfur prosthetic group and is reversible. J Clin Invest 78:790–797

Drapier JC & Hibbs JB (1988) Differentiation of murine macrophages to express nonspecific cytotoxicity for tumor cells results in L-arginine-dependent inhibition of mitochondrial iron-sulfur enzymes in the macrophage effector cells. J Immunol 140:2829–2838

Fleming I, Gray GA, Julou-Schaeffer G, Parratt JR & Stoclet JC (1990) Incubation with endotoxin activates the L-arginine pathway in vascular tissue. Biochem Biophys Res Commun 171:562–568

Förstermann U, Ishii K, Gorsky LD & Murad F (1989) The cytosol of N1E-115 neuroblastoma cells synthesizes an EDRF-like substance that relaxes rabbit aorta. Naunyn-Schmiedebergs Arch Pharmacol 340:771–774

Förstermann U, Pollock JS, Schmidt HHHW, Heller M & Murad F (1991) Calmodulin-dependent endothelium-derived relaxing factor nitric oxide synthase activity is present in the particulate and cytosolic fractions of bovine aortic endothelial cells. Proc Natl Acad Sci USA 88:1788–1792

Förstermann U, Pollock JS, Schmidt HHHW, Mitchell JA, Kohlhaas KL, Heller M & Murad F (1991) Characterization and purification of particulate EDRF-synthase from bovine aortic endothelial cells. FASEB J 5(Part III), A1728 (Abstr)

Furchgott RF (1988) Studies on relaxation of rabbit aorta by sodium nitrite: the basis for the proposal that the acid-activatable inhibitory factor from bovine retractor penis is inorganic nitrite and the endothelium-derived relaxing factor is nitric oxide. In: Vanhoutte PM (ed) Mechanisms of vasodilation, vol 4. Raven Press, New York, pp 401–414

Furchgott RF & Zawadzki JV (1980) The obligatory role of endothelial cells in the relaxation of arterial smooth muscle by acetylcholine. Nature (London) 288:373–376

Gally JA, Montague PR, Reeke Jr GN & Edelman GM (1990) The NO hypothesis: possible effects of a short-lived, rapidly diffusible signal in the development and function of the nervous system. Proc Natl Acad Sci USA 87: 3547–3551

Garthwaite J, Charles SL & Chess-Williams R (1988) Endothelium-derived relaxing factor release on activation of NMDA receptors suggests role as intercellular messenger in the brain. Nature (London) 336:385–388

Garthwaite J, Southam E & Anderton M (1989) A kainate receptor linked to nitric oxide synthesis from arginine. J Neurochem 53:1952–1954

Hauschildt S, Lückhoff A, Mülsch A, Kohler J, Bessler W & Busse R (1990) Induction and activity of NO synthase in bone marrow-derived macrophages are independent of calcium. Biochem J 270:351–356

Hibbs JB, Taitor RE & Vavrin Z (1987) Macrophage cytotoxycity: role for I-arginine deiminase and imino nitrogen oxidation to nitrite. Science 235:473–476

Ignarro LJ & Kadowitz PJ (1985) The pharmacological and the physiological role of cyclic GMP in vascular smooth muscle relaxation. Annu Rev Pharmacol Toxicol 25:171–191

Ignarro LJ, Byrns RE & Wood KS (1988) Biochemical and pharmacological properties of endothelium-derived relaxing factor and its similarity to nitric oxide radical. In: Vanhoutte PM (ed) Mechanisms of vasodilatation, vol 4. Raven Press, New York, pp 427–35

Iyengar R, Stuehr DJ & Marlette MA (1987) Macrophage synthesis of nitrite, nitrate, and N-nitrosamines: precursors and role of the respiratory burst. Proc Natl Acad Sci USA 84:6369–6373

Kilbourn RG & Belloni P (1990) Endothelial cell production of nitrogen oxides in response to interferon-gamma in combination with tumor necrosis factor, interleukin-1, or endotoxin. J Natl Cancer Inst 82:772–776

Knowles RG, Palacios M, Palmer RMJ & Moncada S (1989) Formation of nitric oxide from L-arginine in the central nervous system – a transduction mechanism for stimulation of the soluble guanylate cyclase. Proc Natl Acad Sci USA 86:5159–5162

Kwon NS, Nathan CF & Stuehr DJ (1989) Reduced biopterin as a cofactor in the generation of nitrogen oxides by murine macrophages. J Biol Chem 264:20496–20501

Lückhoff A & Busse R (1986) Increased free calcium in endothelial cells under stimulation with adenine nucleotides. J Cell Physiol 126:414–420

Lückhoff A & Busse R (1990) Calcium influx into endothelial cells and formation of EDRF is controlled by the membrane potential. Pflügers Arch 416:305–311

Lückhoff A, Pohl U, Mülsch A & Busse R (1988) Differential role of extra- and intracellular calcium in the release of EDRF and prostacyclin from cultured endothelial cells. Br J Pharmacol 95:189–196

Marletta MA, Yoon PS, Iyengar R, Leaf CD & Wishnok JS (1988) Macrophage oxidation of L-arginine to nitrite and nitrate: nitric oxide is an intermediate. Biochemistry 27:8706–8711

Martin W, White DG & Henderson AH (1988) Endothelium-derived relaxing factor and atriopeptin II elevate cyclic GMP levels in pig aortic endothelial cells. Br J Pharmacol 93:229–239

Mayer B, John M & Bohme E (1990) Purification of a Ca^{-2+}/calmodulin-dependent nitric oxide synthase from porcine cerebellum-cofactor-role of tetrahydrobiopterin. FEBS Lett 277:215–219

Melkumyants AM, Balashov SA, Veselova ES & Khayutin VM (1987) Continuous control of the lumen of feline conduit arteries by blood flow rate. Cardiovasc Res 21:863–870

Moncada S & Higgs EA (1990) Nitric oxide from L-arginine: a bioregulatory system. Elsevier, Amsterdam

Mülsch A & Busse R (1990) NG-nitro-L-arginine (N5-[imino(nitroamino)methyl]-L-ornithine) impairs endothelium-dependent dilations by inhibiting cytosolic nitric oxide synthesis from L-arginine. Naunyn-Schmiedebergs Arch Pharmacol 341:143–147

Mülsch A & Busse R (1991) Nitric oxide synthase in native and cultured endothelial cells: calcium/calmodulin and tetrahydrobiopterin are cofactors. J Cardiovasc Pharmacol 17 (Suppl 3):S52–S56

Mülsch A, Mordvintcev P, Vanin A & Busse R (1991) Detection of NO from L-arginine in endothelial cells by electron spin resonance. FASEB J 5:A 1016

Myers PR, Minor RL, Guerra R, Bates JN & Harrison DG (1990) Vasorelaxant properties of the endothelium-derived relaxing factor more closely resemble S-nitrosocysteine than nitric oxide. Nature (London) 345:161–163

Olesen SP, Clapham DE & Davies PF (1988) Haemodynamic shear stress activates a K^+ current in vascular endothelial cells. Nature (London) 331:168–170

Palacios M, Knowles RG, Palmer RMJ & Moncada S (1989) Nitric oxide from L-arginine stimulates the soluble guanylate cyclase in adrenal glands. Biochem Biophys Res Commun 165:802–809

Palmer RMJ, Ferrige AG & Moncada S (1987) Nitric oxide release accounts for the biological activity of endothelium-derived relaxing factor. Nature (London) 327:524–526

Palmer RMJ, Rees DD, Ashton DS & Moncada S (1988) L-Arginine is the physiological precursor for the formation of nitric oxide in the endothelium-dependent relaxation. Biochem Biophys Res Commun 153:1251–1256

Pellat C, Henry Y & Drapier JC (1990) IFN-gamma-activated macrophages–detection by electron paramagnetic resonance of complexes between L-arginine-derived nitric oxide and non-heme iron proteins. Biochem Biophys Res Commun 166:119–125

Pfeilschifter J & Vosbeck K (1991) Transforming growth factor β_2 inhibits interleukin 1beta-induction and tumour necrosis factor alpha-induction of nitric oxide synthase in rat renal mesangial cells. Biochem Biophys Res Commun 175:372–379

Pohl U & Busse R (1988) Reduced nutritional blood flow in autoperfused rabbit hindlimbs following inhibition of endothelial vasomotor function. In: Halpern W, Pegram B, Brayden J, Mackey K, McLaughlin M, Osol G (eds) Resistance arteries. Perinatology Press, Ithaca, New York, pp 10–16

Pohl U, Busse R, Kuon E & Bassenge E (1986a) Pulsatile perfusion stimulates the release of endothelial autacoids. J Appl Cardiol 1:215–235

Pohl U, Holtz J, Busse R & Bassenge E (1986b) Crucial role of endothelium in the vasodilator response to increased flow in vivo. Hypertension 8:37–44

Pohl U, Lamontagne D, Busse R & Bassenge E (1990) EDRF reduces coronary autoregulatory responses to increases in perfusion pressure. Arch Int Pharmacodyn Ther 305:273

Radomski MW, Palmer RMJ & Moncada S (1990a) An L-arginine nitric oxide pathway present in human platelets regulates aggregation. Proc Natl Acad Sci USA 87:5193–5197

Radomski MW, Palmer RMJ & Moncada S (1990b) Glucocorticoids inhibit the expression of an inducible, but not the constitutive, nitric oxide synthase in vascular endothelial cells. Proc Natl Acad Sci USA 87:10043–10047

Rapoport RM & Murad F (1983) Agonist-induced endothelium-dependent relaxation in rat thoracic aorta may be mediated through cGMP. Circ Res 52:352–357

Rees DD, Palmer RMJ, Hodson HF & Moncada S (1989) A specific inhibitor from nitric oxide formation from L-arginine attenuates endothelium-dependent relaxation. Br J Pharmacol 96:418–424

Schmidt HHHW, Nau H, Wittfoht W, Gerlach J, Prescher KE, Klein MM, Niroomand F & Böhme E (1988) Arginine is a physiological precursor of endothelium-derived nitric oxide. Eur J Pharmacol 154:213–216

Schmidt HHHW, Zernikow B, Baeblich S & Bohme E (1990) Basal and stimulated formation and release of L-arginine-derived nitrogen oxides from cultured endothelial cells. J Pharmacol Exp Ther 254:591–597

Schmidt HHHW, Pollock JS, Nakane M, Gorsky LD, Förstermann U & Murad F (1991) Purification of a soluble isoform of guanylyl cyclase-activating-factor synthase. Proc Natl Acad Sci USA 88:365–369

Schröder H & Schrör K (1989) Cyclic GMP stimulation by vasopressin in LLC-PK1 kidney epithelial cells is L-arginine-dependent. Naunyn-Schmiedebergs Arch Pharmacol 340:475–477

Shibuki K & Okada D (1991) Endogenous nitric oxide release required for long-term synaptic depression in the cerebellum. Nature (London) 349:326–328

Smiesko V, Kozik J & Dolezel S (1985) Role of endothelium in the control of arterial diameter by blood flow. Blood Vessels 22:247–251

Tayeh MA & Marletta MA (1989) Macrophage oxidation of L-arginine to nitric oxide, nitrite, and nitrate-tetrahydrobiopterin is required as a cofactor. J Biol Chem 264:19654–19658

Tesfamarian B & Cohen RA (1988) Inhibition of adrenergic vasoconstriction by endothelial shear stress. Circ Res 63:720–725

Thomas G & Ramwell P (1986) Induction of vascular relaxation by hydroperoxides. Biochem Biophys Res Commun 139:102–108

Thurm U (1983) Biophysics of mechanoreception. Mechano-electric transduction. In: Hoppe W, Lohmann W, Markl H, Ziegler H (eds) Biophysics, Springer, Berlin Heidelberg New York, pp 666–671

Wright CD, Mülsch A, Busse & Osswald H (1989) Generation of nitric oxide by human neutrophils. Biochem Biophys Res Commun 160:813–819

Molecular Aspects of the Chemistry and Biology of Endotoxin

E. Th. RIETSCHEL[1], T. KIRIKAE[1], W. FEIST[1], H. LOPPNOW[1], P. ZABEL[1], L. BRADE[1], A. J. ULMER[1], H. BRADE[1], U. SEYDEL[1], U. ZÄHRINGER[1], M. SCHLAAK[1], H. D. FLAD[1], and U. SCHADE[1]

1 Introduction

Endotoxins are integral constituents of the outer membrane of gram-negative bacteria such as the Enterobacteriaceae, Neisseriaceae, and Chlamydiaceae (Lüderitz et al. 1982). They participate in various physiological membrane functions essential for bacterial growth and survival. Endotoxins also represent the main heat-stable O-antigens of the bacteria and, thus, identify the multiplicity of serotypes. Finally, endotoxins are endowed with an overwhelming spectrum of biological activities, expressed either after injection into experimental animals, or in vitro. In fact, endotoxins have been recognized as playing an important role in the pathogenesis and manifestations of gram-negative infection, in general, and of septic shock, in particular. Thus, endotoxins are among the most potent agents capable of inducing local or generalized inflammatory reactions in both humans and experimental animals.

In the past, much emphasis has been placed on identifying chemical structures and structural principles of endotoxins, among other reasons, with the aim of elucidating the substructures which determine their various biological effects. This has recently been achieved for a number of activities. These results, in turn, have stimulated approaches to the chemical synthesis of biologically active endotoxin substructures and chemical analogs (for literature, cf. Rietschel et al. 1988). At the same time, the mechanisms of biological action of endotoxin are being studied at the molecular level, providing a rational basis for strategies aiming at the immunological or pharmacological prevention and treatment of endotoxicosis (Morrison and Ryan 1987; Pollack 1990).

The present chapter will summarize our knowledge of the chemical structure of bioactive endotoxin domains, the principal mechanisms involved in the biological activity of endotoxin and recent findings concerning the suppression of inflammatory endotoxic manifestations both in vitro and in vivo.

[1] Forschungsinstitut Borstel, Institut für Experimentelle Biologie und Medizin, Parkallee 1–40, D-2061 Borstel, FRG

42. Colloquium Mosbach 1991
Molecular Aspects of Inflammation

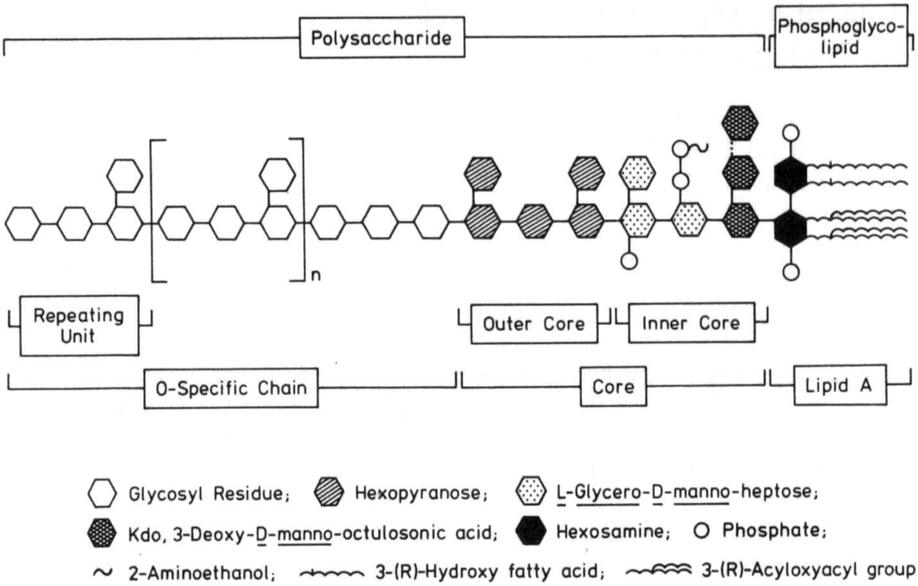

Fig. 1. Architecture and schematic structure of an enterobacterial lipopolysaccharide

2 Chemistry of Endotoxin

2.1 General Structure

Endotoxins represent a characteristic attribute of gram-negative bacteria and form a large, unique class of macromolecules (for reviews, see Lüderitz et al. 1982; Brade et al. 1988b; Rietschel et al. 1990). Various extraction procedures for their isolation have been described, most universally applicable being the phenol/water method (Westphal et al. 1952) and, for R-form bacteria, the phenol/chloroform/petroleum ether procedure (Galanos et al. 1969). Consecutive fractionation steps have been elucidated leading to highly purified endotoxin preparations.

Detailed chemical studies have revealed that, in spite of the great compositional variations among endotoxins derived from different bacterial serotypes, endotoxins conform to a common structural principle. They consist of a hydrophilic heteropolysaccharide and a covalently bound lipid component and chemically, therefore, endotoxins represent lipopolysaccharides (LPS). In the case of Enterobacteriaceae, the molecule can be subdivided into three structurally distinct segments, i.e., a lipid component, called lipid A, a core oligosaccharide, and a long heteropolysaccharide chain, the O-specific chain, which is generally composed of a sequence of identical oligosaccharides, the repeating units (Fig. 1). The core region and lipid A component usually carry polar residues such as phosphoryl, pyrophosphoryl, and 2-aminoethyl(pyro)phosphate groups. The three regions are under separate genetic control and distinct biosynthetic pathways are involved in their assembly (Mäkelä and Stocker 1984).

2.2 Polysaccharide Component

The O-specific chain represents a polymer of oligosaccharides, the so-called repeating units. In an endotoxin preparation, various LPS species are present which differ in the number of repeating units (for literature, cf. Lüderitz et al. 1982; Rietschel et al. 1988). As can be demonstrated by polyacrylamide gel electrophoresis, in a typical enterobacterial LPS preparation molecules with 20 to 30 repeating units predominate, but also molecules with only 1 and with up to 50 repeating units are present. The nature and type of linkage or substitution of the individual monosaccharide residues within a repeating unit are characteristic and unique for a given LPS and the parental bacterial strain. Since the structure of the O-specific chain is serotype-specific, it exhibits, among different serotypes or species, an enormous compositional and structural variability. Although not important in the context of this study, it should be mentioned that a variety of nonenterobacterial wild-type strains of pathogenic gram-negative bacteria such as *Acinetobacter*, *Bordetella*, *Campylobacter*, *Bacteroides*, *Neisseria*, *Haemophilus*, and *Chlamydia* express LPS which lack O-specific chains with repeating units (Hitchcock et al. 1986; Griffis et al. 1988). In their general architecture and behavior in polyacrylamide gel electrophoresis, these LPS resemble those of enterobacterial rough-(R)-mutants, although in their fine structure they are clearly different.

The core region of enterobacterial LPS consists of a hetero-oligosaccharide which can be formally divided into the O-chain-proximal outer core and the lipid A-proximal inner core. The structural variability of the core within the various species of an enterobacterial genus is limited. Thus, in *Salmonella* only one core type (Ra core) exists for all serotypes and in *Escherichia coli* so far five core types (R1 to R4 and K-12) have been described for over 100 different serotypes (for literature, cf. Jann and Jann 1984). Structural differences between these core types are mainly recognized in the outer region, which contains the common hexoses *D*-glucose (Glc), *D*-galactose (Gal) and *N*-acetyl-*D*-glucosamine (GlcNAc) in the pyranosidic ring form and, therefore, is also termed the hexose region. The various enterobacterial outer core types differ by the structural arrangement of these hexoses.

The inner core region of LPS is characterized by the unusual sugars heptose (Hep), present mainly in the *L-glycero-D-manno*, though occasionally in the *D-glycero-D-manno* configuration, and notably 3-deoxy-*D-manno*-octulosonic acid (2-keto-3-deoxyoctonic acid, Kdo). LPS of Enterobacteriaceae and the majority of other bacteria studied (including those which lack the O-specific chain) contain Hep, whereas all LPS, independent of their bacterial origin, harbour at least one α-bound pyranosidic or furanosidic Kdo residue (or a derivative thereof) with a free carboxyl group which occupies the lipid A-proximal position of the reducing terminus of the inner core region (Fig. 1). Thus, the inner core region appears to be structurally rather conserved (Brade et al. 1988b; Rietschel et al. 1988).

Based on biological and immunochemical investigations, the O-specific chain, core oligosaccharide, and lipid A region of LPS have been ascribed different activities and functions. Thus, it is known that the determinants being responsible for the O-antigenic properties of LPS are located in the O-specific chain. Further, this segment plays an important role in the resistance of bacteria against humoral and cellular components of the host defense system and, thus, in bacterial virulence (Mäkelä et al. 1990). The core oligosaccharide, on the other hand, is recognized by R-specific antibodies, and the inner core has been implicated in playing a role in the immunostimulatory activity of endotoxin (for literature, see Rietschel et al. 1988). Finally, a large body of experimental evidence indicated that lipid A is the carrier of the endotoxic properties of LPS, i.e., the induction of fever and shock (Galanos et al. 1977; Lüderitz et al. 1982; Rietschel et al. 1982).

With the aim to understand, at the molecular level, the interaction of the toxic principle of LPS with host components and the subsequent intracellular and extracellular events culminating in irreversible shock, we have investigated the primary structure and the conformation of lipid A. These analyses represented the basis for the chemical synthesis of lipid A partial structures, which proved to be of great importance in the analysis of structure-activity relationships, i.e., the elucidation of those structural parameters of lipid A which are essential for endotoxicity.

2.3 Lipid A

Lipid A represents the covalently linked lipid component of LPS. Enzymes which can cleave the bond between the polysaccharide and lipid region are not known. However, the ketosidic linkage between the Kdo (i.e., polysaccharide) and lipid A regions is particularly susceptible to acid-catalyzed hydrolysis and, hence, polysaccharide-deprived free lipid A can be prepared by treatment of LPS with mild acid.

The general principles governing the lipid A architecture have been elucidated using LPS of *Salmonella minnesota* and *Escherichia coli* rough (Re) mutants. In Fig. 2 the primary structure of *E. coli* lipid A is shown (Rietschel et al. 1984). In the following, *E. coli* lipid A will serve as a prototype structure in relation to which other lipid A derived from various entero- or nonenterobacterial LPS are discussed in order to illustrate some of the structural parameters which may be subject to structural variation. The predominant component of *E. coli* lipid A (Fig. 2) is composed of a β-glucosaminyl-(1–6)-α-*D*-glucosamine disaccharide, which carries two phosphoryl groups, one in position 4' (of the distal glucosaminyl residue, GlcN II) and one in position 1 (of the reducing glucosaminyl residue, GlcN I). This hydrophilic lipid A backbone is acylated by four (primary) acyl groups, i.e., four (*R*)-3-hydroxytetradecanoic acid residues [14:0(3-OH)] at positions 2, 3, 2', and 3', and it contains two unsubstituted hydroxyl groups at positions 4 and 6'. The latter primary hydroxyl group is only free in polysaccharide-deprived free lipid A since, in LPS, it represents the attachment site of Kdo, i.e., of the polysaccharide component (Zähringer et al. 1985). The hydroxyl groups of the two 14:0(3-OH) residues bound to GlcN II at positions 2' and 3' carry, as secondary acyl groups, dodecanoic (12:0) and tetradecanoic acid (14:0), respectively, whereas the two 14:0(3-OH) groups at GlcN I are not 3-O-acylated.

In addition to the main structure shown in Fig. 2, *E. coli* and other enterobacterial lipid A preparations contain molecules with possess the phosphorylated glucosamine backbone but a smaller number of acyl groups. Such acyl-deficient structures are particularly found in enterobacterial S-form LPS (Jiao et al. 1989). The phosphoryl

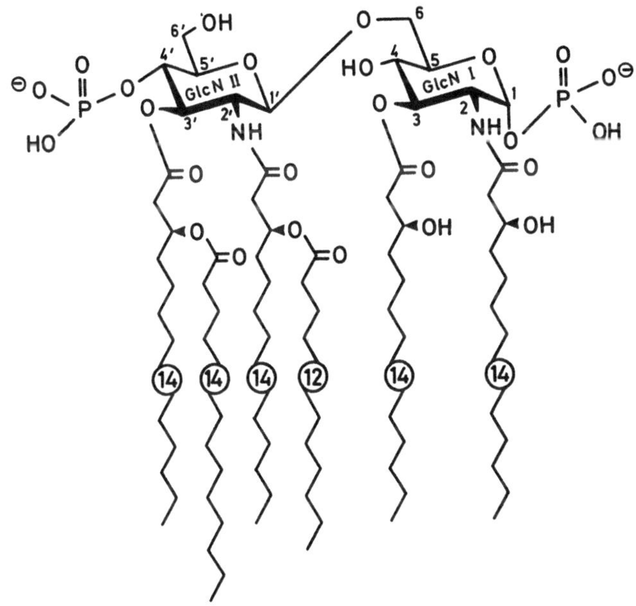

Fig. 2. Primary structure of the lipid A component of *Escherichia coli* LPS. The hydroxyl group in position 6' represents the attachment point of Kdo, i.e., the inner core region of LPS. The *numbers* in the acyl chain signify the numbers of carbon atoms in fatty acids. The 3-hydroxytetradecanoyl residues possess the (*R*)-configuration. In bacterial *E. coli* lipid A, the glycosidic phosphate group is nonstoichiometrically substituted by phosphate

residue in positions 1 may be substituted with phosphate in nonstoichiometric amounts indicating intrinsic heterogeneity of lipid A. The same structure as shown for *E.coli* lipid A has been identified in *S. typhimurium* (Qureshi et al. 1982), and this type of lipid A is also present in other enterobacterial and nonenterobacterial genera such as *Haemophilus* (Helander et al. 1988) and *Providencia* (Mayer et al. 1989). Hexaacyl *E. coli* lipid A has been chemically synthesized and shown to be identical in all chemical and physical properties to its purified bacterial counterpart (Galanos et al. 1985; see below).

Lipid A of nonenterobacterial LPS differs in structure from *E. coli* in several parameters, including e.g., the nature, number, chain length, and location of fatty acids, the absence of the nonglycosidic phosphate group at GlcN II, and the positional replacement of GlcN by 2,3-diamino-2,3-dideoxy-*D*-glucose (GlcN3N). To illustrate such variations, Fig. 3 shows the chemical structure of the major components of lipid A of *Chromobacterium violaceum* (Fig. 3A), *Pseudomonas aeruginosa* (Fig. 3B), *Bacteroides fragilis* (Fig. 3C), and *Campylobacter jejuni* (Fig. 3D).

In *E. coli* hexaacyl lipid A, GlcN carries a total of four, and GlcN I two acyl groups. Thus, the acylation pattern over the two GlcN residues is asymmetric [4 + 2]. In contrast, a symmetrical fatty acid distribution [3 + 3] is recognized in hexaacyl *Chromobacterium violaceum* lipid A where GlcN I and GlcN II each carry three fatty acids (Fig. 3A). It should be mentioned that the amide-linked 12:0 (3-OH) of GlcN II is 3-O-acylated by either 12:0 or (*S*)-2-hydroxydodecanoic acid (12:0(2-

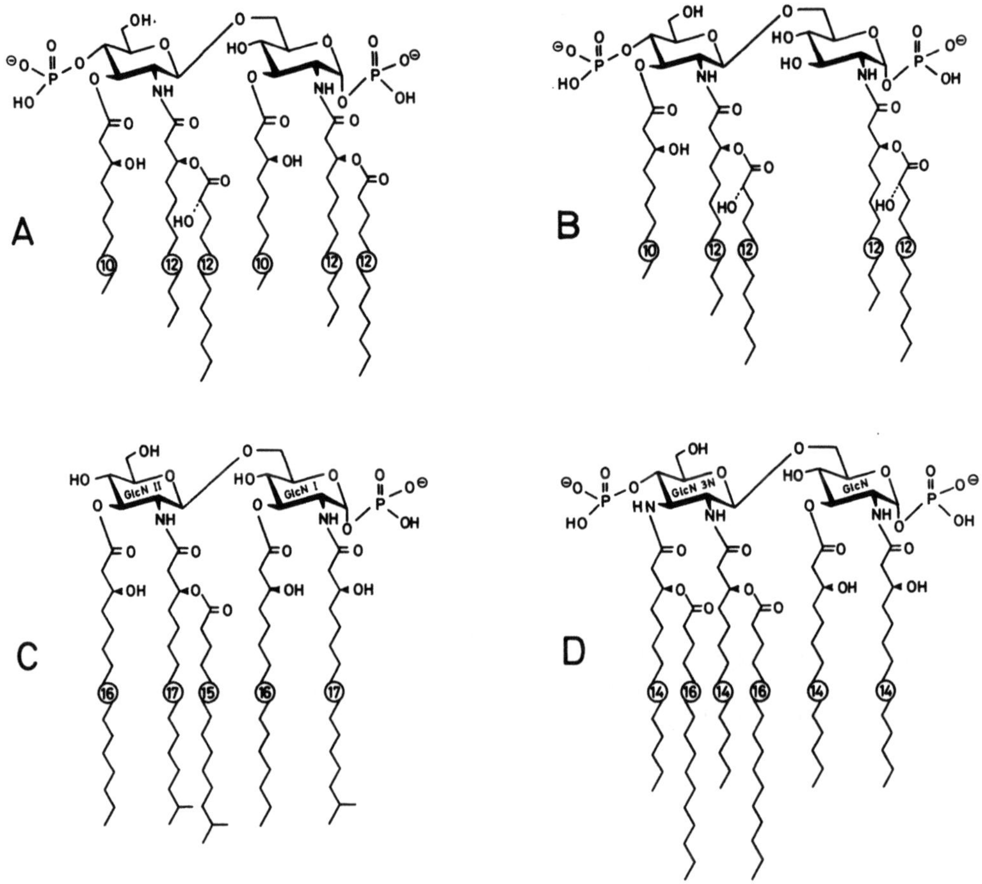

Fig. 3. Chemical structure of the main components of lipid A of **A** *Chromobacterium violaceum*; **B** *Pseudomonas aeruginosa*; **C** *Bacteroides fragilis* and **D** *Campylobacter jejuni*. For a detailed discussion of these lipid A structures, cf. text; see also legend to Fig. 2

OH)) yielding two molecular species as indicated by the dashed line (Hase and Rietschel 1977; Wollenweber et al. 1984). Lipid A of *C. violaceum* further differs from that of *E. coli* in the chain length of the fatty acids present (Figs. 2 and 3A). This type of lipid A has also been identified in *Neisseria gonorrhoeae* (Takayama et al. 1986) and *N. meningitidis* (Kulshin et al., submitted).

In the main species of *Pseudomonas aeruginosa* lipid A (Fig. 3B), in contrast to *E. coli* lipid A, only a total of five fatty acids is detected since the primary acyl residue at position 3 of GlcN is lacking (Kulshin et al. 1991). This prominent species, comprising approximately 75% of a *P. aeruginosa* lipid A preparation, harbors four structural forms of which all possess the same backbone with three 3-hydroxy fatty acids attached to positions 2,2', and 3' (Fig. 3B). These structural forms merely differ from each other by the 3-O-acylation of each of the two amide-linked 12:0(3-OH) residues by either 12:0 or 12:0(2-OH), as indicated by the dashed lines in Fig. 3B.

In the major component of *Bacteroides fragilis* lipid A (Fig. 3C) also only five, comparatively long-chain and, in part, isobranched fatty acids are recognized (Weintraub et al. 1989), but in location different to that of *P. aeruginosa* lipid A. The most striking difference to *E. coli* and the other lipid A structures shown, however, is that in *B. fragilis* lipid A the hydroxyl group at position 4' of GlcN II is free, i.e. the nonglycosidic phosphate group is absent.

Campylobacter jejuni lipid A (Fig. 3D) harbors, like *E. coli* lipid A, six acyl groups in an asymmetrical [4 + 2] distribution (Moran et al. 1991). However, it significantly differs from *E. coli* lipid A and the other structures shown in that GlcN3N represents the distal backbone hexosamine group, positionally replacing the GlcN II residue of *E. coli* lipid A. The hexosamine disaccharide is β(1-6)-linked and the location of the phosphoryl and acyl groups is similar to that in *E. coli* lipid A. However, in *C. jejuni* lipid A, three amide-bound primary acyl groups are revealed, whereas in *E. coli* only two fatty acids are present in the amide linkage.

In other lipid A preparations, further compositional and architectural differences as compared to the *E. coli* model structure were revealed. To name only a few examples, in nontoxic *Rhodopseudomonas viridis* lipid A, an acylated GlcN3N monosaccharide forms the lipid A backbone (Mayer et al. 1989), in *Rhodobacter sphaeroides* 3-hydroxy fatty acids are, in part, replaced by 3-oxo fatty acids (Mayer et al. 1989), and in *Rhizobium trifolii* 27-hydroxyoctaeicosanoic acid, which is believed to be bound to *D*-glucosaminuronic acid (Hollingsworth and Lill-Elghanian 1989), represents the major lipid A acyl group (for further structural variations, see Rietschel et al. 1982, 1984).

Thus, lipid A of distinct bacterial origin share basic structural elements but may differ in some chemical features. Nevertheless, as these analyses have shown, endotoxically active lipid A of different bacterial origin are structurally closely related. Characteristic and common to them is the presence of a bisphosphorylated β(1-6)-linked disaccharide of *gluco*-configurated *D*-hexosamines. This structure has not been identified so far in other natural compounds and, hence, it is unique to lipid A. The lipid A backbone carries, in general, up to four mole equivalents of primary (*R*)-3-hydroxy fatty acids, two of which occupy amino functions and two of which are linked to the backbone hydroxyl groups at positions 3 and 3'. Both amide- and ester-bound (*R*)-3-hydroxy fatty acids are, in part, substituted at their 3-hydroxyl groups by (secondary) saturated and nonhydroxylated fatty acids. Such (*R*)-3-acyloxyacyl residues are found in taxonomically distant groups of gram-negative bacteria and they are also characteristic for lipid A. In summary, therefore, lipid A expresses, as compared to the other LPS regions, a rather low structural variability.

2.4 Synthetic Lipid A

Based on the results of chemical analysis, lipid A has been chemically synthesized (Kusumoto et al. 1989, 1990). The first fully synthetic lipid A molecule (preparation 506 or LA-15-PP) corresponds in structure to *E. coli* lipid A (Fig. 2). Also, other lipid A and lipid A partial structures were prepared which all contain a β(1-6)-linked *D*-glucosamine disaccharide but which differ in the acylation and phosphorylation pat-

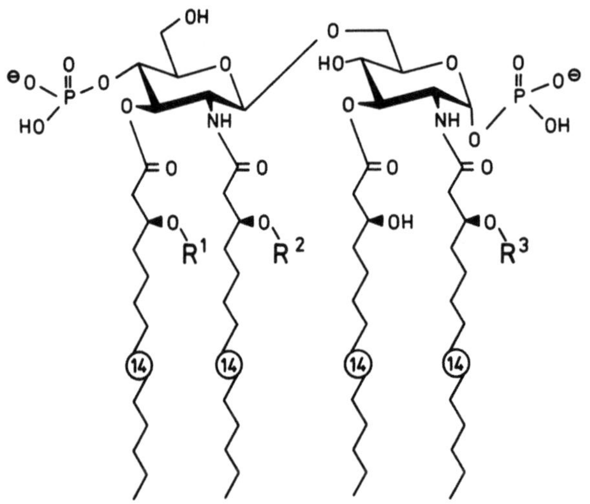

Designation of Preparation		Number of Fatty Acids	Nature of		
Bacterial	Synthetic		R^1	R^2	R^3
Precursor Ia	LA-14-PP or 406	4	H	H	H
Precursor Ib	LA-20-PP	5	H	H	16:0
E.coli Lipid A	LA-15-PP or 506	6	14:0	12:0	H
S.minnesota Lipid A	LA-16-PP or 516	7	14:0	12:0	16:0

Fig. 4. Chemical structure of some synthetic lipid A and partial structures. For details, cf. legend of Fig. 1. *12:0* Dodecanoic acid; *14:0* tetradecanoic acid; *16:0* hexadecanoic acid

tern. These preparations include the heptaacyl species of *S. minnesota* lipid A (compound 516 or LA-16-PP), the pentaacyl precursor Ib (LA-20-PP), and the tetraacyl precursor Ia (406 or LA-14-PP), the structure of which is shown in Fig. 4. In addition, the 1-dephospho (e.g., compound 504 or LA-15-PH) and the 4'-dephospho (505 or LA-15-HP) partial structures of *E. coli* lipid A and of acyl-deficient compounds such as precursor Ia were prepared. Also, lipid A disaccharide analogs, such as the pentaacyl compound 316 with an acylation pattern different from that of lipid A, have been chemically synthesized (for literature, see Kusumoto et al. 1989, 1990; Stütz et al. 1990). Further, a great number of monosaccharide partial structures with different acylation and phosphorylation patterns have been prepared by several groups (for literature, see Stütz et al. 1990). These compounds include synthetic counterparts of lipid A-related bacterial products such as lipid X and lipid Y, as well as other partial structures and analogs corresponding to either the reducing or the distal GlcN unit of the lipid A backbone.

3 Biology of Endotoxin

3.1 Lipid A, the Endotoxic principle of LPS

For a long time, and based on a large body of experimental evidence, it had been postulated that the lipid A component represents the endotoxic principle, being responsible for the induction of the manifold pathophysiological effects of LPS (Westphal and Lüderitz 1954; Galanos et al. 1977; Lüderitz et al. 1982; Rietschel et al. 1982). The successful chemical synthesis of lipid A and partial structures then offered the possibility to confirm this concept and to verify the structural proposals made for lipid A. In fact, biological assays showed that chemically synthesized *E. coli* lipid A (compound 506) expresses, with similar doses, the same spectrum of endotoxic effects as bacterial (*E. coli*) free lipid A and endotoxin (Galanos et al. 1985). Thus, lipid A represents the lethal, leukopenic, pyrogenic, i.e., the endotoxic principle of LPS, its endotoxic properties being embedded in a molecule with the structure shown in Fig. 2.

E. coli lipid A possesses a complex structure and the question arose as to which of its constituents is essential for, or contributed to, bioactivity. To answer this, various lipid A partial structures were analyzed in vivo for endotoxic activity. The results of such studies, with emphasis on the fever-producing capacity, i.e., pyrogenicity, performed by our and other groups can be summarized as follows (for literature, see Rietschel et al. 1987, 1988, 1990). Hexosamine monosaccharide compounds largely lack in vivo endotoxic activity [pyrogenicity (rabbit), lethal toxicity (mouse)]. Glucosamine disaccharide structures carrying two phosphate groups and seven, five or four fatty acids exhibit in vivo pyrogenic activity which, however, is weaker than that of *E. coli* lipid A (containing six fatty acids). Compounds with only two fatty acids are completely inactive. Glucosamine disaccharide preparations carrying only one phosphate group (at either position 1 or 4', see Fig. 2) are also less pyrogenic than *E. coli* lipid A (containing two phosphoryl residues). Most interestingly, none of the synthetic compounds so far available exhibits stronger pyrogenicity and endotoxic activity, in general, than natural or synthetic *E. coli* type lipid A.

According to present knowledge, pyrogenicity, i.e., endotoxic activity, is optimally expressed by a molecule containing two β(1-6)-interlinked D-*gluco*-configurated pyranosidic hexosamine residues, two phosphoryl groups, and six fatty acids (saturated and, in part, 3-hydroxylated) including 3-acyloxyacyl groups in a defined location as it is present, e.g., in *E. coli* lipid A. Partial structures lacking one constituent, or molecules with a different distribution of constituents, are endotoxically less or not active. Thus, slight modifications at any site of the *E. coli* lipid A architecture result in a significant change in biological activity, showing that endotoxicity is not dependent on one single lipid A constituent or a toxophore group. It appears, rather, that it is unique molecular arrangement of lipid A and, therefore, a unique supramolecular conformation (endotoxic conformation), which allows the expression of endotoxic activity (Rietschel et al. 1987).

3.2 Interaction of Endotoxin or Lipid A with Host Target Cells

After entering the circulation of higher organisms, endotoxins interact with a number of humoral and cellular components of the host. Thus, LPS has been shown to bind to serum proteins such as high and low density lipoproteins, serum albumin, complement components, a 28-kDa factor, and perhaps, most importantly, to an acute phase protein termed lipopolysaccharide-binding protein (LBP) (for literature, see Morrison 1990; Schumann et al. 1990). Some of these proteins probably prevent the random attachment of the amphiphilic LPS molecule to cells and tissues. They rather appear to mediate a more specific transport of LPS to organs involved in endotoxin accumulation, i.e., the liver and, to some extent, the spleen and lung. At these body sites and in the circulation, endotoxins interact with cells, and this event is believed to play a decisive role in the mechanism of endotoxin action. Endothelial cells, granulocytes, platelets, B-lymphocytes, and, in particular, monocytes and macrophages have been identified as primary endotoxin target cells. The result of this interaction may be either cellular differentiation and proliferation, or the formation and secretion of lipid and peptide mediator molecules, which are endowed with potent biologic activity (see Sect. 3.3).

The initial and perhaps central event which triggers the reaction cascade, resulting in endotoxic effects, is the interaction of endotoxin, via its lipid A component, with the membrane of mononuclear phagocytes resulting in the production and release of bioactive mediators. It is possible that lipid A, by way of its hydrophobic region, intercalates into the lipid matrix of the mammalian cell in a nonspecific way. More recently, however, it has been discussed that on the target cell particular binding molecules are exposed which are able to specifically recognize either lipid A (endotoxin) in the active conformation or lipid A/protein complexes. The presence of specific lipid A binding proteins on the surface of host macrophages such as the 80-kDa LPS-binding protein (Morrison 1989) and the macrophage surface protein CD 14, which binds LPS/LBP complexes (Wright et al. 1990), has been experimentally demonstrated (see also Golenbrock et al. 1990; Tahri-Jouti et al. 1990).

Recently, we have found that on murine lymphocytes (pre-B cell lines, B- and T-cell lines) a 40-kDa protein (45 kDa under nonreducing conditions) is present which interacts specifically with endotoxin or free lipid A. In a murine macrophage-like cell line (J774.1), however, two lipid A binding proteins, one having an Mr of 40 kDa and the other of 80 kDa (45 and 95 kDa, respectively, under nonreducing conditions) were detected (Kirikae et al. 1991). These two proteins were also identified in peritoneal macrophages of the mouse strain C3H/HeN (T. Kirikae, unpubl. results). The presence of two distinct LPS-binding proteins on macrophages may be related to the extreme sensitivity of this cell type to LPS-induced activation. Thus, macrophages respond with enhanced phagocytosis capacity (Schade 1986) and mediator production (Loppnow et al. 1989) to LPS or lipid A doses in the picogram to low nanogram range, whereas microgram amounts of LPS are required to induce proliferation and differentiation, e.g., of B-lymphocytes (Brade et al. 1988a). It is thus conceivable that macrophages dispose of a high affinity two-component receptor system (40 and 80 kDa), which can be triggered by very small LPS doses, whereas B-lymphocytes possess only one protein component of the receptor system and, therefore, require higher amounts of LPS for activation. At present, this concept is hypothetical as it is not known whether the recognition proteins identified are functional receptor molecules which, after binding of endotoxin or lipid A, deliver an activation signal. In addition, it remains to be investigated whether further steps (such as internalization of endotoxin or lipid A into the cell interior and/or binding to a cytosolic receptor), perhaps in addition to activation of a cell surface receptor, are required to initiate a biological response.

3.3 Endotoxin-Induced Monokine Production

Biological activity of endotoxin or lipid A expressed in vivo is dependent on the presence of LPS-reactive mononuclear cells (Freudenberg et al. 1986), and it is mediated by bioactive molecules, produced and released by these cells. Among such mediators, tumor necrosis factor α (TNF) and interleukin 1 (IL-1), together with interleukin 6 (IL-6) figure prominently. These secondary, hormone-like proteins can induce many of the typical endotoxin effects such as fever and shock by acting independently, in sequence, and, most importantly, synergistically (Vogel and Hogan 1990).

We hypothesized that the structure-activity relationships revealed in the in vivo pyrogenicity system may also be recognized by in vitro systems involving the production and release of such bioactive peptide mediators like IL-1 and TNF, which are involved in the generation of fever. Therefore, the structural prerequisites of endotoxins with respect to their capacity to induce IL-1 and TNF production in human peripheral monocytes were examined (Feist et al. 1989; Loppnow et al. 1989). In these studies, various S- and R-form LPS as well as bacterial and synthetic free lipid A and lipid A partial structures were employed. Some of the results obtained with TNF in numerous dose-response experiments with cells of different human low- and high-responder donors are shown in Fig. 5.

In inducing about 1 U/ml of TNF, S-, and notably R-form LPS was found to be very potent (concentration range 1 pg to 1 ng/ml) as was synthetic *E. coli* lipid A (preparation 506), the latter, however, exhibiting lower activity (concentration range 1 to 100 ng/ml). Synthetic *S. minnesota* heptaacyl lipid A (preparation 516) and the monophosphoryl partial structures 504 (and 505, not shown) were significantly less active (concentration range 10 ng/ml to 10 μg/ml). The tetraacyl precursor Ia (compound 406), the bisacyl compound 606 (not shown), and the monosaccharide structures 401 (lipid X) and 410 were completely inactive (Fig. 5). The same hierarchy of LPS and partial structures was established for their IL-1 inducing capacity in human peripheral monocytes (Loppnow et al. 1989). Again, LPS was a very potent inducer of IL-1 release and free lipid A was less active, whereas lipid A partial structures showed greatly reduced or, like precursor Ia (compound 406), no activity. On stimulation of cells with precursor Ia and lipid X, also no intracellular IL-1 could be detected.

In summary, the rules of structure-activity relationships established in these in vitro analyses parallel those deduced from in vivo assay systems such as pyrogenicity. Thus, full activity is expressed by a lipid A molecule corresponding in structure to *E. coli* lipid A. It is most remarkable that precursor Ia, which differs from *E. coli* lipid

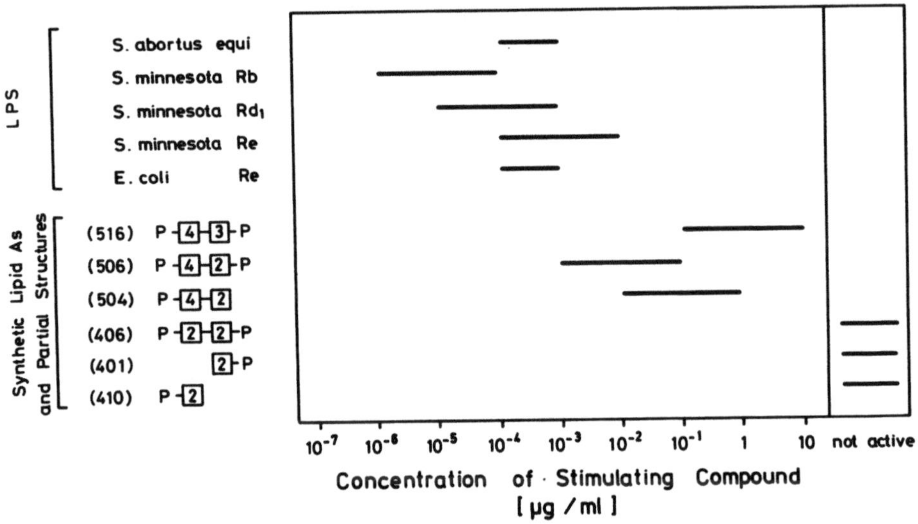

Fig. 5. Tumor necrosis factor α (TNF)-inducing capacity of LPS, lipid A, and partial structures. Shown is the minimal concentration of compounds required for the induction of > 1 U/ml of released TNFα, and in a schematic representation, the compounds used for stimulation of cells. The ***boxes*** represent GlcN residues and the ***numbers*** in the schematic formulas indicate the number of fatty acids bound to either GlcN I or GlcN II

A only by the absence of two acyl groups (12:0 and 14:0), thus lacking 3-acyloxyacyl groups, is completely inactive in stimulating human mononuclear cells. This points to an extreme structural specificity and the importance of the acylation pattern of lipid A for the expression of its endotoxic activity (cf. Sect. 3.4).

As Fig. 5 shows, S- and R-form LPS is significantly more active in inducing cytokine production than free lipid A. This is shown in more detail in Fig. 6, which illustrates that the minimal concentration of free lipid A, necessary for the stimulation of TNF production by human mononuclear cells, is about 100-fold higher than the minimal bioactive concentration of LPS. Figure 6 also shows that the amount of TNF produced after stimulation of cells with free lipid A is significantly lower than that detectable after LPS, e.g., Re LPS (Fig. 5), which differs chemically from lipid A only by the presence of two Kdo residues. It, therefore, appears that Kdo or the Kdo-containing inner core is capable of upregulating lipid A bioactivity, as was also observed with other endotoxin activities (for literature, cf. Rietschel et al. 1988). It is possible that this effect is based on the fact that, in addition to lipid A, also the Kdo-containing inner core may interact with a specific receptor on mononuclear cells indicating the necessity of two signals for an optimal cellular response (Cavaillon and Haeffner-Cavaillon 1990). On the other hand, the charged Kdo groups are likely to change the physicochemical properties of lipid A, and they may contribute to a peculiar conformation required for full bioactivity of lipid A. It remains to be studied wether these or other factors are responsible for the modulating effect of the Kdo-containing oligosaccharide on the bioactivity of lipid A.

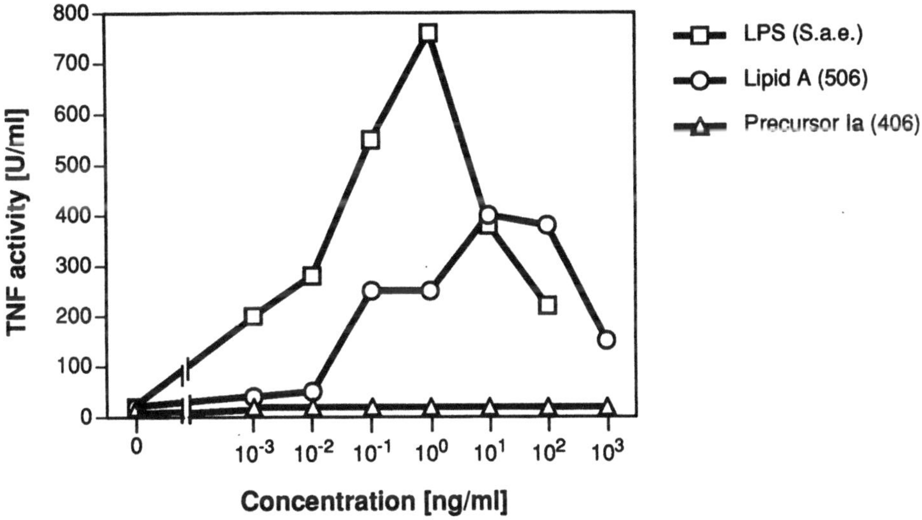

Fig. 6. Induction of TNF production in human peripheral monocytes by LPS (*S. abortus equi*), synthetic *E. coli* lipid A (506), and precursor Ia (406). Shown is the dependency of TNF amounts in cellular supernatants on the dose of stimulating compound. Precursor Ia (406) does not cause TNF production. TNF levels were determined by bioassay

3.4 Inhibition of Mediator Induction by Lipid A Partial Structures

It has previously been shown that partially deacylated LPS preparations are capable of inhibiting LPS-induced adherence of granulocytes to vascular endothelium (Pohlman et al. 1987). We have asked the question wether oligoacyl partial structures of lipid A such as precursor Ia (compound 406), which by themselves are inactive (cf. Sect. 3.3 and Fig. 5), are able to inhibit LPS- or lipid A-induced monokine production in human peripheral monocytes. It was found that the tetraacyl compound 406, when added to cells either 1 h before or simultaneously with LPS, indeed, inhibited in a dose-dependent manner the monokine production by human monocytes induced by LPS but not by *Staphylococcus epidermidis*, lipoprotein, or BCG (Loppnow et al. 1989; Wang et al. 1990). Figure 7 shows the dependency of the inhibition of LPS-induced TNF production on the amount of precursor Ia. Synthetic precursor Ia (406) did not exhibit its inhibitory effect by suppressing the release of monokines. Rather, as was shown by Northern blot analysis and as documented in Fig. 8, monokine production was inhibited at an early stage. i.e., precursor Ia suppressed the LPS-induced formation of mRNA specific for TNF and IL-1 (Feist et al., in press). Similar results were reported for the bacterial precursor Ia isolated from *S. typhimurium* (Kovach et al. 1990) and also with nontoxic lipid A species of *Rhodobacter sphaeroides* (Takayama et al. 1989) and *Rh. capsulatus* (Loppnow et al. 1990).

These findings raise the question as to the mechanism of the inhibitory action of lipid A partial structures such as precursor Ia (compound 406). At present, we favor the view that the inhibitory action of precursor Ia is based on a competitive binding to a specific LPS receptor, i.e., that precursor Ia acts via an antagonistic mechanism.

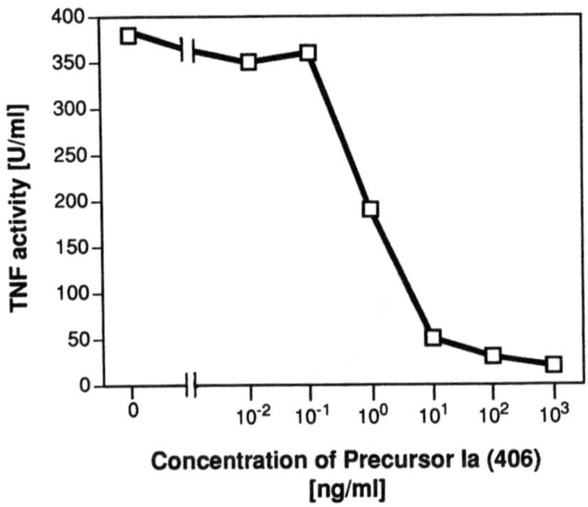

Fig. 7. Inhibition of LPS-induced TNF production by precursor Ia (406). Shown is the dependency of the degree of inhibition on the amount of precursor Ia (406) added. Human peripheral monocytes were stimulated with 1 ng/ml of S. abortus equi LPS, and precursor Ia (406) was added to cells 1 h before LPS. TNF activity was assessed by bioassay

This assumption is based, in part, on the observation that the inhibitory effect of precursor Ia is LPS- or lipid A-specific, i.e., monokine (TNF and IL-1) production induced by phorbolmyristate acetate, *S. epidermidis*, *S. aureus*, BCG, or a synthetic lipopeptide is not affected by compound 406. Further, as shown in Fig. 9, the binding of (radiolabeled) Re-mutant LPS to mouse macrophages is equally well inhibited by LPS and compound 406 (Kirikae et al. 1991). Compound 406 also strongly competes with binding of LPS to human monocytes (W. Feist, T. Kirikae, U. Schade, A. J. Ulmer, unpubl. results). These and other considerations lead us to believe that compound 406, lipid A, and LPS bind to the same cellular receptor site, but that compound 406, in contrast to lipid A and LPS, is not able to activate it, i.e., to trigger signal-transducing mechanisms.

Presently, we are studying the structural prerequisites of the capacity of compounds, which are structurally related to lipid A, to antagonize endotoxin effects. As discussed in Section 3.3, endotoxic (agonistic) activity of lipid A-related preparations containing the bisphosphorylated hexosamine disaccharide strongly depends on the number of fatty acids present. Thus, hexaacyl *E. coli* lipid A (compound 506) is a strong mediator-inducing agonist, pentaacyl precursor Ib (LA-20-PP) is less potent, whereas the tetraacyl precursor Ia (406) is completely inactive. The latter compound, on the other hand, possesses potent antagonistic properties, whereas a synthetic compound with only two fatty acids (compound LA-19-PP or 606) is a weaker antagonist. Thus, a preparation with four fatty acids (like preparation 406) lacks endotoxicity but exhibits optimal antagonistic activity. This was established for reactivity with *human* mononuclear cells. In the *murine* system, however, the tetraacyl compound 406 expresses, as compared to hexaacyl *E. coli* lipid A (506), significant, although reduced, endotoxicity (Galanos et al. 1984). Therefore, in the human and murine systems different structural rules appear to govern the expression of agonistic or antagonistic activity. It was found that LPS of *Rh. capsulatus*, harboring also only four fatty acids (like precursor Ia), is endotoxically nonactive in the mouse (and human) system and that it exhibits strong antagonistic activity (Loppnow et al. 1990). Tetraacyl compound 406 and tetraacyl *Rh. capsulatus* lipid A only differ in the nature of their fatty

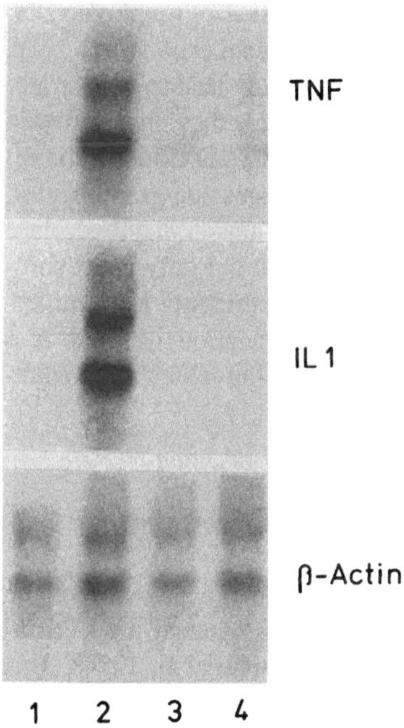

Fig. 8. LPS-induced expression of mRNA specific for TNF and IL-1, and its inhibition by precursor Ia (406). *Lane 1* Control cells without LPS; *lane 2* cells stimulated with LPS (1 ng/ml); *lane 3* cells treated with precursor Ia (100 ng/ml); *lane 4* cells first treated for 1 h with precursor Ia (100 ng/ml), followed by LPS (1 ng/ml)

acids: compound 406 contains four 14:0(3-OH) residues, whereas *Rh. capsulatus* harbors two amide-bound fatty acids with 14 carbon atoms (in the form of the 3-oxo derivative) but, unlike compound 406, the two ester-linked acyl residues have only ten carbon atoms [10:0(3-OH)]. The difference in activity of these tetraacyl compounds in the murine system, therefore, may depend on either the different *nature* or

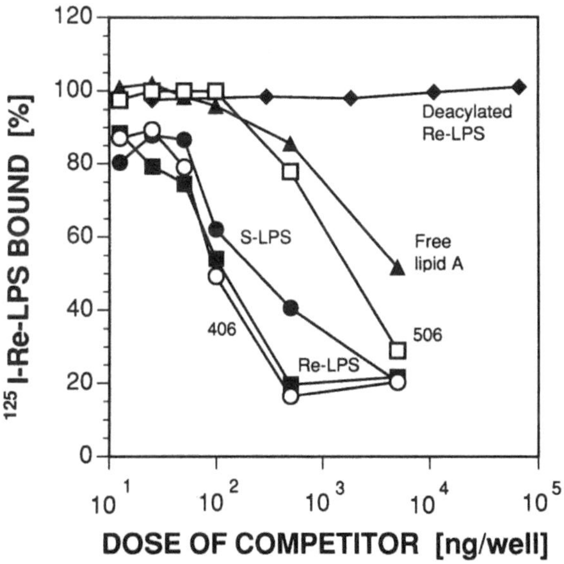

Fig. 9. Inhibition of binding of radiolabeled *S. minnesota* Re LPS to mouse peritoneal macrophages by LPS, free lipid A, and partial structures. Cells were incubated with [^{125}I]-labeled Re LPS (67 ng/ml, S. minnesota R595) at 0–2 °C in the absence or presence of different competitors in varying doses. The degree of inhibition of binding is expressed in % of optimal binding of [^{125}I]-Re LPS. Deacylated Re LPS does not compete with LPS binding, whereas Re LPS and precursor Ia (406) are very efficient inhibitors

on the different *chain length of the fatty acids* present. Recent studies employing *Rh. sphaeroides* lipid A render the latter alternative very likely (Qureshi et al. 1991).

Studies along these lines appear to be important since they may lead to new strategies to pharmacologically control endotoxin effects, and since they may provide deeper insight into the interaction of endotoxins with the host. In this context it is remarkable that on biodegradation of *E. coli* lipid A by macrophages and granulocytes, two fatty acids (12:0 and 14:0) are removed by the enzyme 3-acyloxyacylhydrolase yielding a structural counterpart of precursor Ia. This compound not only lacks stimulatory activity, i.e., endotoxicity for human cells but rather represents a potent endotoxin antagonist. Thus, it appears that on enzymatic degradation of lipid A a molecule (precursor Ia) is formed which not only is detoxified but which even counteracts endotoxin bioactivity (Riedo et al. 1990).

3.5 Regulation of Endotoxin-Induced Monokine Formation in Macrophages

It has previously been observed that inhibitors of the lipoxygenase pathway of the arachidonic acid cascade, but not cyclooxygenase inhibitors, suppressed the LPS-induced formation of IL-1 in human peripheral monocytes (Dinarello et al. 1984). This finding suggested that an arachidonic acid metabolite, formed via the lipoxygenase pathway, is involved in the production of IL-1. Despite several studies, the nature of the lipoxygenase product involved in monokine formation remained unknown (for literature, see Schade et al. 1989a). In more recent studies, we found that also the LPS-induced formation of TNF in murine macrophages is blocked by lipoxygenase inhibitors (Schade et al. 1989b). Moreover, lipoxygenase inhibitors suppressed the LPS-induced production of TNF *in vivo* as well as LPS-induced lethality in mice (Keppler et al. 1987; Schade et al. 1989b). Most interestingly, however, TNF production in murine macrophages was not blocked by a specific inhibitor (MK 886) of the enzyme 5-lipoxygenase, which is involved in the transformation of arachidonic acid to 5-hydroperoxyeicosatetraenoic acid (5-HETE) and, further, to the leukotrienes LTA_4 and LTB_4, as well as LTC_4 and LTD_4 (Schade et al. 1991). This result suggested that the lipoxygenase product, which plays a role in TNF production, was not derived from arachidonic acid. In subsequent studies, we were unable to detect lipoxygenase products in supernatants of LPS-activated mouse macrophages. [The reported production of LTC_4 in macrophages (Lüderitz et al. 1989) proceeds only under special experimental conditions and with R-mutant LPS]. However, in alkaline hydrolysates of LPS-activated macrophages, the compound (*S*)-13-hydroxyoctadecadienoic acid [(*S*)-13-hydroxylinoleic acid, 13-HODD] was detected, being present in approximately tenfold higher amounts in activated as in control cells (Schade et al. 1987). This compound is derived from linoleic (and not from arachidonic) acid and is not secreted by macrophages into the surrounding medium as are prostaglandins or LTs. Rather, 13-HODD is present within cells in a bound form as a constituent of phospholipids, located in position C2 of phosphatidylethanolamine, -serin, and -inositol (I. Burmeister and U. Schade, unpubl. results). In summary, these data suggested that LPS induces, in macrophages, the activation of a 15-lipoxygenase, which is involved in the transformation of bound or free linoleic acid to 13-HODD, as is shown in Fig. 10.

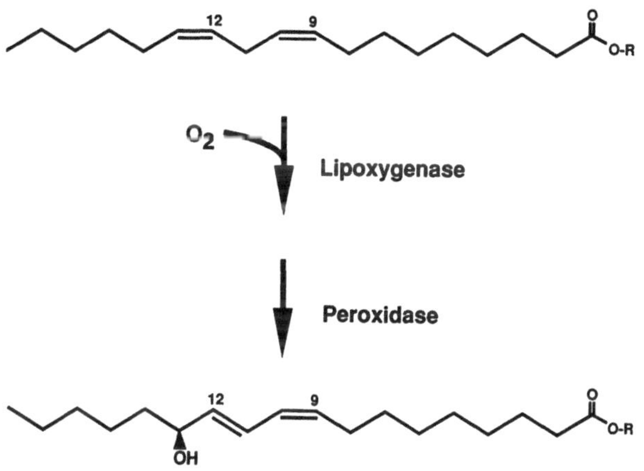

Fig. 10. Lipoxygenase-catalyzed oxidation of linoleic acid to (S)-13-hydroxylinoleic acid [(S)-13-HODD]. It is not known wether free (R = H) or bound linoleic acid serves as substrate for the 15-lipoxygenase

The notion that inhibition of the 15-lipoxygenase blocks the formation of TNF indicates that the 15-lipoxygenase product 13-HODD plays a role in the activation of murine macrophage i.e., the production and release of TNF (and IL-1). So far, direct proof for this hypothesis is lacking, but several lines of evidence tend to support it. Thus, 13-HODD, added to macrophages induces, like LPS, enhanced phagocytic capacity. LTC_4 and LTB_4, on the other hand, are inactive in this system. Further, 13-HODD, in doses of 10^{-7} to 10^{-5} M, does not induce TNF formation in murine macrophages. If, however, 13-HODD is added to macrophages together with substimulatory amounts of LPS, it induces the formation of TNF to a similar extend as do optimal doses of LPS. Finally, the TNF production of murine macrophages, which had been blocked by lipoxygenase inhibitors, can be fully restored by the addition of 13-HODD (U. Schade, unpubl. data). In a similar way, the inhibition of the LPS-induced expression of TNF-specific mRNA in macrophages by lipoxygenase blockers can be fully reversed by the addition of 13-HODD. These data are compatible with the concept that (*S*)-13-HODD mediates an essential step in the LPS-induced activation and TNF production of macrophages. It should be noted, however, that in addition to 13-HODD, a further signal(s), being provided by LPS, seem(s) to be required for optimal cellular response.

3.6 Endotoxin-Induced TNF Production in Humans and Its Inhibition by Pentoxifylline

After injection of endotoxin into experimental animals such as mice and rabbits, an initial leukopenia, associated with reduced capillary blood flow, is observed. Both phenomena are related and largely based on the fact that polymorphonuclear neutrophils adhere *via* specific adhesion molecules to endothelial cells and, thus, disappear from the circulation. Poor capillary flow and neutrophil-dependent injury lead to cellular anoxia, lactic acidosis, and ultimately to life-threatening conditions and shock. Thus, neutrophil adherence to the endothelial vessel lining – a phenomenon induced by TNF – greatly contributes to the deleterious effects of endotoxins *in vivo*. Therefore, drugs which interfere with neutrophil adherence or which improve capillary blood flow are expected to counteract LPS-induced inflammatory effects.

The xanthine derivative pentoxifylline [POF, oxpentifyllin; 3, 7-dimethyl-1-(5-oxohexyl)-xanthine] is known to reduce blood viscosity, and it is widely used clinically in the treatment of various types of vascular insufficiencies. We and others have, therefore, studied its effect in different animal models and it was found that POF protected from increased pulmonary vascular permeability and sequestration of neutrophils into the lung (Welsh et al. 1988) as well as from LPS-induced lethal shock (for literature, see Schönharting and Schade 1989; Schade 1990). The protective effect of POF was dose-dependent and observed even when the drug was given up to 4h after endotoxin. These data suggested that POF prevents endotoxin effects and, therefore, may be of benefit in gram-negative septic shock and related disorders.

In order to evaluate the pharmacological effect of POF in humans in more detail, a study was performed on ten healthy volunteers under controlled conditions of endotoxinemia (Zabel et al. 1989a). It was first shown that a bolus injection of endotoxin (100 ng of *S. abortus equi* LPS) caused a transient increase of circulating TNF (Fig. 11) and IL-6 levels. The levels of IL-6 were maximal 3h after endotoxin injection and they correlated well with the rise and decline of body temperature values in time. Three weeks later, these volunteers again received endotoxin (100 ng), but at this time POF was also infused (500 mg over 4h). The most notable result was that, due to POF administration, there was no significant rise in circulating TNF levels (Fig. 11). In contrast, IL-6 levels were comparable to those seen in the first part of the study, and again rose and declined in parallel with body temperature. Furthermore, the endotoxin-induced initial leukocytopenia, caused by adherence of activated leukocytes in the microcirculation, was counteracted by POF (Zabel et al. 1989b). In this and other studies up to now, a total of 22 healthy volunteers have received different doses of endotoxin (30, 50, or 100 ng), their circulating leukocytes being counted at different time intervals before and after LPS administration. Interestingly, the initial leukocytopenia (2h post-endotoxin) was most pronounced after low LPS dosage, whereas with higher LPS doses leukocytosis was increasingly dominant, superimposing the initial drop in leukocyte counts. Our combined results show that treatment with POF

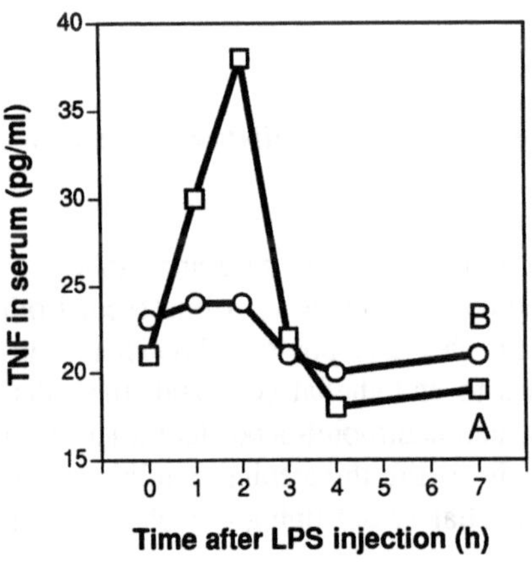

Fig. 11. Effect of pentoxifylline on circulating TNF levels in healthy volunteers after administration of *S. abortus equi* LPS. *A* Healthy human volunteers received LPS (100 ng), and TNF levels in serum were determined (□–□). *B* LPS was administered to volunteers together with pentoxifylline and serum TNF levels were estimated (○–○). TNF levels were estimated by ELISA

(500 mg over 4h) interferes with the activation of neutrophils by LPS, inhibiting the initial decrease in circulating leukocytes, whereas the IL-6 mediated leukocytosis (4h post-endotoxin) remains unaffected.

In studies addressing the mechanism by which POF exerts its beneficial effects, it was found that this pharmaceutical drug reduced, in macrophage cultures *in vitro*, the formation of TNF-specific mRNA (Strieter et al. 1982), and (in mice) the production of circulating TNF (Schade 1990). It was also found that POF selectively inhibits the formation of TNF but not IL-6, most likely by inhibiting phosphodiesterase activity and, thus, causing accumulation of intracellular cAMP (Tannenbaum and Hamilton 1989). In addition to blocking TNF production, POF also interferes with TNF activity. Thus, POF has been shown to induce prostacyclin (PGI_2) synthesis in endothelial cells (Schade 1989). It is, therefore, likely that POF also counteracts the TNF-mediated adherence of neutrophils to vascular endothelium by inducing the production of PGI_2.

In summary, it appears that protection against endotoxin shock provided by POF is based on two independent mechanisms: POF selectively inhibits TNF formation without affecting the production of other cytokines, such as IL-6. In addition, POF counteracts TNF activity such as TNF-mediated adhesion of leukocytes to endothelial cells. Our data show that the very encouraging findings with POF in animal shock models may also apply to human beings. Although controlled endotoxinaemia in healthy volunteers differs from septic shock and related conditions, it appears worthwhile and promising to investigate whether POF exhibits beneficial effects also in such clinical settings.

4 Concluding Remarks

Great progress has been made in recent years in our understanding of the pathogenesis of local and generalized inflammatory reactions induced by gram-negative bacteria and, in particular, of the role of endotoxins in such processes. Thus lipid A was identified as the endotoxic principle of LPS, its primary structure was elucidated, it was chemically synthesized, and quantitative relationships between its structure and bioactivity were established. In addition, the *in vivo* mechanisms of endotoxin involving mononuclear cells and monocyte-derived bioactive peptide and lipid mediators as well as activated oxygen and nitroxide species, and the intracellular regulation of mediator production have been studied in some detail. Finally, the molecular basis of the primary steps of the interaction between endotoxin and the host cell, which involves humoral and cellular endotoxin binding proteins and receptors, and which initiates the cascade of events leading to endotoxic manifestations, is presently being characterized at the molecular level. Some of the pertinent findings have been summarized in this review.

Of present worldwide concern is the recognition that the incidence of gram-negative septicemia, which represents the leading cause of morbidity and mortality in hospitalized patients, is steadily increasing (for literature, see Pollack 1990). Only in the United States of America, 425 000 cases of bacterial septicemia were reported in

1987. In the group of gram-negative bacteremia, the fatality rate is in the order of 30% and, thus, particularly high. This figure has remained unchanged over the past years despite improvements in antibiotics and critical care therapy.

In view of the limitations encountered with the presently available therapeutic possibilities, new strategies must be envisaged in the fight against gram-negative sepsis, in general, and against those bacterial components responsible for many of the manifestations of the septic shock syndrome such as endotoxin. Due to the progress made in the molecular characterization of the bacterial and host cell structures involved in infection-caused inflammation, the mode of interaction of these structures, and its biological consequences, the design of such new strategies is, in fact, realistic today. Thus, worldwide efforts are being made in laboratories of academia and industry to generate monoclonal antiendotoxin antibodies which ideally cross-react with LPS of different bacterial genera and cross-neutralize endotoxin effects. Epitopes for such antibodies are assumed to be localized in the phylogenetically conserved LPS domain comprising the inner core and the lipid A component (see Sects. 2.2 and 2.3). Such antibodies have been developed (Ziegler et al. 1991), but it is not unanimously accepted that they exhibit cross-reactivity and favorable effects (Baumgartner et al. 1990). Obviously, further studies are indicated and mandatory in view of the clinical need for efficient and cross-reactive anti-LPS antibodies. In addition to immunological approaches in the control of endotoxicosis, also pharmacological strategies are now conceivable on a rational basis. As the data presented in this work suggested, three new therapeutical concepts could be envisaged: (1) lipid A partial structures such as precursor Ia (compound 406) are likely to antagonize LPS activity at the cellular receptor level; (2) lipoxygenase inhibitors suppress the formation of toxic mediators such as TNF; and, finally, (3) certain pharmaceutical drugs such as pentoxifylline counteract both the production as well as the action of TNF. None of these agents has entered yet a prospective and controlled clinical trial involving patients suffering from severe gram-negative infection and associated inflammatory syndromes. Thus, it is too early to speculate on the therapeutical value of these pharmaceutical drugs. It is, however, important to note that the selection and definition of these agents as potential antiendotoxic and antiinflammatory compounds were made possible through basic research which, initially, had no apparent bearing on clinically relevant topics. Thus, the elucidation of the chemical structure of lipid A was a prerequisite before the synthesis of antagonistic partial structures, such as precursor Ia, could be designed. Likewise, acquisition of knowledge on the molecular biochemistry of macrophage activation involving the metabolism of linoleic acid was essential to select agents of potential therapeutic value such as 15-lipoxygenase inhibitors, which may specifically interfere with TNF production and, hence, generalized inflammatory reactions.

Bacterial endotoxins, because of their complex structure, their significance for bacterial viability, and their important role in gram-negative infection and sepsis, have fascinated for several decades, researchers in the fields of chemistry, physics, biology, immunology, genetics, microbiology, pharmacology, and clinical medicine. It was through the efforts of these scientists that we dispose of deep knowledge on various aspects on the molecular chemistry and biology of endotoxin. It is to be hoped that advantage will be taken of the knowledge acquired in approaching the immunological as well as pharmacological control of endotoxicosis and associated local and generalized inflammatory reactions.

References

Baumgartner JD, Heumann D, Gerain J, Weinbreck P, Grau GE & Glauser MP (1990) Association between protective efficacy of anti-lipopolysaccharide (LPS) antibodies and suppression of LPS-induced tumor necrosis factor α and interleukin 6. Comparison of O side chain-specific antibodies with core LPS antibodies. J Exp Med 171:889–896

Brade L, Bessler W & Brade H (1988a) Mitogenic activities of synthetic *Escherichia coli* lipid A and a synthetic partial structure (tripalmitoyl pentapeptide) of *E. coli* lipoprotein. Infect Immun 56:1382–1384

Brade H, Brade L & Rietschel ETh (1988b) Structure-activity relationships of bacterial lipopolysaccharides (endotoxins). Current and future aspects. Zentralbl Bakt Hyg 268:151–179

Cavaillon JM & Haeffner-Cavaillon N (1990) Signals involved in interleukin 1 synthesis and release by lipopolysaccharide-stimulated monocytes/macrophages. Cytokine 2:313–329

Dinarello CA, Bishai J, Rosenwasser LJ & Coceani F (1984) The influence of lipoxygenase inhibitors on the *in vitro* production of human leukocyte pyrogen and lymphocyte activating factor (interleukin 1). Int J Immunopharmacol 6:43–50

Feist W, Ulmer AJ, Musehold J, Brade H, Kusumoto S & Flad HD (1989) Induction of tumor necrosis factor alpha-release by lipopolysaccharide and defined lipopolysaccharide partial structures. Immunobiology 179:293–307

Feist W, Ulmer AJ, Wang MH, Musehold J, Schlüter C, Gerdes J, Herzbeck H, Brade H, Kusumoto S, Diamantstein T, Rietschel ETh & Flad HD (1991) Modulation of lipopolysaccharide-induced production of tumor necrosis factor, interleukin 1, and interleukin 6 by synthetic precursor Ia of lipid A FEMS Microbiol Immunol, in press

Freudenberg MA, Keppler D & Galanos C (1986) Requirement for lipopolysaccharide-responsive macrophages in D-galactosamine-induced sensitization to endotoxin. Infect Immun 51:891–895

Galanos C, Lüderitz O & Westphal O (1969) A new method for the extraction of R lipopolysaccharides. Eur J Biochem 9:245–249

Galanos C, Lüderitz O, Rietschel ETh & Westphal O (1977) Newer aspects of the chemistry and biology of bacterial lipopolysaccharides, with special reference to their lipid A component. In: Goodwin TW (ed) International review of biochemistry, biochemistry of lipids vol. 14 II. University Park Press, Baltimore pp 239–335

Galanos C, Lehmann V, Lüderitz O, Rietschel ETh, Westphal O, Brade H, Brade L, Freudenberg MA, Hansen-Hagge T, Lüderitz T, McKenzie G, Schade U, Strittmatter W, Tanamoto K, Zähringer U, Imoto M, Yoshimura H, Yamamoto M, Shimamoto T, Kusumoto S & Shiba T (1984) Endotoxic properties of chemically synthesized lipid A part structures: comparison of synthetic lipid A precursor and synthetic analogues with biosynthetic lipid A precursor and free lipid A. Eur J Biochem 140:221–227

Galanos C, Lüderitz O, Rietschel ETh, Westphal O, Brade H, Brade L, Freudenberg MA, Schade U, Imoto M, Yoshimura S, Kusumoto S & Shiba T (1985) Synthetic and natural *Escherichia* coli free lipid A express identical endotoxic activities. Eur J Biochem 148:1–5

Golenbrock DT, Hampton RY, Raetz CRH & Wright SD (1990) Human phagocytes have multiple lipid A-binding sites. Infect Immun 58:4069–4075

Griffis JM, Schneider H, Mandrell RE, Yamasaki R, Jarvis GA, Kim JJ, Gibson BW, Hamadeh R & Apicella MA (1988) Lipooligosaccharides: the principal glycolipids of the neisserial outer membrane. Rev Infect Dis 10:S287–S295

Hase S & Rietschel ETh (1977) The chemical structure of the lipid A component of lipopolysaccharides from *Chromobacterium violaceum* NCTC 9694. Eur J Biochem 75:23–34

Helander I, Lindner B, Brade H, Altmann K, Lindberg BB, Rietschel ETh & Zähringer U (1988) Chemical structure of the lipopolysaccharide of *Haemophilus influenzae* strain I-69 Rd^{-}/b^{+}. Description of a novel deep rough chemotype. Eur J Biochem 177:483–492

Hitchcock P, Leive L, Mäkelä PH, Rietschel ETh, Strittmatter W & Morrison D (1986) Lipopolysaccharide nomenclature – past, present and future. J Bacteriol 166:699–705

Hollingsworth RI & Lill Elghanian DA (1989) Isolation and characterization of the unusual lipopolysaccharide component, 2-amino-2-deoxy-2-N-(27-hydroxyoctacosanoyl)-3-O-(3-hydroxytetradecanoyl)-*gluco*-hexuronic acid, and its de-O-acylation product from the free lipid A of *Rhizobium trifolii* ANU 843. J Biol Chem 264:14039–14042

Jann K & Jann B (1984) Structure and biosynthesis of O-antigens. In: Proctor RA & Rietschel ETH (eds) Handbook of endotoxin, vol 1. Chemistry of endotoxin. Elsevier, Amsterdam, pp 138–186

Jiao B, Freudenberg MA & Galanos C (1989) Characterization of the lipid A component of genuine smooth-form lipopolysaccharide. Eur J Biochem 180:515–518

Keppler D, Hagmann W & Rapp S (1987) Role of leukotrienes in endotoxin action *in vivo*. Rev Infect Dis 9:S580–S584

Kirikae T, Kirikae F, Schade FU, Yoshida M, Kondo S, Hisatsune K, Nishikawa SI & Rietschel ETh (1991) Detection of lipopolysaccharide binding proteins on membranes of murine lymphocyte and macrophage-like cell lines. FEMS Microbiol Immunol (in Press)

Kovach NL, Yee E, Munford RS, Raetz CRH & Harlan JM (1990) Lipid IV_A inhibits synthesis and release of tumor necrosis factor induced by lipopolysaccharide in human whole blood *ex vivo*. J Exp Med 172:77–84

Kusumoto S, Kusunose N, Imoto M, Shimamoto T, Kamikawa T, Takada H, Kotani S, Rietschel ETh & Shiba T (1989) Synthesis and biological function of bacterial endotoxin. Pure Appl Chem 61:461–464

Kusumoto S, Kamikawa T, Kurosawa M, Fukase K, Shiba T, Kusama T, Soga T, Shioya E, Nakayama K, Nakajima H, Osada Y & Ono Y (1990) New results in the synthesis of lipid A and endotoxins. In: Nowotny A, Spitzer JJ & Ziegler EJ (eds) Endotoxin research series vol. 1. Cellular and molecular aspects of endotoxin reactions. Elsevier, Amsterdam, pp 41–50

Kulshin VA, Zähringer U, Lindner B, Jäger KE, Dmitriev B & Rietschel ETh (1991) Structural characterization of the lipid A component of *Pseudonomas aeruginosa* lipopolysaccharides. Eur J Biochem 198:697–704

Kulshin VA, Zähringer U, Lindner B, Frasch C, Dmitriev B & Rietschel ETh Structural analysis of the lipid A component of *Neisseria meningitidis*. J Bacteriol (submitted)

Loppnow H, Brade H, Dürrbaum I, Dinarello CA, Kusumoto S, Rietschel ETh & Flad HD (1989) Interleukin 1 induction-capacity of defined lipopolysaccharide partial structures. J Immunol 142:3229–3238

Loppnow H, Libby P, Freudenberg MA, Krauss JH, Weckesser J & Mayer H (1990) Cytokine induction by lipopolysaccharide (LPS) corresponds to lethal toxicity and is inhibited by nontoxic *Rhodobacter capsulatus* LPS. Infect Immun 58:3743–3750

Lüderitz O, Freudenberg MA, Galanos C, Lehmann V, Rietschel ETh & Shaw DH (1982) Lipopolysaccharides of gram-negative bacteria. In: Razin S & Rottem S (eds) Microbial membrane lipids vol. 17. Academic Press, New York, pp 79–151

Lüderitz TH, Brandenburg K, Seydel U, Roth A, Galanos C & Rietschel ETh (1989) Structural and physiochemical requirements of endotoxins for the activation of arachidonic acid metabolism in mouse peritoneal macrophages *in vitro*. Eur J Biochem 179:11–16

Mäkelä PH & Stocker BAD (1984) Genetics of lipopolysaccharide. In: Proctor RA & Rietschel Th (eds) Handbook of endotoxin vol. 1. Chemistry of endotoxin. Elsevier, Amsterdam, pp 59–137

Mäkelä PH, Hovi M, Saxen H, Muotiala A & Rhen M (1990) Role of LPS in the pathogenesis of salmonellosis. In: Nowotny A, Spitzer JY & Ziegler EJ (eds) Endotoxin research series, vol. 1. Cellular and molecular aspects of endotoxin reactions. Elsevier, Amsterdam, pp 537–546

Mayer H, Bhat RU, Masoud H, Radziejewska-Lebrecht J, Widemann C & Krauss JH (1989) Bacterial lipopolysaccharides. Pure Appl Chem 61:1271–1282

Moran AP, Zähringer U, Seydel U, Scholz D, Stütz P & Rietschel ETh (1991) Structural analysis of the lipid A component of *Campylobacter jejuni* CCUG 10936 (serotype 0:2) lipopolysaccharides. Description of a lipid A containing a hybrid backbone of 2-amino-2-deoxy-*D*-glucose and 2, 3-diamino-2, 3-dideoxy-*D*-glucose. Eur J Biochem 198:459–469

Morrison DC (1989) The case for specific lipopolysaccharide receptors expressed on mammalian cells. Microb Pathogen 7:389–398

Morrison DC (1990) Diversity of mammalian macromolecules which bind to bacterial lipopolysaccharides. In: Nowotny A, Spitzer JY & Ziegler EJ (eds) Endotoxin research series, vol. 1. Cellular and molecular aspects of endotoxin reactions. Elsevier, Amsterdam, pp 183–189

Morrison DC & Ryan JL (1987) Endotoxins and disease mechanisms. Ann Rev Med 38:417–432

Pohlmann TH, Munford RS & Harlan JM (1987) Deacylated lipopolysaccharides inhibits neutrophil adherence to endothelium induced by lipopolysaccharide *in vitro*. J Exp Med 165:1393–1402

Pollack M (1990) New therapeutic strategies in gram-negative sepsis and septic shock based on molecular mechanisms of pathogenesis. In: Mandell, Douglas & Bennet (eds) Principles and practice of infection diseases update 8. Churchill-Livingstone, New York, pp 1–18

Qureshi N, Takayama K & Ribi E (1982) Purification and structural determination of nontoxic lipid A obtained from the lipopolysaccharide of *Salmonella typhimurium*. J Biol Chem 257:11808–11815

Qureshi N, Takayama K, Meyer KC, Kirkland ThN, Bush CA, Chen L, Wang R & Cotter RJ (1991) Chemical reduction of 3-oxo and unsaturated groups in fatty acids of diphosphoryl lipid A from the lipopolysaccharide of *Rhodopseudomonas sphaeroides*. J Biol Chem 266:6532–6538

Riedo FX, Munford RS, Campbell WB, Reisch JS, Chien KR & Gerard RD (1990) Deacylated lipopolysaccharide inhibits plasminogen activator inhibitor-1, prostacyclin, and prostaglandin E_2 induction by lipopolysaccharide but not by tumor necrosis factor-α J Immunol 144:3506–3512

Rietschel ETh, Galanos C, Lüderitz O & Westphal O (1982) Chemical structure, physiological function and biological activity of lipopolysaccharides and their lipid A component. In: Webb D (ed) Immunopharmacology and the regulation of leukocyte function. Marcel Dekker, New York, pp 183–229

Rietschel ETh, Wollenweber HW, Brade H, Zähringer U, Lindner B, Seydel U, Bradaczek H, Barnickel G, Labischinski H & Giesbrecht P (1984) Structure and conformation of the lipid A component of lipopolysaccharides. In: Proctor R & Rietschel ETh (eds) Handbook of endotoxin, vol 1. Chemistry of endotoxin Elsevier, Amsterdam, pp 187-246

Rietschel ETh, Brade H, Brade L, Brandenburg K, Schade U, Seydel U, Zähringer U, Galanos C, Lüderitz O, Westphal O, Labischinski H, Kusomoto S & Shiba T (1987) Lipid A, the endotoxic center of bacterial lipopolysaccharides: relation of chemical structure to biological activity. Prog Clin Biol Res 231:25–53

Rietschel ETh, Brade L, Schade UF, Seydel U, Zähringer U & Brade H (1988) Bacterial endotoxins: properties and structure of biologically active domains. In: Schrinner E, Richmond M, Seibert G & Schwarz U (eds) Surface structures of microorganisms and their interaction with the mammalian host. Verlag Chemie, Weinheim, FRG, pp 1-41

Rietschel ETh, Brade L, Schade U, Seydel U, Zähringer U, Brandenburg K, Helander I, Holst O, Kondo S, Kuhn HM, Lindner B, Röhrscheidt E, Russa R, Labischinski H, Naumann D & Brade H (1990) Bacterial lipopolysaccharides: relationship of structure and conformation to endotoxic activity, serological specificity and biological function. Adv Exp Med Biol 256:81–99

Schade FU (1986) Involvement of lipoxygenase in the activation of mouse peritoneal macrophages by endotoxin. Biochem Biophys Res Commun 138:842–849

Schade FU (1989) The role of prostacyclin in the protective effects of pentoxifylline and other xanthine derivatives in endotoxin action in mice. Eicosanoids 2:183–188

Schade FU (1990) Pentoxifylline increases survival in murine endotoxin shock and decreases formation of tumor necrosis factor. Circulatory Shock 31:171–181

Schade FU, Burmeister I & Engel R (1987) Increased 13-hydroxyoctadecadienoic acid content in lipopolysaccharide-stimulated macrophages Biochem Biophys Res Commun 147:695–700

Schade FU, Burmeister I, Elekes E, Engel R & Wolter DT (1989a) Mononuclear phagocytes and eicosanoids: aspects of their synthesis and biological activities. Blut 59:475–485

Schade FU, Ernst M, Reinke, M & Wolter DT (1989b) Lipoxygenase inhibitors suppress formation of tumor necrosis factor in vitro and in vivo. Biochem Biophys Res Commun 159:748–754

Schade FU, Engel R & Jacobs D (1991) Differential protective activities of site-specific lipoxygenase inhibitors in endotoxic shock and production of tumor necrosis factor. Int J Immunopharmacol, in press

Schönharting MM & Schade FU (1989) The effect of pentoxifylline in septic shock – new pharmacological aspects of an established drug. J Med 20:97–105

Schumann RR, Leong SR, Flaggs GW, Gray PW, Wright SD, Mathison JC, Tobias PS & Ulevitch RJ (1990) Structure and function of lipopolysaccharide binding protein. Science 249:1429–1431

Strieter RM, Rennick DG, Ward PA, Spengler RN, Lynch III JP, Larick S & Kunkel SL (1988) Cellular and molecular regulation of tumor necrosis factor-α production by pentoxifylline. Biochem Biophys Res Commun 155:1230–1236

Stütz PL, Aschauer H, Hildebrandt J, Lam C, Loibner H, Macher I, Scholz D, Schütze E & Vyplel H (1990) Chemical synthesis of endotoxin analogs and some structure activity relationships: In: Nowotny A, Spitzer JY & Ziegler EJ (eds) Endotoxin research series, vol. 1. Cellular and molecular aspects of endotoxin reactions. Elsevier, Amsterdam, pp 129–144

Tahri-Jouti MA, Mondange M, Le Dur A, Auzanneau FI, Charon D, Girard R & Chaby R (1990) Specific binding of lipopolysaccharides to mouse macrophages. II Involvement of distinct lipid A substructures. Mol Immunol 27:763–770

Takayama K, Qureshi N, Hyver K, Honovich J, Cotter RJ, Mascagni P & Schneider H (1986) Characterization of a structural series of lipid A obtained from the lipopolysaccharides of *Neisseria gonorrhoeae*. J Biol Chem 261:10624–10631

Takayama K, Qureshi N, Beutler B & Kirkland TN (1989) Diphosphoryl lipid A from *Rhodopseudomonas sphaeroides* ATCC 17023 blocks induction of cachectin in macrophages by lipopolysaccharide. Infect Immun 57:1336–1338

Tannenbaum CS & Hamilton TA (1989) Lipopolysaccharide-induced gene expression in murine peritoneal macrophages is selectively suppressed by agents that elevate intracellular cAMP. J Immunol 142:1274–1280

Vogel SN & Hogan MM (1990) The role of cytokines in endotoxin-mediated host responses. In: Oppenheim JJ & Shevack EM (eds) Immunopharmacology – the role of cells and cytokines in immunity and inflammation. Oxford University Press, New York, pp 238–258

Wang MH, Feist W, Herzbeck H, Brade H, Kusumoto S, Rietschel ETh, Flad HD & Ulmer AJ (1990) Suppressive effect of lipid A partial structures on lipopolysaccharide or lipid A-induced release of interleukin 1 by human monocytes. FEMS Microb Immunol 64:179–186

Weintraub A, Zähringer U, Wollenweber HW, Seydel U & Rietschel ETh (1989) Structural characterization of the lipid A component of *Bacteroides fragilis* strain NCTC 9343 lipopolysaccharide. Eur Biochem 183: 425–431

Welsh CH, Lien D, Worthen GS & Weil JV (1988) Pentoxifylline decreases endotoxin-induced pulmonary neutrophil suquestration and extravascular protein accumulation in the dog. Am Rev Respir Dis 138:1106–1114

Westphal O & Lüderitz O (1954) Chemische Erforschung von Lipopolysacchariden Gram-negativer Bakterien. Angew Chem 66:407–417

Westphal O, Lüderitz O & Bister F (1952) Über die Extraktion von Bakterien mit Phenol/Wasser. Z Naturforsch B 7B:148–155

Wollenweber HW, Seydel U, Lindner B, Lüderitz O & Rietschel ETh (1984) Nature and location of amide-bound (R)-3-acyloxyacyl groups in lipid A of lipopolysaccharides from various gram-negative bacteria. Eur J Biochem 145:265–272

Wright SD, Ramos RA, Tobias PS, Ulevitch RJ & Mathison JC (1990) CD14, a receptor for complexes of lipopolysaccharide (LPS) and LPS binding protein. Science 249:1431–1433

Zabel P, Wolter DT, Schönharting MM & Schade FU (1989a) Oxpentifylline in endotoxaemia Lancet 1474-1477

Zabel P & Schade FU (1989b) Endotoxin-induced sticking of leukocytes in the microcirculation and its reversal by pentoxifylline. Clin Hemorheol 9:A534

Zähringer U, Lindner B, Seydel U, Rietschel ETh, Naoki H, Unger FM, Imoto M, Kusumoto S & Shiba T (1985) Structure of de-O-acylated lipopolysaccharide from the *Escherichia coli* Re mutant strain F515. Tetrahedron Lett 26:6321–6324

Ziegler EJ, Fisher CJ, Sprung CL, Straube RC, Sadoff JC, Foulke GE, Wortel CH, Fink MP, Dellinger RP, Teng NNH, Allen IE, Berger HJ, Knatterud GL, LoBuglio AF, Smith CR & the HA-1A Sepsis Study Group (1991) Treatment of gram-negative bacteremia and septic shock with HA-1A human monoclonal antibody against endotoxin, A randomized, double-blind, placebo-controlled trial. N Engl J Med 324:429–436

Bacterial Exotoxins and Acute Lung Failure

W. SEEGER[1], F. GRIMMINGER[1], D. WALMRATH[1], N. SUTTORP[1], and S. BHAKDI[2]

1 Introduction

The adult respiratory distress syndrome (ARDS), developing under conditions of sepsis, shock, and severe polytrauma, is characterized by increased pulmonary artery pressure with concomitant perfusion inhomogeneities and increased permeability of lung endothelial and epithelial membranes (Fig. 1). The permeability increase results in the formation of protein-rich interstitial and alveolar edema. Ventilation perfusion mismatch, shunt flow, and diffusion impairment cause severe disturbances of gas exchange.

Bacterial sepsis is the most consistent factor associated with the development of ARDS [1–3]. Moreover, acute diffuse lung injury of different etiology is often complicated by nosocomial pneumonia [4, 5]. In experimental models, septic lung failure is mimicked by the application of lipopolysaccharide components of gram-negative bacteria (endotoxins), and abundant detailed information is available on the mecha-

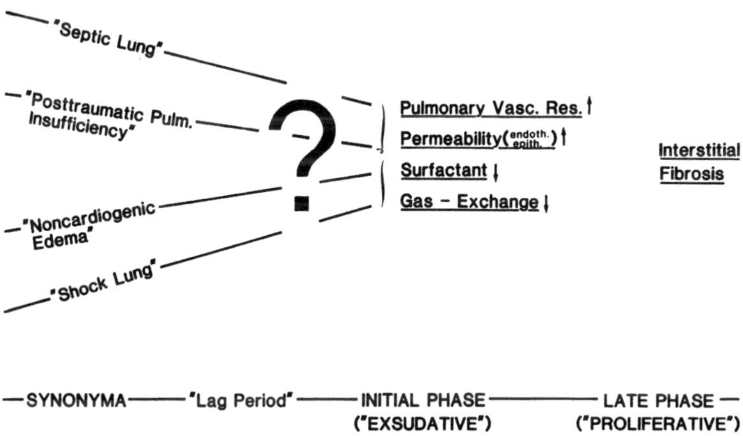

Fig. 1. Basic features of ARDS. A variety of catastrophic events can lead to the acute respiratory distress syndrome of the adult (ARDS). Its central pathophysiological characteristics include an increase in pulmonary vascular resistance with maldistribution of blood flow, increased permeability of the capillary endothelial and the alveolar epithelial membranes, disturbances of the alveolar surfactant system, and severe deterioration of gas exchange. The initial exudative phase may progress to a late proliferative phase with marked interstitial fibrosis within a few weeks. Underlying pathophysiological events are still unknown to a large extent

[1] Klinikum der Justus-Liebig-Universität Medizinische Klinik I, 6300 Gießen, FRG
[2] Klinikum der Gutenberg-Universität, Med. Mikrobiologie, 6500 Mainz, FRG

42. Colloquium Mosbach 1991
Molecular Aspects of Inflammation

nisms of lung injury by these agents. There is evidence for a critical role of circulating inflammatory cells and activation of several humoral mediator systems [complement, arachidonic acid (AA) metabolites, reactive oxygen species, proteases] has been documented [6, 7].

Only very little information is available on abnormalities inflicted by "nonendotoxin toxins" in the pulmonary circulation. A variety of bacteria are capable of liberating membrane-perturbating exotoxins. These include *Staphylococcus aureus, Pseudomonas aeruginosa, Escherichia coli, Morganella morganii, Proteus* spp., *Pasteurella haemolytica,* and others. Several of these bacterial agents have been shown to generate discrete transmembrane aqueous pores in cellular membranes of erythrocytes, but also of nucleated target cells. We hypothesized that these toxins might induce marked inflammatory mediator generation, thereby eliciting pulmonary circulatory disorders and disturbances of gas exchange. The subsequent elucidation of studies from our laboratory will concentrate on the microcirculatory effects of *Escherichia coli* hemolysin.

Certain strains of *Escherichia coli* produce a proteinaceous exotoxin, designated hemolysin (Hly), which was originally characterized by its lytic effect on erythrocytes [8–11]. Secretion of active hemolysin requires the expression of four genes (hlyA, hlyB, hlyC, hlyD), constituting the hly operon [12, 13]. The hlyA gene encodes the 110-kDa hemolysin protein, which is activated by the hlyC gene product. Secretion of HlyA occurs via a transport mechanism involving the HlyB and HlyD proteins. Approximately 50% of *E. coli* isolates causing extraintestinal infections in man produce Hly; this agent represents one of the most prevalent bacterial exotoxins encountered in humans [8, 10, 11, 14–16]. Its pathogenetic relevance has been established in animal models by using genetically engineered bacterial strains [8, 14, 17]. An analogous role in human infections has been inferred from the high association of hemolysin production with disease, including pyelonephritis and septicemia [15, 16].

The primary structure of Hly has been deduced from the nucleotide sequence [18]. The membrane permeabilizing capacities of Hly were first demonstrated in erythrocytes, artificial lipid membranes, and neutrophils [9, 19–21]. Unlike the pores formed by membrane insertion of oligomeric aggregates of gram-positive exotoxins, such as staphylococcal alpha toxin, membrane permeabilization by Hly appears to be exerted by monomers. A high affinity calcium-binding domain, required for toxin membrane insertion, has been identified [22–24].

2 Vasoconstrictor Potency of Hly in the Pulmonary Circulation – The Role of Prostanoid Generation

To assess this issue, isolated hemolysin as well as Hly liberating viable *E. coli* bacteria were infused in a model of isolated, perfused, and ventilated rabbit lungs. The model is illustrated in Fig. 2. In the first study [25], direct intravascular application of purified Hly was performed. Due to a glycerol density gradient centrifugation step, the contaminating lipopolysaccharide concentration of the Hly preparation was reduced to about 3 ng LPS/μg protein. One hemolytic unit of these bacterial products

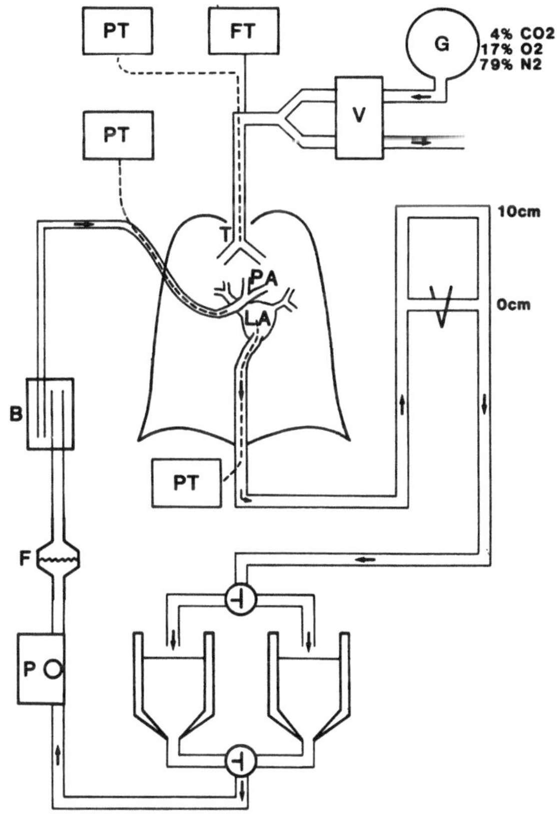

Fig. 2. Schematic representation of the isolated perfused lung model. Lungs are perfused with Krebs Henseleit buffer, including albumin or hydroxyethylamylopectin as oncotic substance in a recirculating system. The use of two alternate perfusate reservoirs allows the repeated exchange of perfusion fluid. The perfusion circuit includes a roller pump (*P*), a filter system (*F*), as well as a bubble trap (*B*). The 0 cm venous effluent can be clamped to induce a sudden venous pressure elevation of 10 cm H_2O, for evaluation of hydrostatic step-induced fluid filtration, which allows the calculation of the capillary filtration coefficient. The lung is ventilated (*V*) with a CO_2-containing gas mixture for maintenance of physiological pH and gas partial pressures. Pressure transducers in the pulmonary artery (*PA*) catheter, in the trachea, and in the left atrium (*LA*) allow the continuous monitoring of perfusion pressures and ventilation pressure. The lung is freely suspended from a force transducer (*FT*) for continuous registration of organ weight. Perfusate aliquots can be taken repeatedly for analysis of inflammatory mediators. The bronchoalveolar space is accessible by bronchoalveolar lavage

approximates 100 ng protein. Low concentrations of the toxin in the perfusate (0.05–5 hemolytic u/ml) caused a dose- and time-dependent release of potassium, thromboxane A_2, and prostaglandin I_2, but not of lactate dehydrogenase, into the recirculating medium (Figs. 3–6). Concomitantly, a dose-dependent liberation of the prostanoids into the bronchoalveolar space was noted. These events were paralleled by pulmonary hypertension, and together with studies using, different inhibitors indicated that the vasoconstrictor response was mediated predominantly by pulmonary

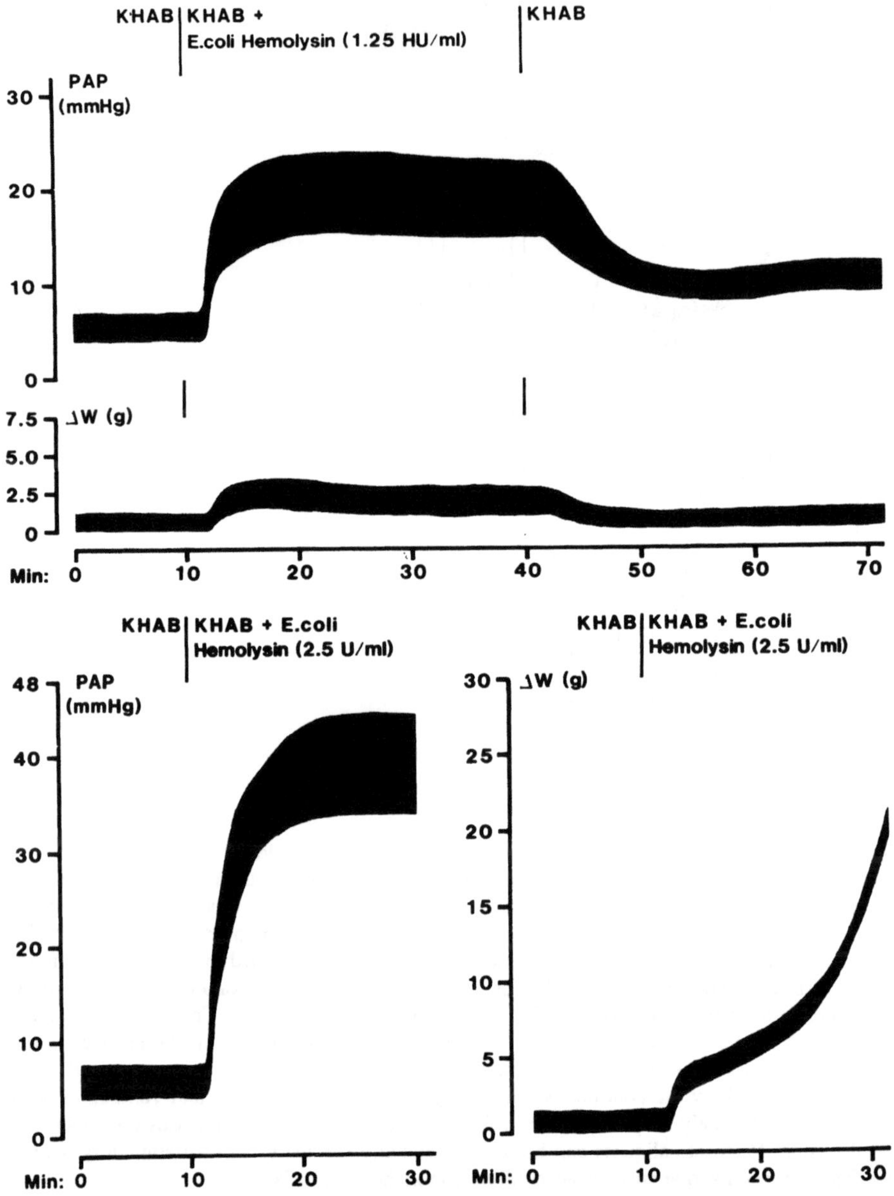

Fig. 3. *E. coli* hemolysin-evoked pressor response and edema formation in perfused rabbit lungs. The pulmonary artery pressure (*PAP*) and changes in lung weight (Δ *W*) in response to 1.25 and 2.5 HU/ml Hly are given. Experiments were performed in Krebs Henseleit buffer in the presence of albumin (*KHAB*). *Vertical lines* indicate exchange of perfusion fluid. In the absence of albumin, corresponding features of immediate pressure rise and delayed onset of lung edema formation are evoked by ≈ 1/50 of the exotoxin dose. [25]

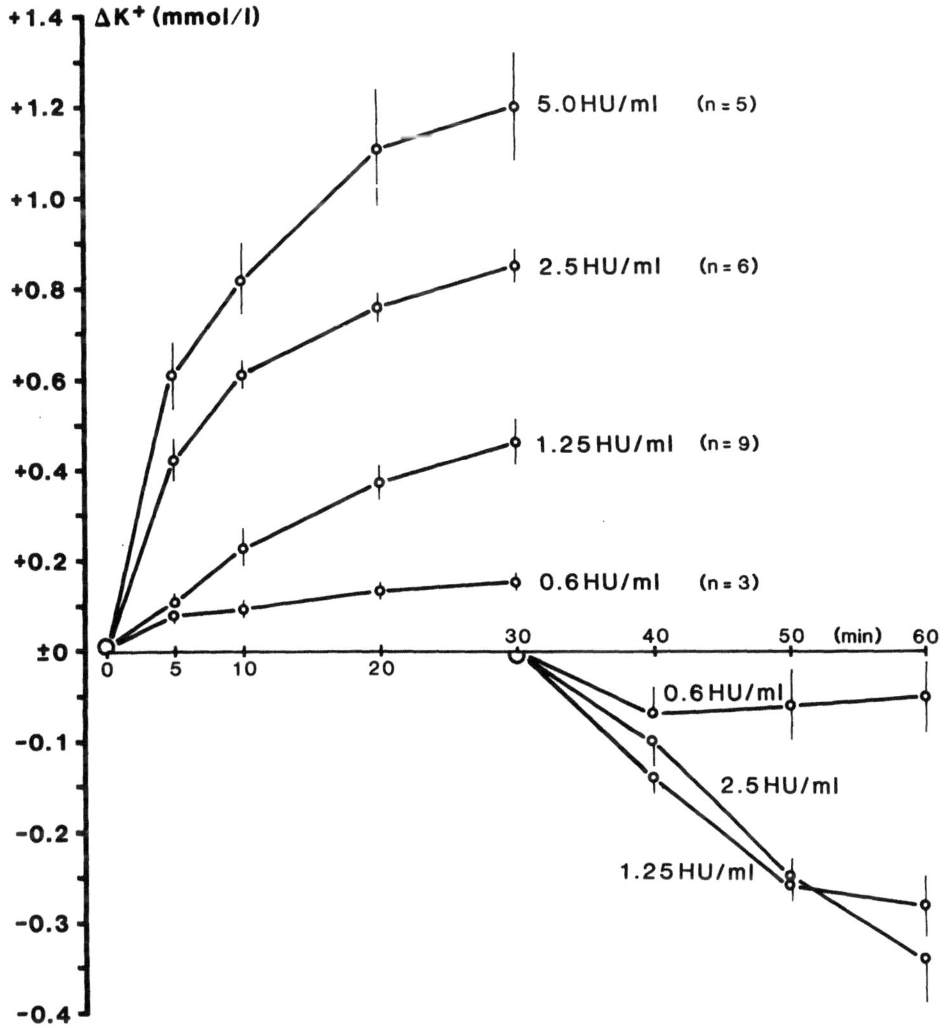

Fig. 4. Dose dependence and time course of hemolysin-induced potassium release: partial reuptake after exchange of perfusion fluid. The Krebs Henseleit buffer fluid contained 5.2 mM potassium before being recirculated. During a 30-min period of lung exposure to Hly, a time- and dose-dependent increase in the perfusate potassium concentration occurred (Δ *K*). Next, the perfusate was exchanged by a toxin-free buffer fluid, again containing 5.2 mM potassium. In the posttoxin phase, a decrease in the potassium concentration in the recirculating buffer fluid was noted, reflecting partial reuptake of this cation by lung cells after toxin washout. This second (posttoxin) phase could not be performed in lungs exposed to 5 HU Hly/ml because of too rapid edema formation even in the presence of a cyclooxygenase inhibitor. The potassium release indicates primary membrane attack by the exotoxin, independent of inflammatory mediator generation. The partial reuptake of K^+ after exchange of perfusate may suggest repair mechanisms. [25]

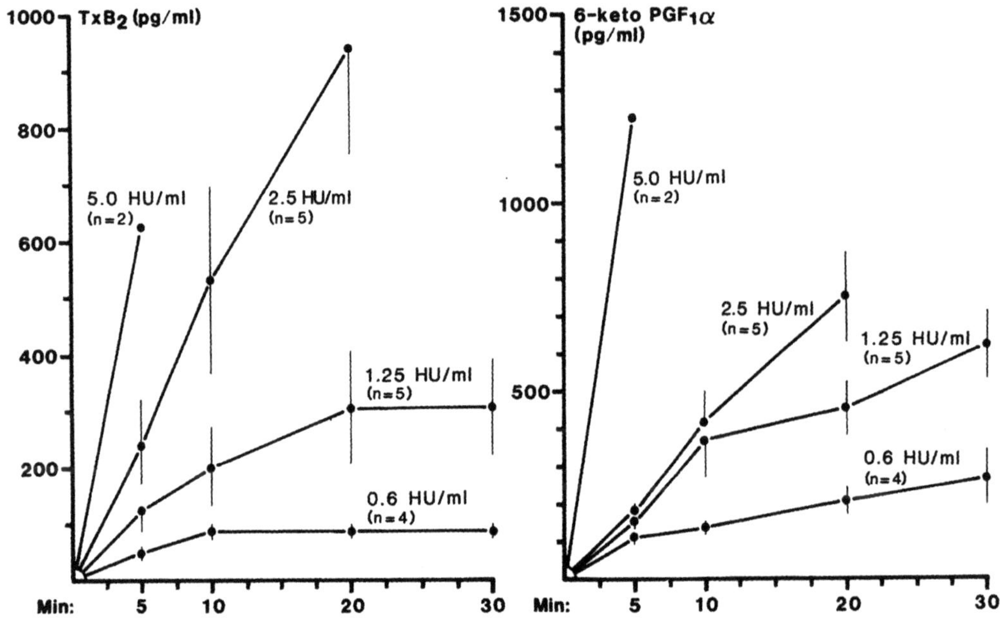

Fig. 5. Dose dependence and time course of *E. coli* hemolysin-induced prostanoid release into the perfusion fluid. The concentrations of TXB_2 and 6-keto-$PGF_{1\alpha}$, measured in aliquots of the recirculating buffer fluid, at different times after administration of the various doses of the hemolysin are given. In the studies with application of 2.5 and 5 HU/ml, the experiments had to be stopped before termination of the 30-min period because of excessive edema formation. [25]

thromboxane generation (Fig. 6). In addition, Hly elicited a protracted, dose-dependent increase in the lung capillary filtration coefficient, which was independent of the prostanoid-mediated pressor response and resulted in severe pulmonary edema formation.

In the second study [26], viable Hly-forming *E. coli* bacteria were infused in isolated rabbit lungs. In control experiments an *E. coli* strain was used which releases an inactive form of the hemolysin (lack of the HlyC gene). Thus, 10^4–10^7 Hly-forming *E. coli*/ml perfusate caused a dose- and time-dependent appearance of hemolysin in the recirculating buffer fluid, again accompanied by the release of potassium, thromboxane A_2, and prostaglandin I_2 into the buffer fluid. Concomitantly, marked pulmonary hypertension developed (Fig. 7). Inhibitor studies, using cyclooxygenase inhibitors, thromboxane synthetase inhibitors, and a thromboxane receptor antagonist, again indicated that the pressor response was predominantly mediated via pulmonary thromboxane generation. Administration of hemolysin-forming *E. coli* additionally caused a protracted, dose-dependent increase in the lung capillary filtration coefficient, followed by severe edema formation (Fig. 8). The HlyC-lacking *E. coli* strain completely failed to provoke the described biophysical and biochemical responses up to 10^7/ml bacteria. The addition of an admixture of rabbit plasma, which contains putative hemolysin-binding lipoproteins and naturally occurring antibodies, to the perfusate did not suppress the Hly-related biophysical and biochemical responses.

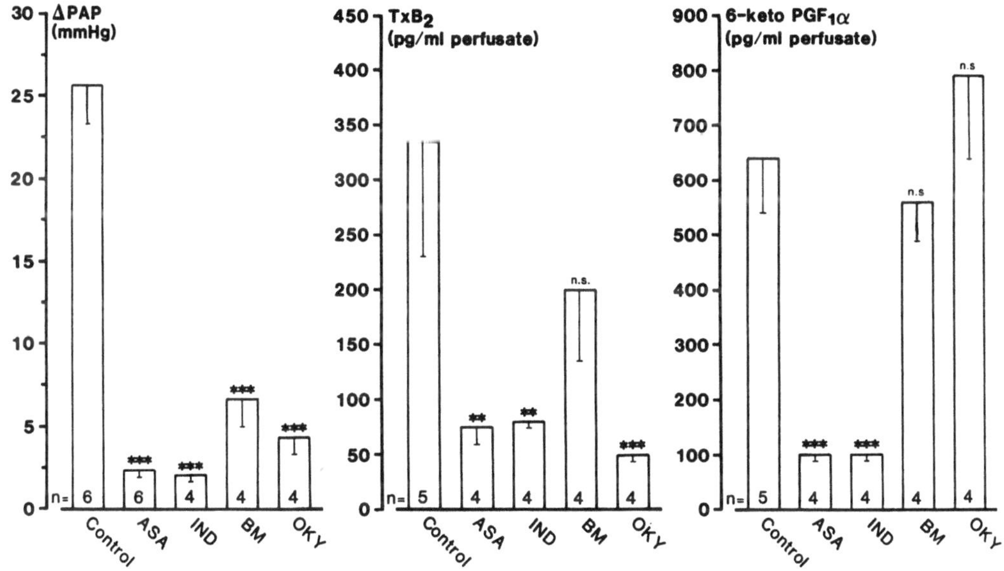

Fig. 6. Influence of different AA cascade inhibitors on *E. coli* hemolysin-induced prostanoid generation and pressor response. The lungs were exposed to 1.25 HU hemolysin/ml perfusate in the absence (control) or presence of different inhibitors of the lung AA system. The maximum values of PAP rise (Δ *PAP*) as well as TXB_2 and 6-keto-$PGF_{1\alpha}$ concentrations in the perfusion fluid are given. Data of the different inhibitor groups are compared with the control group by one-way analysis of variance. **, $p < 0.01$; ***, $p < 0.001$; *ASA* acetylsalicylic acid, 100 μM; *IND* indomethacin, 35 μM; *BM* BM 13.177, 10 μM; *OKY* OKY-046, 2.5 μM. [25]

These two studies strongly suggest that Hly is a potent vasoconstrictor in the pulmonary vascular bed. Dose-dependent release of potassium in the absence of significant LDH liberation indicates toxin-induced membrane permeabilization as a primary exotoxin effect also in this intact organ model. Secondary generation of the potent vasoconstrictive agent thromboxane A_2 occurs, which surpasses the vasodilatory effect of the simultaneously generated prostaglandin I_2 and which is predominantly responsible for the toxin-evoked pulmonary hypertension. In addition to the pressor response, prostanoid-independent vascular leakage ensues, which results in the formation of protein-rich interstitial and alveolar edema. Putative, underlying events of the toxin-induced vascular leakage will be discussed in the following sections. A high sensitivity of the pulmonary vasculature toward hemolysin was found even in the presence of plasma with putative, Hly-inactivating capacity (antibodies, lipoproteins). This finding supports the notion that rapid exotoxin binding to target cells competes with toxin neutralization. Furthermore, it is conceivable that a direct transfer of the exotoxin from target cell-adherent bacteria occurs. In such a "hidden microenvironment", plasma inactivators may be virtually ineffective. The pulmonary target cells responsible for the predominant thromboxane generation in the intact rabbit vasculature have not yet been conclusively identified. Intravascular resident, interstitial, and intraalveolar macrophages are likely candidates. Additional cell types such as alveolar epithelial cells type II and interstitial fibroblasts may, however, also be involved.

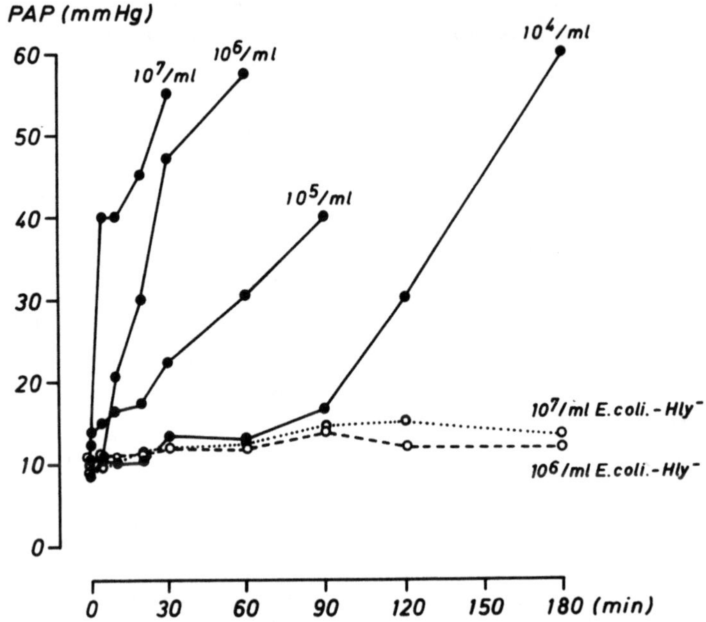

Fig. 7. Time and dose dependency of pulmonary artery pressure (*PAP*) rise evoked by Hly-forming *E. coli*. The experiments were performed in Krebs Henseleit albumin buffer-perfused lungs. Time was set to zero when the hemolysin-forming *E. coli* were administered to the buffer medium at the given final concentrations. Mean values of two to three experiments at each concentration are given. As controls, 10^6/ml and 10^7/ml bacteria of a genetically related *E. coli* strain, which secretes an inactive toxin, were used. [26]

The lung microvascular endothelium is the most likely candidate for the Hly-evoked prostaglandin I_2 generation.

3 Ventilation Perfusion Mismatch, Resulting from Exotoxin-Induced "Inflammatory" Vasoconstriction

The Hly-elicited vasoconstriction results in an inhomogeneous narrowing of the pulmonary vascular bed [27]. As a consequence, the profile of the perfusion distribution is markedly affected, resulting in severe ventilation perfusion mismatch in otherwise healthy lungs. This marked effect on perfusion distribution is exemplified in Fig. 9. Such a mismatch results in severe gas exchange disturbances even in the absence of lung edema formation and atelectasis.

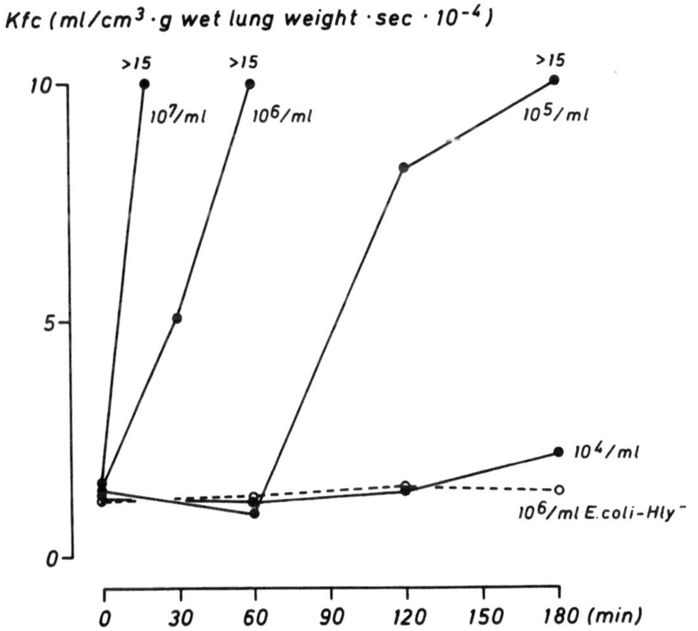

Fig. 8. Time and dose dependency of *E. coli* (Hly+)-evoked increase in the capillary filtration coefficient (*Kfc*). The experiments were performed in Krebs Henseleit albumin buffer-perfused lungs in the presence of 500 μM ASA, in order to block any major Hly-induced pressor response. Baseline Kfc-data (0') were determined directly before administration of the hemolysin-forming *E. coli* to the buffer medium at the given final concentrations. Mean values of two to three experiments at each concentration are given. As a control, 10^6/ml *E. coli* of the non-hemolysin-forming strain was used. [26]

4 Morphological Examination of Hemolysin-Exposed Rabbit Lungs

In a preceding electron microscope examination of staphylococcal α toxin-exposed rabbit lungs, pulmonary hypertension and vascular leakage were found to be paralleled by severe dose- and time-dependent structural changes. In particular, increasing electron density of microvascular endothelial cell nuclei and subsequently of their cytoplasma was noted, followed by detachment of these cells from their mutual endo-epithelial basal lamina [28]. Edema was localized in the blood-gas exchange area, in contrast to hydrostatically induced lung fluid accumulation. A recent electron- microscope examination of perfusion-fixed lungs after Hly incubation did, however, display a different morphological picture. Interstitial edema formation was again noted, but this was predominantly located in areas remote from the blood-gas exchange barrier. The endothelial cells displayed normal electron density, and were not detached from the fused basal lamina. These findings suggest alterations of functional endothelial sieving properties in Hly-treated lungs, rather than gross endothelial cell damage. There were, however, markedly increased numbers endothelial plasmalemmal vesicles, which may suggest intercompartmental fluid shifts. Such transport pro-

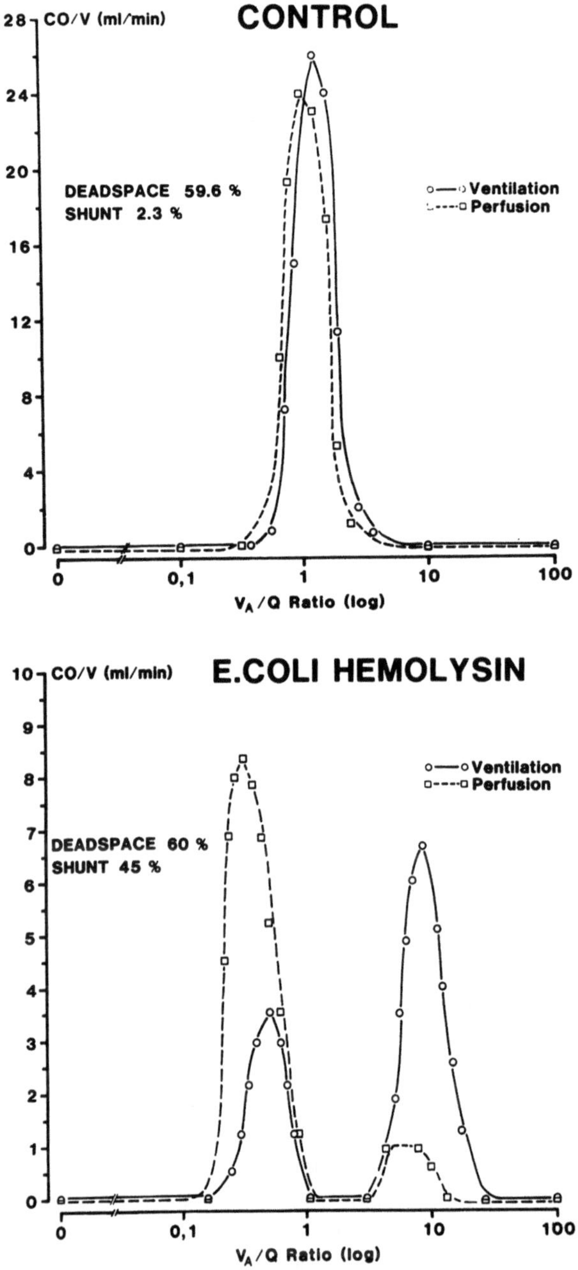

Fig. 9. Hemolysin-induced ventilation-perfusion mismatch. Ventilation (V_A)-perfusion (Q) ratios of the different lung compartments were assessed by the multiple inert gas elimination technique as described in [27]. Under baseline conditions, both ventilation and perfusion ("cardiac output", CO) are distributed nearly quantitatively to areas with V_A/Q ratios approximating 1. In the presence of hemolysin, mismatch of ventilation and perfusion occurs, and shunt flow dramatically increases.

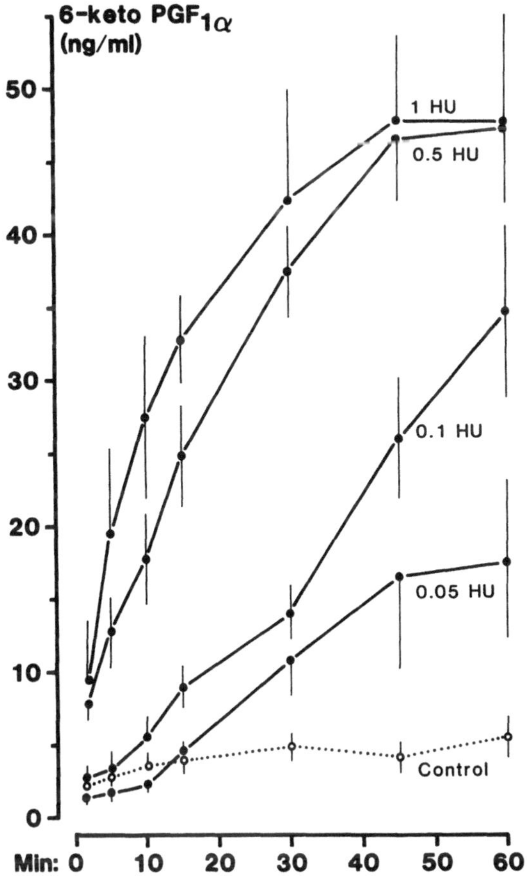

Fig. 10. Time- and dose-dependent PGI_2 synthesis evoked by Hly in cultured pulmonary artery endothelial cells. Cells were incubated with 0, 0.05, 0.1, 0.5, and 1 HU of Hly per ml for 1 to 60 min, and then the medium was taken for determination of PGI_2 levels (measured radioimmunologically as 6-keto-$PGF_{1\alpha}$). Data presented are means ± SE from seven separate experiments. [30]

cesses, when directed from the abluminal to the luminal side of the microvessels, have been suggested as a second-line defense against pulmonary edema formation in lungs with hydrostatically induced edema.

5 Hemolysin Effects on Endothelial Cell Monolayers in Vitro, Prostaglandin I_2 Generation, and Interendothelial Permeability Increase

These experiments were performed in pig pulmonary artery endothelial cell monolayers in culture [30]. To assess the monolayer permeability characteristics, the cells were grown on polycarbonate filter membranes, which were positioned in a modified chemotaxis chamber for measurement of the transendothelial flux of water and labeled dextran. In concentration ranges between 0.05 and 1 hemolytic u/ml, Hly evoked dose- and time-dependent generation of prostaglandin I_2 (Fig. 10). At concentrations between 1 and 5 hemolytic u/ml, the hemolysin attack caused a time- and dose-dependent increase in the permeability characteristics of the endothelial cell

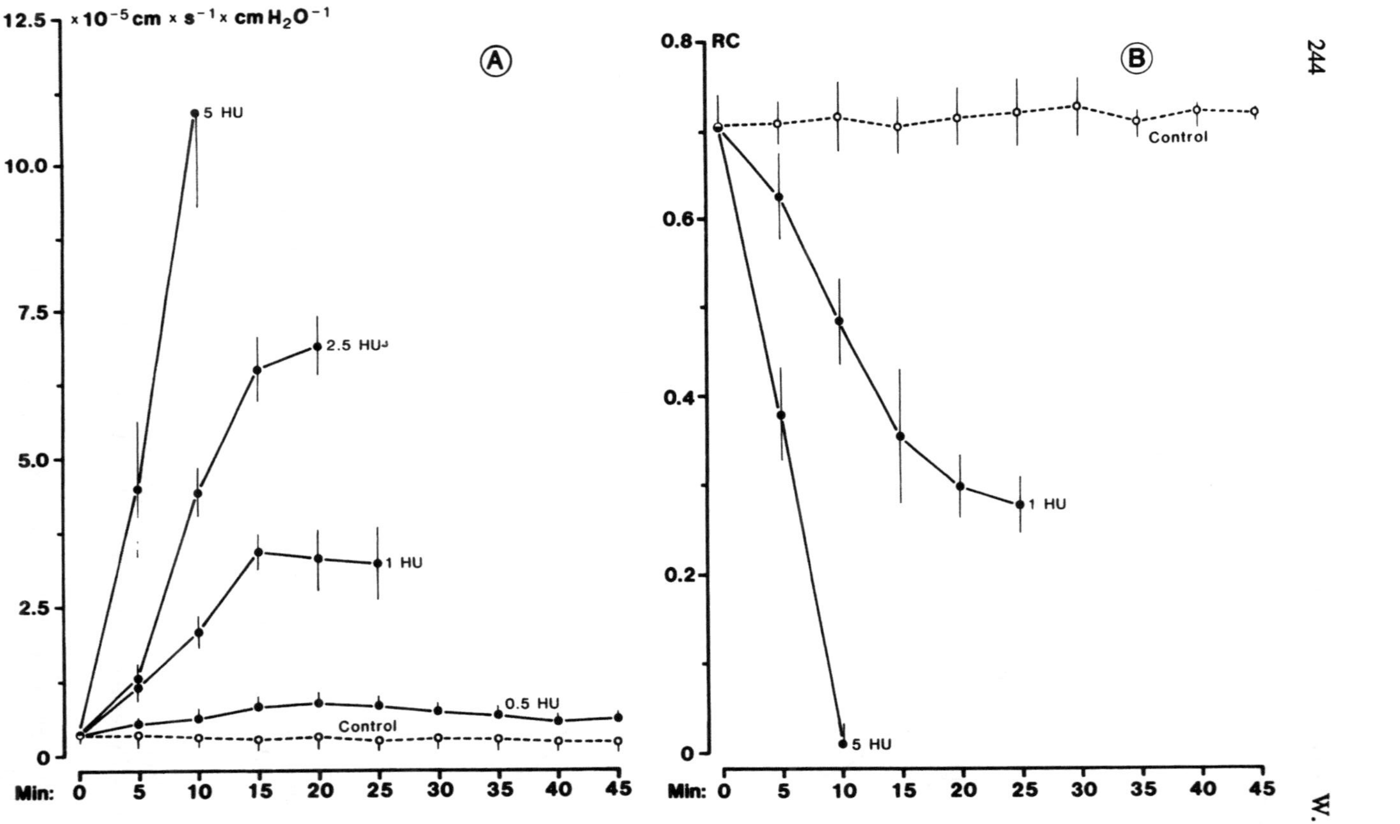

Fig. 11. Time- and dose-dependent increase in permeability of endothelial cell monolayers caused by Hly. Confluent cell monolayers on polycarbonate filter membranes were mounted in a modified chemotaxis chamber and exposed to a continuous hydrostatic pressure of 10 cm H_2O. After a steady state was reached, 0, 0.5, 1, 2.5, or 5 HU of Hly per ml was added. Hly induced a time- and dose-dependent increase in the hydraulic conductivity (*A*) and a decrease in the reflection coefficient (*RE*) for dextran (*B*), indicating severe damage to the barrier function of the endothelial cell monolayers. Data presented are means ± SE from ten separate experiments. [30]

monolayer (Fig. 11). The hydraulic conductivity increased approximately 30-fold, and the reflection coefficient for large molecules (dextran) dropped from 0.71 to < 0.05, indicating toxin-induced loss of endothelial barrier function. All such biochemical and biophysical alterations occurred below the threshold of overt cell damage, i.e., in the absence of significant LDH release. The alterations in endothelial monolayer permeability were accompanied by cell retraction and interendothelial gap formation, as assessed by rhodamin-labeled phalloidin staining of the endothelial microfilament network.

6 Hemolysin as a Potent Inductor of Arachidonic Acid Lipoxygenase Metabolism

In addition to its induction of prostanoid generation in perfused rabbit lungs and in cultured pulmonary artery endothelial cells, Hly was noted to be a potent inductor of arachidonic acid lipoxygenase metabolism. In perfused lungs, subcytolytic concentrations of the exotoxin caused a rapid release of large quantities of cysteinyl leukotrienes and leukotriene B_4 into the intravascular space (Fig. 12; [31]). Similar quantities have hitherto been elicited only by high concentrations of the artificial calcium ionophore A23187. Moreover, a marked liberation of 5-HETE into the buffer medium occurred, whereas LTB_4 represented the predominant compound in the bronchoalveolar lavage fluid. The hemolysin-induced burst of leukotriene and HETE generation preceded the onset of vascular leakage in this model. Similarly, large amounts of arachidonic acid lipoxygenase products were elicited by infusion of viable hemolysin liberating *E. coli* bacteria in the perfused lung model (Fig. 13; [32]). Used in a concentration range between 10^4 and 10^7 *E. coli*/ml, predominant cysteinyl LT formation was again elicited. In particular, rapid generation of LTC_4 occurred, with subsequent metabolic conversion to LTD_4 and LTE_4 in the lung vascular compartment. Concomitantly, significant amounts of LTB_4 were elicited by the hemolysin liberating *E. coli* bacteria. In accordance with the physiological measurements, corresponding amounts of nonhemolysin forming *E. coli* bacteria (lacking the HlyC gene) again failed to induce substantial arachidonic acid lipoxygenase metabolism. The leukotriene induction by in situ liberation of hemolysin was not inhibited in the presence of plasma with toxin-neutralizing capacity.

Pre-application of human granulocytes, which sequestered in the lung microvasculature, caused a severalfold increase in leukotriene generation upon subsequent challenge with hemolysin forming *E. coli* bacteria. This finding deserves special attention, since granulocytes have been noted to be preferred Hly targets within the various circulating blood cells [33]. The first observation in this context addressed release of granule constituents from and killing capacity of hemolysin-exposed blood leukocytes. In these experiments, Hly concentrations far below those causing erythrocyte lyses were noted to induce the release of elastase and other granule constituents and to cause loss of phagocytosing capacity in human neutrophil granulocytes. Extending these experiments to neutrophil arachidonic acid metabolism, we recently noted that subhemolytic doses of this exotoxin are potent inductors of the 5-lipoxygenase metabolism in these cells [34]. In the absence of exogenous arachidonic acid

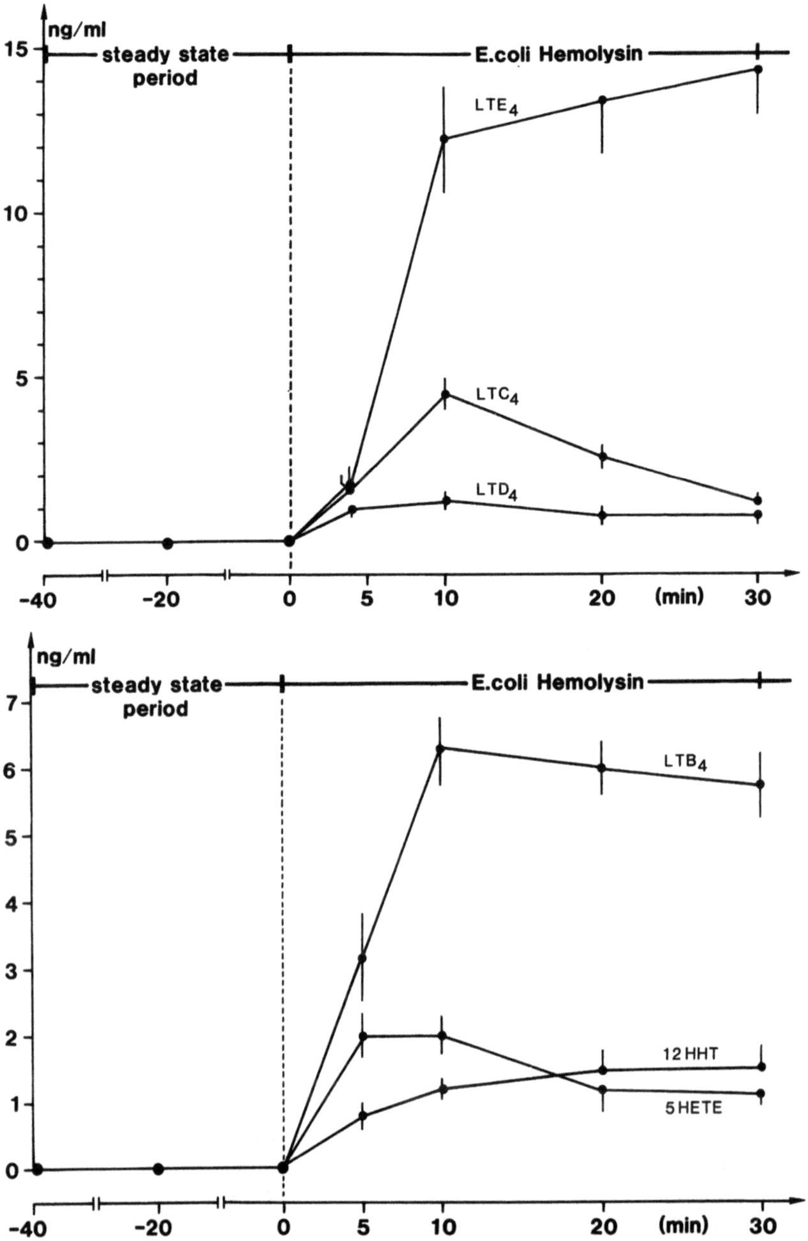

Fig. 12. Time courses of leukotriene LT and hydroxyeicosatetraenoic acid (HETE) release into the lung perfusion fluid after application of 2.5 HU of *E. coli* hemolysin per ml. All values are calculated from the UV absorbance of corresponding peaks in RP- and SP-HPLC procedures, corrected for the respective recoveries of the different compounds in the overall extraction and detection procedure, and are given in ng/ml perfusate (mean ± SE; n = 5 individual lung preparations). After an initial steady-state period of 40 min and the subsequent total exchange of perfusate (*dotted lines*), time was set at zero and hemolysin was administered. *12 HHT* 12-hydroxyheptadecatrienoic acid. [31]

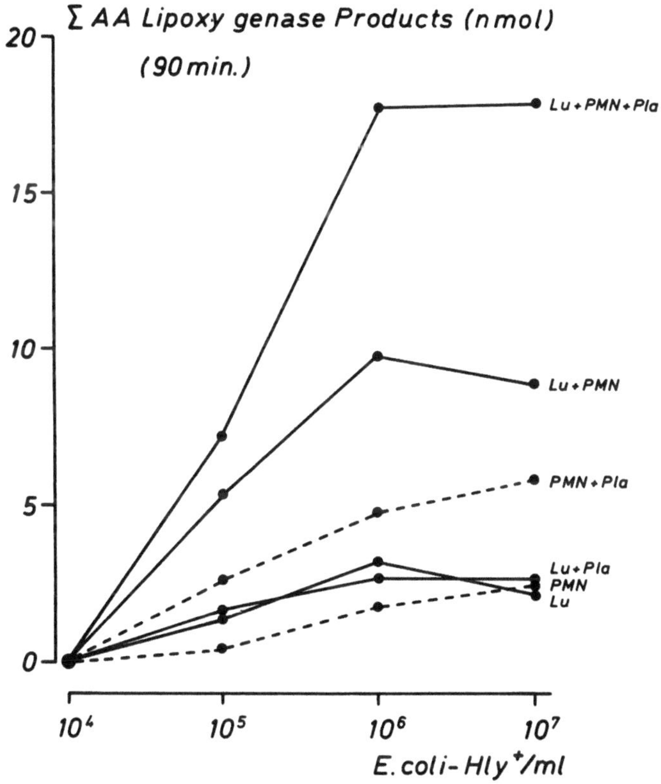

Fig. 13. Time-dependent liberation of AA lipoxygenase products into the buffer medium of lungs and/or PMN challenged with 10^6/ml hemolysin-forming *E. coli*. Lungs (*Lu*) were perfused with 200 ml Krebs Henseleit albumin buffer in the absence or presence of rabbit plasma (*Pla*; 15% v/v) and human PMN (2×10^8). Total amounts of the different AA lipoxygenase products in the recirculating perfusate volume are summed. In accompanying invitro experiments, 5×10^6 PMN were incubated with 10^6/ml *E. coli* Hly^+ in a buffer volume of 5 ml; total amounts of eicosanoids measured in these studies were corrected (factor 40) to correspond to the buffer volume in the perfused lung experiments. In the isolated lung experiments, cysteinyl LTs represented the predominant compounds. Total lipoxygenase product formation is amplified in the presence of PMN (cooperative LT synthesis) and again in the presence of plasma (phagocytosis-related stimulation in addition to the Hly-elicited events. [32]

supply, Hly (0.01–10 hemolytic u/ml) induced moderate generation of LTB_4 and its omega oxidation products (Fig. 14). Simultaneous supply with free arachidonic acid during hemolysin challenge evoked large quantities of metabolites, dependent on both Hly (maximum at 0.1 HU/ml) and free arachidonic acid concentration (maximum at 10 μM). In parallel, large amounts of 5-HETE and nonenzymatic LTA_4 hydrolysis products appeared. The total amounts evoked by subhemolytic doses of Hly in the presence of exogenous arachidonic acid supply approximate those elicited by the calcium ionophore A23187 under optimum conditions. These studies thus demonstrate potent activation of the 5-lipoxygenase pathway in the Hly-exposed neutrophils. This feature predisposes neutrophils for cooperative leukotriene synthesis under conditions

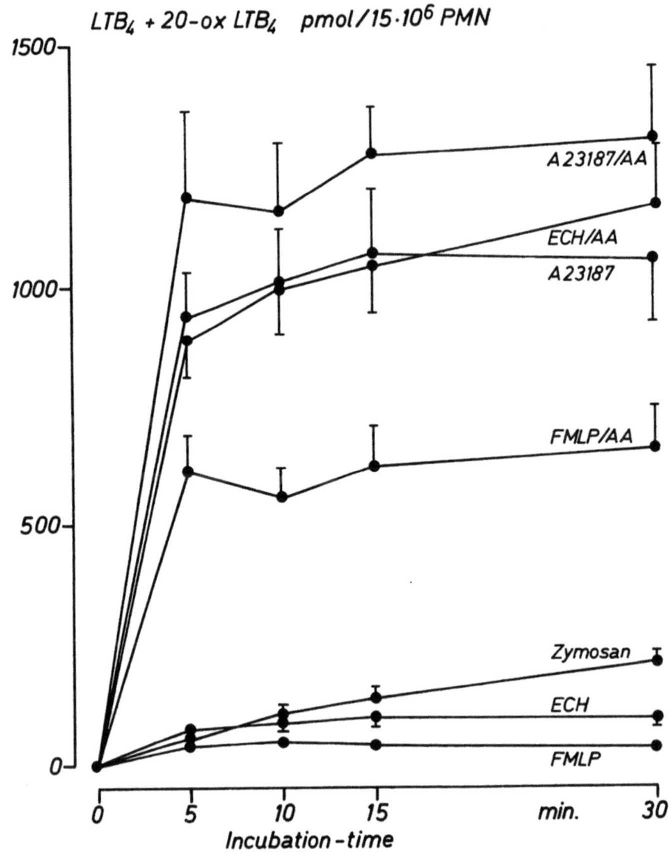

Fig. 14. Comparison of hemolysin (*ECH*)-evoked leukotriene generation in neutrophils (PMN) with the PMN response to, the calcium ionophore A23187, FMLP, and zymosan. 1.5×10^7 PMN were incubated with Hly (0.1 HU/ml), A23187 (1 µM), FMLP (1 µM) or opsonized zymosan (2 mg/Ml) in the absence or presence of free arachidonic acid (10 µM) for various time periods. LTB_4, 20-OH-LTB_4, and 20-COOH-LTB_4 are comprised of LTB_4 + 20 ox LTB_4. Mean ± SEM of five independent experiments are given. [34]

of Hly challenge as outlined in Fig. 15. Activated neutrophils, when investigated in vitro, release the unstable leukotriene intermediate LTA_4 into the extracellular space, with subsequent rapid degradation to nonenzymatic hydrolysis products. Under conditions of cell-to-cell contact with endothelial cells, however this LTA_4 is transferred into the endothelium as acceptor cell, and is further processed to cysteinyl leukotrienes by the glutathione S-transferase of this latter cell. The enzymatic equipments of the neutrophil "feeder" cell and the endothelial "acceptor" cell thus cooperate to generate cysteinyl leukotrienes, which are not generated by each single cell type, as the neutrophils are lacking the glutathione S-transferase, and the endothelial cells are lacking the 5-lipoxygenase step. In addition, under circumstances of direct cell-to-cell contact with endothelial cells, the granulocyte 5-lipoxygenase pathway is diverted from the generation of 5-HETE to an increased formation of LTA_4 from

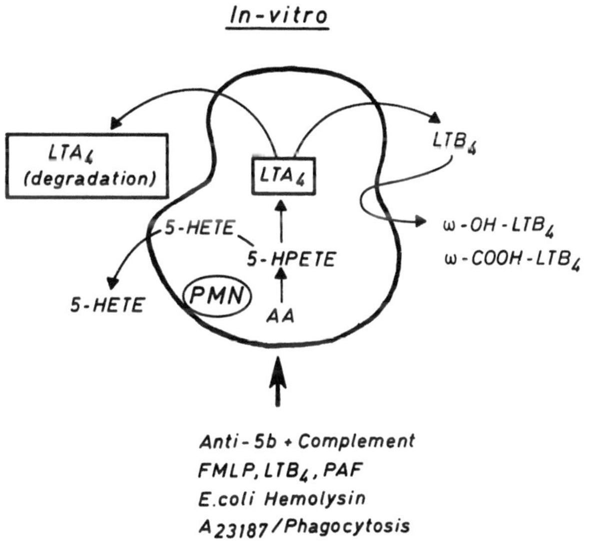

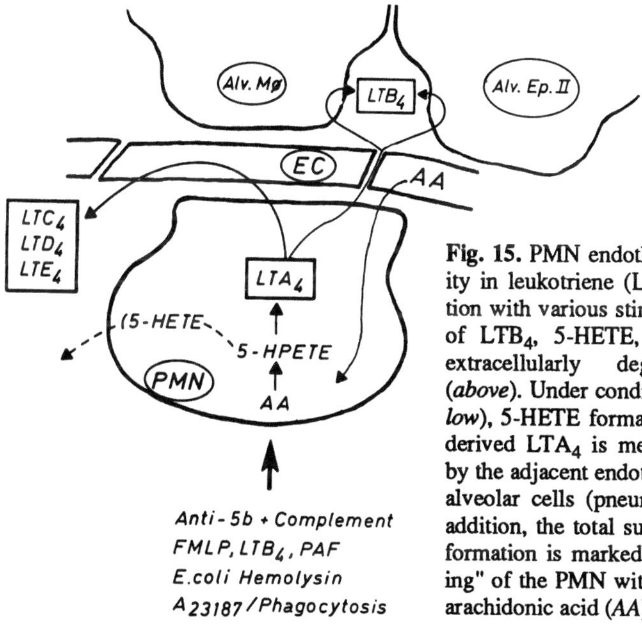

Fig. 15. PMN endothelial cell (*EC*) cooperativity in leukotriene (LT) synthesis. PMN activation with various stimuli results in the secretion of LTB_4, 5-HETE, and intact LTA_4, which extracellularly degrades nonenzymatically (*above*). Under conditions of PMN sticking (*below*), 5-HETE formation is reduced, and PMN-derived LTA_4 is metabolized to cysteinyl-LTs by the adjacent endothelial cells and to LTB_4 by alveolar cells (pneumocytes, macrophages). In addition, the total sum of lipoxygenase product formation is markedly increased by "backfeeding" of the PMN with endothelium-derived free arachidonic acid (*AA*)

their mutual precursor 5-hydroperoxyeicosatetraenoic acid. Moreover, under these conditions, there is a "feedback" mechanism, which provides the neutrophil with arachidonic acid. Experiments with coincubation of PMN and endothelial cells demonstrated transfer of arachidonic acid from the endothelial cell to the neutrophil upon mutual cell activation. Both mechanisms add to a severalfold amplification of leukotriene generation.

This concept of cooperative leukotriene generation can also be demonstrated in blood-free perfused rabbit lungs, when these organs are preloaded with neutrophils and subsequently stimulated with subhemolytic doses of Hly ([32]; Fig. 13). In the presence of neutrophils, there is an amplification of 5-lipoxygenase metabolism, which is predominantly caused by increased generation of cysteinyl leukotrienes and, to a minor extent, LTB_4. Concomitantly, 5-HETE- and LTA_4-hydrolysis product release decreases as compared to Hly neutrophil stimulation in vitro. This pattern of metabolite generation is in accordance with previous studies, in which selective stimulation of sequestered neutrophils in the lung vasculature [35, 36] as well as direct intravascular application of freshly prepared LTA_4 demonstrated neutrophil-endothelial cell cooperativity in leukotriene metabolism.

In summary, Hly evokes marked leukotriene generation in cell-free perfused lungs, and it is a potent inductor of 5-lipoxygenase metabolism in isolated neutrophils. Under conditions of lung microcirculatory neutrophil entrapment, which occurs, e.g., during shock and sepsis, a severalfold amplification of leukotriene generation is noted. The underlying mechanism include neutrophil-endothelial cell LTA_4 transfer and "backfeeding" of neutrophils with endothelial-derived arachidonic acid under conditions of mutual stimulation by the exotoxin. The amounts of leukotrienes elicited under these circumstances surpass those elicited by optimum concentrations of the artificial calcium ionophore A23187. The potency in leukotriene generation is not diminished in the presence of plasma.

7 Hemolysin-Evoked Cytokin Generation in Monocytes

Incubation of human monocytes with subcytocidal doses of Hly (0.1 – 2 HU/ml) induced rapid release of large amounts of interleukin 1ß within 60–120 min [38]. IL-1ß release occurred in the presence of actinomycin-D and cycloheximide, and was thus probably due to processing and export of the intracellular IL-1ß precursor. Correspondingly, IL-1ß release was induced by incubation of Hly producing *E. coli* with cultured human monocytes at ratios of only 0.3 to 3 colony-forming units per monocyte. In contrast, nontoxin producers were ineffective with respect to interleukin generation. This feature of Hly efficacy again parallels staphylococcal α-toxin properties, as the bacterial agent from the gram-positive rod was similarly found to induce the release of large amounts of IL-1ß from cultured human monocytes at subcytolytic doses [39]. In addition, release of tumor necrosis factor α was observed in these experiments. Taken together, these results demonstrate that IL-1ß, possibly tumor necrosis factor, and possibly a variety of further cytokines not yet experimentally addressed belong to the battery of inflammatory mediators which can be elicited by Hly. Preferred target cells in this respect are evidently monocytes, however, this finding may extend also to a variety of tissue macrophages.

8 Signal Transduction Events in Hly-Challenged Target Cells

Two concepts have emerged to explain the different cell activation phenomena in Hly-exposed targets.

1. Calcium flux through toxin-created nonphysiological bypass gates. Prostanoid generation and vasoconstriction in cell-free perfused rabbit lungs as well as prostacyclin generation and cell retraction phenomena in cultured endothelial cells are strictly dependent upon extracellular calcium. In addition, sieving experiments in the cultured endothelial cells indicated the generation of small transmembrane hydrophilic pores with an effective diameter of approximately 2 nm in the presence of Hly. An enhanced passive influx of ^{45}Ca into the hemolysin-treated cells was directly demonstrated. These data together suggest that these pores may act as nonphysiological calcium gates, allowing extra- to intracellular calcium flux along the steep calcium gradient, thus initiating different calcium-dependent cellular processes. These processes include stimulation of phospholipolytic activities in the endothelial cells and putatively in toxin-exposed, thromboxane-generating lung macrophages, calcium-dependent cell retraction phenomena with generation of intercellular gaps in the endothelial cells, and possible processing of intracellular interleukin 1ß precursor in monocytes. A corresponding role of subcytolytic amounts of exotoxin-created pores has also been suggested for α-toxin-attacked cells [40]. Moreover, such a signal transduction mechanism may exist for activation of intracellular processes in nucleated cells exposed to subcytolytic doses of the terminal complement complex. Such toxin-related calcium gating is thus distinctly different from the action of a calcium ionphore such as A23187 or from the calcium shift following receptor occupancy and activation of receptor-operated events.

2. Induction of phosphoinositol hydrolysis and related metabolic responses. A different signal transduction pathway in response to Hly challenge was recently noted in human granulocytes [41]. In these cells, Hly provoked rapid secretion of elastase and myeloperoxidase, generation of superoxide, and synthesis of platelet-activating factor (PAF) and lyso-PAF. Concomitantly, marked phosphatidylinositol hydrolysis with sequential appearance of the inositol phosphates IP3, IP2, and IP1 and the formation of diacylglycerol occurred. The metabolic responses displayed distinct bell-shaped dose dependencies, with maximum events noted at low toxin concentrations of 0.1 – 0.5 HU/ml.

Phosphatidylinositol hydrolysis and metabolic responses elicited by Hly exceeded those evoked by optimum concentrations of FMLP, PAF, LTB_4, A23187, and staphylococcal α-toxin. The toxin-induced effects were sensitive to pertussis toxin, thus indicating enrollment of G-proteins. These data suggest that the marked capacity of low doses of Hly to elicit degranulation, respiratory burst, and lipid-mediator generation in human neutrophils probably involves signal transduction via the preformed phosphatidylinositol hydrolysis pathway. Further studies will be necessary to elucidate whether this finding is restricted to exotoxin-exposed neutrophils, or whether other target cells may similarly employ the phosphatidylinositol hydrolysis related signal transduction pathway in addition to toxin pore-related nonphysiological calcium bypass phenomena.

9 Summary

In isolated, ventilated, and blood-free perfused rabbit lungs, pathophysiological characteristics of the acute respiratory distress syndrome can be fully reproduced by the admixture of low doses of *E. coli* hemolysin. Correspondingly, ARDS events are mimicked by intravenous infusion of viable Hly-forming *E. coli*, but not of isogenic non-Hly formers in this model. The exotoxin-induced vasoconstrictor responses are mediated predominantly via induction of pulmonary thromboxane generation. This vasoconstrictor response results in severe maldistribution of perfusion, with subsequent ventilation perfusion mismatch. In addition, Hly exposure causes delayed onset, but severe lung vascular permeability increase. This vascular leakage can be reproduced in Hly-exposed endothelial cell monolayers in culture. Markedly increased hydraulic conductivity and decreased protein-reflecting properties represent the underlying events. The induction of vascular leakage is independent of the Hly-evoked prostanoid generation. It may well be related to the burst of arachidonic acid lipoxygenase metabolites, generated in cell-free perfused lungs and, in particular, elicited under conditions of neutrophil entrapment in the lung vasculature due to neutrophil-endothelial cell cooperativity in arachidonic acid lipoxygenase metabolism. In addition, generation of cytokines, hitherto only incompletely addressed, may add to the hemolysin-induced severe loss of lung microvascular integrity. Finally, endothelial cell retraction phenomena, directly related to Hly-induced transmembrane calcium fluxes in these cells, may add to the vascular leakage. Stimulus transmission pathways in the Hly-exposed target cells appeared to include toxin pore-related nonphysiological calcium bypass phenomena as well as – in neutrophils – potent induction of the phosphatidylinositol hydrolysis signal transduction pathway with related metabolic responses. Overall, *E. coli* emerges as a potent inductor of a diversity of inflammatory mediators in multiple target cells. Inflammatory mediator release, particularly elicited by subcytocidal concentrations of the bacterial exotoxin, must be expected to contribute substantially to microcirculatory disturbances and tissue injury in hemolysin-exposed organs.

References

1. Fowler AA, Hamman RF, Zerbe GO, Benson KN & Hyers TM (1985) Adult respiratory distress syndrome. Prognosis after onset. Am Rev Respir Dis 132:427–478
2. Montgomery AB, Stager MA, Carrico CJ & Hudson LD (1985) Causes of mortality in patients with the adult respiratory distress syndrome. Am Rev Respir Dis 132:485–489
3. Petty TL (1985) Indicators of risk, course, and prognosis in adult respiratory distress syndrome (ARDS). Am Rev Respir Dis 132:471–472

4. Andrews CP, Coalson JJ, Smith JD & Johanson WG (1981) Diagnosis of nosocomial pneumonia in acute, diffuse lung injury. Chest 80:254–258
5. Bodey GP, Jadea L & Elting I (1985) *Pseudomonas bacteremia.* Arch Intern Med 145:1621–1629
6. Brigham KL & Meyrick B (1984) Interactions of granulocytes with the lung. Circ Res 54:623–635
7. Brigham KL & Meyrick B (1977) Endotoxin and lung injury. Am Rev Respir Dis 133:913–927
8. Cavalierei S, Bohach GA & Snyder IS (1984) *Escherichia coli* α-hemolysin: characteristics and probable role in pathgenecity. Microbiol Rev 1(48):326–343
9. Bhakdi S, Mackman N, Nicaud J-M & Holland IB (1986) *Escherichia coli* hemolysin may damage target cell membranes by generating transmembrane pores. Infect Immun 52:63–69
10. Bhakdi S & Tranum-Jensen J (1987) Damage to mammalian cells by proteins that form transmembrane pores. Rev Physiol Biochem Pharmacol 107:148–223
11. Bhakdi S & Tranum-Jensen J (1988) Damage to cell membranes by pore-forming bacteria cytolysins. Progr Allerg 40:1–43
12. Mackman N & Hollan IB (1984) Secretion of a 107 K dalton polypetide onto the medium from a hemolytic *E. coli* K12 strain. Mol Gen Genet 193:312–315
13. Welch RA & Pellet S (1988) Transcriptional organization of the *Escherichia coli* hemolysin genes. J Bacteriol 170:1622–1630
14. Welch RA, Dellinger EP, Minshew B & Falkow S (1981) Hemolysin contributes to virulence of extra-intestinal *E. coli* infections. Nature (London) 294:665–667
15. Welch RA & Falkow S (1984) Characterization of *Escherichia coli* hemolysin conferring quantitative differences in virulence. Infect Immun 43:156–160
16. Fünfstück R, Tschäpe H, Stein G, Kunath H, Bergner M & Wessel G (1986) Virulence properties of *Escherichia coli* strains in patients with chronic pyelonephritis. Infection 14:145–150
17. Hacker J, Hughes C, Hof H & Goebel W (1983) Cloned hemolysin genes from *Escherichia coli* that cause urinary tract infection determine different levels of toxicity in mice. Infect Immun 42:57–63
18. Flemlee T, Pellet S & Welch RA (1985) Nucleotide sequence of an *E. coli* chromosomal hemolysin. J Bacteriol 163:94–105
19. Menestrina G, Mackman N, Holland IB & Bhakdi S (1987) *E. coli* hemolysin forms voltage-dependent ion channels in lipid membranes. Biochim Biophys Acta 905:109–117
20. Menestrina G (1988) *Escherichia coli* hemolysin permeabilizes small unilamellar vesicles loaded with calcium by a single-hit mechanism. FEBS Lett. 232:217–224
21. Bhakdi S, Greulich S, Muhly M, Eberspächer B, Becker H, Thiele A & Hugo F (1989) Potent leucocidal action of *Escherichia coli* hemolysin mediated by permeabilization of target cell membranes. J Exp Med 169:737–754
22. Boehm DF, Welch RA & Snyder IS (1990) Calcium is required for binding of *Escherichia coli* hemolysin (Hlya) to erythrocyte membrane. Infect Immun 58:1951–1958
23. Ludwig A, Jarchau T, Benz R & Goebel W (1988) The repeat domain of *E. coli* hemolysin is responsible for its Ca-dependent binding to erythrocytes. Mol Gen Genet 214:553–561
24. Boehm DF, Welch RA & Snyder IS (1990) Domains of *Escherichia coli* hemolysin (HlyA) involved in binding of calcium and erythrocyte membranes. Infect Immun 58:1959–1964
25. Seeger W, Walter H, Suttorp N & Bhakdi S (1989) Thromboxane-mediated hypertension and vascular leakage evoked by low doses of *Escherichia coli* hemolysin in rabbit lungs. J Clin Invest 84:220–227
26. Seeger W, Obernitz R, Thomas M, Walmrath D, Holland IB, Grimminger F, Eberspächer B, Hugo F, Suttorp N & Bhakdi S (1991) Lung vascular injury after administration of viable hemolysin-forming *Escherichia coli* in isolated rabbit lungs. Am Rev Respir Dis (in press)

27. Walmrath E, König R, Ernst C, Brückner H, Grimminger F & Seeger W (1991) Ventilation-perfusion relationships in isolated bloodfree perfused rabbit lungs. Am Rev Respir Dis (in press)
28. Seeger W, Birkemeyer RG, Ermert L, Suttorp N, Bhakdi S & Duncker HR (1990) Staphylococcal alpha-toxin induced vascular leakage in rabbit lungs. Lab Invest 63:341–349
29. Ermert L, Rousseau S, Schütte H, Birkemeyer RG, Grimminger F, Bhakdi S, Duncker HR & Seeger W, Induction of severe vascular leakage by low doses of *Escherichia coli* hemolysin in perfused rabbit lungs. (submitted)
30. Suttorp N, Flöer B, Seeger W, Schnittler H & Bhakdi S (1990) Effects of *E. coli* hemolysin on endothelial cell function. Infect Immun 58:3796–3801
31. Grimminger F, Walmrath D, Birkemeyer RG, Bhakdi S & Seeger W (1990) Burst of leukotriene- and HETE-generation elicited by low dose *Escherichia coli* hemolysin in rabbit lungs. Infect Immun 58:2659–2663
32. Grimminger F, Thomas M, Obernitz R, Walmrath D, Bhakdi S & Seeger W (1990) Inflammatory lipid mediator generation elicited by viable hemolysin-forming *Escherichia coli* in lung vasculature. J Exp Med 172:1115–1125
33. Bhakdi S, Greulich S, Muhly M, Eberspächer B, Becker H, Thiele A & Hugo F (1989) Potent leucocidal action of *Escherichia coli* hemolysin mediated by permeabilization of target cell membranes. J Exp Med 169:737–754
34. Grimminger F, Scholz C, Bhakdi S & Seeger W (1991) Subhemolytic doses of *Escherichia coli* hemolysin evoke large quantities of 4- and 5-series leukotrienes in human neutrophils in dependence on exogenous fatty acid supply. J Biol Chem (in press)
35. Grimminger F, Menger M, Becker G & Seeger W (1988) Potentiation of leukotriene generation following sequestration of neutrophils in isolated lungs. Blood 72:1687–1692
36. Grimminger F, Kreusler B, Schneider U, Becker G & Seeger W (1990) Influence of microvascular adherence on neutrophil leukotriene generation – evidence for cooperative eicosanoid synthesis. J Immunol 144:1866–1872
37. Seeger W, Suttorp N, Hellwig A & Bhakdi S (1986) Noncytolytic terminal complement complexes may serve as calcium gates to elicit leukotriene generation in human polymorphonuclear leukocytes. J Immunol 137:1286–1293
38. Bhakdi S, Greulich S, Muhly M, Korom S & Schmidt G (1990) Effects of *E. coli* hemolysin on human monocytes: cytocidal action and stimulation of interleukin-1 release. J Clin Invest 85:1746–1753
39. Bhakdi S, Muhly M, Korom S & Hugo F (1989) Release of interleukin-1-beta associated with potent cytocidal action of staphylococcal alpha-toxin on human monocytes. Infect Immun 57:3512–3519
40. Seeger W & Suttorp N (1988) Role of membrane lipids in the pulmonary vascular abnormalities caused by bacterial toxins. Am Rev Respir Dis 136:462–466
41. Grimminger F, Sibelius U, Bhakdi S, Suttorp N & Seeger W, *Escherichia coli* hemolysin-induced secretory responses and PAF generation in human neutrophils is related to phosphoinositide hydrolysis In press: J Clin Invent

Pathophysiological Aspects of Polytrauma, Shock and Organ Failure

H. REDL[1] and G. SCHLAG[1]

1 Trauma – Hemorrhage

1.1 Introduction

Multiorgan failure (MOF) is a complication of the posttraumatic or postoperative course secondary to shock-related events. The lethality of MOF by far exceeds 50%. In the clinical setting the injury severity is a major criterion for the onset of MOF. Additional patient-related factors such as antiinflammatory activity, the individual potential to produce different amounts of inflammatory factors, and the sensitivity of target cells are decisive for the patient's outcome. These factors are in a delicate balance, which can lead to either wound healing and recovery or to the development of organ damage and finally to organ failure. The development of organ failure (OF) may occur within the first days after trauma (early OF) or after 1 or 2 weeks (late OF) in ICU patients. The reaction that leads to organ damage and OF is a generalized inflammatory event (Goris et al. 1985). This inflammation may be nonbacterial in the early stage of OF, and may be aggravated by the influence of bacterial factors during the development of late OF (Fig. 1). The nonbacterial inflammatory reaction is induced by the initial injury and is usually based on four factors:

1. tissue ischemia/hypoxia,
2. reperfusion events with generation of oxygen radicals,
3. stimulation of phagocytes by complement split products or other humoral inflammatory factors,
4. initial direct tissue damage with necrosis.

These four primary trigger mechanisms cause either local inflammation at the site of injury or, if severity is beyond a certain threshold, a so-called whole-body inflammation may ensue.

The target organs of MOF may vary from one patient to another as to time sequence and severity but usually involve lung and liver. The sequence of OF is not only influenced by the patient's individual reaction, but also by modern intensive care procedures, as therapeutic modalities can mask a single organ failure for a certain period of time.

[1] Ludwig Boltzmann Institute for Experimental and Clinical Traumatology, Vienna, Austria

42. Colloquium Mosbach 1991
Molecular Aspects of Inflammation

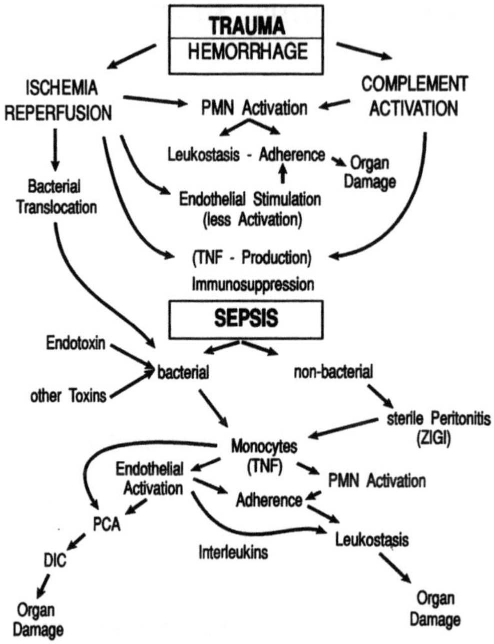

Fig. 1. Working hypothesis of trauma- and sepsis-induced inflammatory reactions involved in the development of organ failure. *TNF* tumor necrosis factor; *PCA* procoagulant activity; *DIC* disseminated intravascular coagulation; *ZIGI* zymosan-induced peritonitis model

1.2 Ischemia-Reperfusion and Complement Activation

According to McCord (1985), hemorrhagic shock may be viewed as "whole-body ischemia". The consequence of acute ischemia is inhibition of cellular ATP synthesis by substrate (glucose) and oxygen deprivation. There is an increase in the levels of the ATP degradation products adenosine, inosine, and hypoxanthine (Saugstad et al. 1977). An involvement of oxygen radicals was first demonstrated by Granger et al. (1981) in studies of small intestine ischemia. Tissue damaged by oxidative processes may leak sufficient xanthine oxidase (XO) into the systemic circulation to mediate injury to the microvasculature of unaffected tissue (Yokoyama et al. 1988). Xanthine oxidase may be detected in the sera of patients with pathologic conditions such as ARDS (Grum et al. 1987), extremity ischemia (tourniquet)/reperfusion (Friedl et al. 1990), or ischemia/reperfusion of the liver (Yokoyama et al. 1990). Increased substrate levels (hypoxanthine) were found during hypovolemic-traumatic shock (Redl et al. 1988).

Radical action could also be due to the generation of various oxidation products by activated leukocytes. For the shock situation, at least with complement/neutrophil-induced lung injury, ·OH is thought to be the decisive radical species (Ward et al. 1986).

Two events are most prominent during polytrauma: (1) the previously mentioned hemorrhage and ischemia caused by the accompanying hypovolemia, and (2) complement activation. Hypovolemia leads to malperfusion, especially of skin and muscle tissue, and, what could be more important, to malperfusion of the splanchnic area (see Sect. 2.1). In baboon studies polytrauma was simulated by hypovolemic-traumatic shock (including bone fracture, soft tissue trauma, and hemorrhage to about 50% of total blood volume with 70% reduction of cardiac output and a mean arterial pressure around 40 mm Hg) (Schlag et al. 1991). As a result, the mesenteric artery flow dropped from 100 to 17 ml/min within 1 h. The reduction of splanchnic flow was immediately followed by a gut response in the form of anaerobic metabolism, which led to a reduction of gut wall pH from 7.36 to 6.96 (6.56 after 3 h) and finally to bacterial translocation.

Traumatic tissue damage leads to the unmasking of different cellular components, which might be the reason for the massive complement activation seen after polytrauma. Similar conclusions have been drawn from experiments in which homogenized muscles tissue was infused into experimental animals (Heidemann et al. 1978). Equally, ischemic tissue causes massive complement activation, resulting in elevated levels of the anaphylatoxins C3a and C5a (Bengtson et al. 1987). The important reactions during complement activation are, probably the release of C3a, C5a, and the formation of terminal complement complexes, which in turn can activate and damage cells. C3a is most useful for monitoring this complement activation. Initially high levels of circulating C3a have been demonstrated to be related to the subsequent development of organ failure and mortality (Zilow et al. 1989). C3a levels were also seen to be significantly higher in patients with early organ failure and early mortality after severe polytrauma (unpubl. data).

1.3 Granulocyte Activation

Anaphylatoxins can activate WBC in vitro. The question is whether phagocytes are also activated in vivo during shock and trauma. Evidence can be obtained from (1) the in vitro (ex vivo) activation capacity of the cells, and (2) from products released into the plasma, e.g., proteinases, during the activation process.

A good example is elastase (also see Jochum et al., this Vol.) liberated by PMN, measured as the elastase α-1-proteinase inhibitor complex in plasma (Neumann et al. 1984). There is a positive correlation between trauma severity and elastase plasma levels in patients (Nuytinck et al. 1986). In the baboon hypovolemic-traumatic shock model (Pretorius et al. 1987) we found a significant increase in plasma elastase levels both at the end of the shock period and after retransfusion.

There is also a significant correlation between sepsis and PMN-elastase plasma levels and the severity of multiorgan failure (Nuytinck et. al. 1986), which indicates a link between the activation of granulocytes and the development of organ failure (Nuytinck et al. 1986; Lang et al. 1989). These findings are also relevant for the role of oxygen radicals in organ failure, as identification of circulating elastase does not necessarily indicate elastase-induced organ failure but suggests the involvement of PMN.

The involvement of PMN in shock-related injuries has been confirmed by PMN depletion. Experimental studies with pharmacologically induced leukopenia have shown that shock-related organ damage, e.g., in the lung, can be prevented (Schraufstätter et al. 1984).

Activated complement components stimulate neutrophils to perform the respiratory burst reaction with the formation of reactive oxygen species and the release of cytotoxic proteases. These complement components also induce an upregulation of neutrophil adherence molecules, especially of CD18/CD11b (Mac-1) (Harlan et al. 1985), which is both an important complement receptor (CR3) involved in phagocytosis, and one of the ligands of the endothelial adherence molecule ICAM-1, and is thus involved in the endothelial-leukocyte interaction. Besides neutrophils, monocytes are target cells for C5a, which could induce the formation of monokines (Scholz et al. 1990), the respiratory burst, and the release of oxygen radicals and enzymes. Complement activation as well as ischemia/reperfusion could trigger neutrophil activation, the mechanism of which is not entirely clear.

1.4 Endothelial Stimulation

One hypothesis assumes that activated endothelial cells might play an important role in this ischemia-derived activation. This is based on the observation that hydrogen peroxide can induce PAF formation on the endothelial surface (Lewis et al. 1988), which upon contact with neutrophils could lead to neutrophil activation via the PAF receptors. PAF formation on the endothelial cell (EC) surface is part of the reactions which, according to Pober (1990), are defined as "endothelial stimulation," probably an initial event in the sequence of generalized inflammatory reactions (Fig. 1). This endothelial stimulation differs from endothelial activation by the nature of EC stimuli, by the reaction time (minutes) and by the fact that no protein synthesis must be induced. EC stimulation involves the transport of the adherence molecule GMP-140, which is located in Weibel-Palade bodies (WPB), to the surface of EC, which leads to fusion of the WPB with the plasma membrane so that GMP-140 molecules can interact with neutrophils (Geng et al. 1990).

It is proposed that both GMP-140 together with PAF on the EC and the PAF receptor on the neutrophils together with the Lewis-X structure are responsible for early adherence of neutrophils to endothelial cells. This initial contact between EC and PMN seems to be independent of neutrophil activation. Presumably, PMN are only activated through contact with PAF/PAF-R. In vivo evidence for this hypothesis has been obtained in splanchnic shock experiments in rats. Leukocyte adherence was noted upon reperfusion as judged by vital microscopy, and was attenuated by SOD, and especially by PAF antagonists in this model (Bühren et al. 1991; Marzi et al. 1991).

1.5 Leukostasis

Leukostasis may occur as a result of this interaction and of trauma/reperfusion – complement-related PMN activation. This has been demonstrated in several studies using different methods including labeled neutrophils, histochemical localization, and

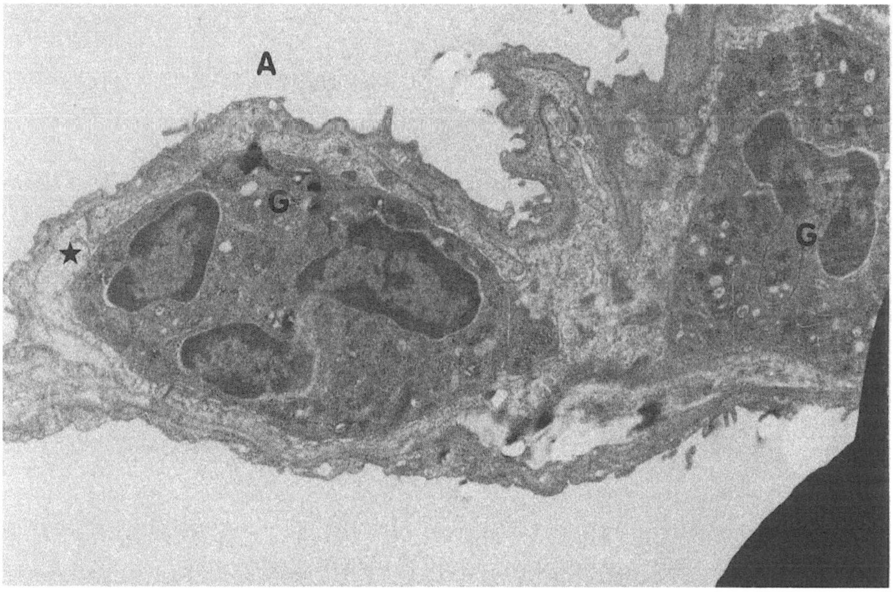

Fig. 2. Typical leukostasis occurring after trauma (or sepsis) in a baboon model. *G* Granulocytes; *A* alveolar Space; * endothelial swelling

morphometry as well as myeloperoxidase tissue measurements. These studies indicate that there is shock/trauma-induced leukostasis, which is evident both in experimental animals as well as in the clinical situation (Fig. 2). In addition to endothelial-neutrophil interaction, PMN-PMN interaction (granulocyte aggregation), aggravated by the involvement of platelets (products) (Redl et al. 1983), is responsible for this event. There is evidence that endothelial cells tend to be damaged, with edema or bleb formation (Schlag et al. 1975), parallel to leukostasis (Fig. 3). As a result, edema – specifically of the lung – may occur and is then known as "posttraumatic early lung failure."

1.6 Cytokine Formation

Recently, the formation of anaphylatoxins and cytokines such as TNF or IL-6 has been observed in the initial phase of hemorrhage and trauma (Ayala et al. 1990). This early cytokine formation is either triggered by complement activation products or by ischemia (as was shown, e.g., for the liver by Colletti et al. 1990). An additional stimulus could be bacterial toxin liberation during bacterial translocation (see below). One interesting aspect of TNF action is the possible link between the observed overactivation of certain immune responses, e.g., PMN activation, and the reactions resulting in immunosuppression. As demonstrated by Chaudry et al. (1990), released TNF could be responsible for the reduced antigen presentation by Kupffer cells after hemorrhagic shock. The observed immunosuppression, which has mostly been demonstrated by in vitro test systems, but also in vivo (skin test) (Puyana et al. 1988),

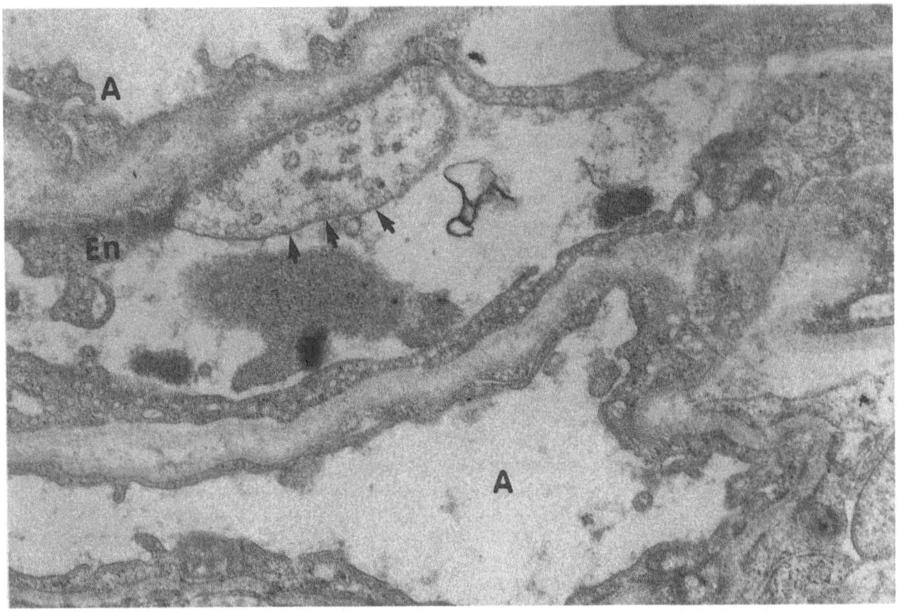

Fig. 3. Endothelial (*En*) swelling (*arrows*) in a lung capillary after polytrauma in a baboon. *A* Alveolar space (Pretorius et al. 1987)

could enhance the subsequent development of bacterial infection and sepsis. This immunosuppression is also due to reduced IL-2 formation, overproduction of PGE2, and the formation of immunosuppressive peptides (Ozkan et al. 1988). This subject has been reviewed extensively, e.g., by Faist et al. (1988) and Chaudry et al. (1990).

2 Sepsis

2.1 Bacterial Translocation (BT) – Toxin Liberation

Endotoxemia or bacteremia occurs posttrauma or after major surgery (Rush et al. 1988) and is frequently without a septic focus. Based on the work of Fine et al. (1959), Meakins and Marshall (1986) have stated that the gut is the source of endotoxin and the "motor" of multiorgan failure (sepsis-like syndrome) during bacterial translocation. The breakdown of the intestinal mucosa is due to hypoxia, reperfusion injury, and granulocyte action. Endotoxin perpetuates this process by a positive feedback loop (Deitch et al. 1987).

Beside BT, the septic focus is obviously a source of bacterial product release. Among the major release products are endotoxins (Rietschel et al., this Vol.) and pore-forming exotoxins (see Seeger et al., this Vol.). In addition to these toxins, the bacterial wall component peptidoglycan, when released, has similar cell-activating properties. In a sheep model, peptidoglycan has been shown to induce a permeability increase in the lung comparable to LPS-induced elevation, although peptidoglycan requires much higher concentrations (Redl et al. 1989).

2.2 Monitoring of Endotoxin Plasma Levels

The release of endotoxin can be monitored in the in vivo setting by measuring plasma endotoxin levels. The occurrence of LPS in the plasma could be a predictive factor.

However, there are several obstacles to the limulus LPS measurement since:

1. The test is not absolutely specific.
2. Contamination can easily occur during blood sampling and plasma preparation.
3. Endotoxin leaking into the circulation is quickly bound to plasma proteins and blood cells. Therefore, the kinetics of infused endotoxin is characterized by a very short half-life (Freudenberg et al. 1984).
   To overcome obstacle (2) specific LPS blood-sampling tubes have been developed to avoid LPS binding to vial walls and contamination due to the pyrogen content of the vials and to permit storage of samples after plasma and blood cell separation in the freezer until the plasma samples are analyzed.

Cytokine formation is probably the major event following the release of bacterial products, especially by monocytes, but also by, e.g., fibroblasts. In addition, LPS activates the complement system via the alternative pathway and, at least in vitro, endothelial cells (see below).

2.3 Nonbacterial Inflammation

Aside from bacterial stimuli, sepsis is also caused by nonbacterial events (Goris et al. 1985). In both instances, sepsis may be characterized by leukocytosis, temperature, thrombocytopenia, and the need of myocardial support. This has been most effectively demonstrated by the group around Goris, who has set up a model of short- and/or long-term organ failure with sterile peritonitis induced by zymosan suspension in paraffin oil (Goris et al. 1986). This model has been extensively studied. However, it is currently not clear which inflammatory reactions are decisive for the development of organ failure, since it has been shown that even in pathogen-free rats late organ failure occurs. On the other hand, there is evidence that at least in "normal" rats part of the inflammatory reaction could be related to bacterial translocation products and TNF (Von Asmuth et al. 1990).

2.4 Involvement of Cytokines

The toxin-induced formation of TNF and the mechanism involved in this reaction have been described in the chapter by Rietschel et al. (this Vol.) and Wendel et al. (this Vol.). Although studies (Tracey et al. 1987; Schlag et. al. 1991) have clearly indicated that TNF blockade leads to decreased mortality, it is difficult to prove the formation and occurrence of TNF in plasma in the clinical setting since the release kinetics do not permit reliable monitoring. In contrast, the monocyte/macrophage activation can be nicely monitored by measuring neopterin plasma levels, as neopterin production is more protracted than TNF formation. Therefore, neopterin has proved to be a valuable monitoring parameter in the intensive care situation (Strohmaier et al. 1987), but has no known mediator role. In contrast, IL-6 acts as an important mediator, e.g., of acute-phase reactions in the liver (Heinrich et al. 1990), and is thus a valid parameter in the clinical setting as its release characteristics are suited for monitoring.

Besides IL-6, the proinflammatory cytokines IL-1 (see Dinarello, this Vol.) and IL-8 (see Baggiolini, this Vol.) are thought to be of central importance in the pathophysiology of sepsis. Similar to TNF, the key role of IL-1 in organ failure has been supported by recent studies, where sepsis-related organ failure was prevented by the IL-1 receptor antagonist (Ohlsson et al. 1990). The release kinetics of IL-8 during sepsis are similar to those of IL-1 and IL-6 as observed in baboon studies (Redl et al. 1991a). Antibody interference with IL-8 action is currently being investigated to study the potential link between monocytes and PMN activation. The question remains whether interference at the level of IL-1 or IL-8 is a useful approach, since blockade of TNF by an antibody (or e.g., by a TNF-binding protein) (Himmler et al. 1991) significantly diminishes the production of IL-1, IL-6, and IL-8 (Redl et al. 1991a).

2.5 Endothelial Activation

A distinct sequence of events occurs if the EC are activated by LPS or by cytokines. The in vitro properties of the endothelium, inducible by LPS and cytokines, include cytokine expression, procoagulant activity, immunological functions, and increased adhesiveness for leukocytes due to expression of adherence molecules. These events are within the definition of "endothelial activation (Pober et al. 1990). We can now demonstrate that a (de novo) expression of adhesion molecules occurs in vivo under septic conditions in subprimate animal models by using two antibodies to the ELAM-1 structure (Redl et al. 1991c), which seems to be TNF-dependent (Redl et al. 1991b).

Endothelial leukocyte adhesion molecules (ELAM-1) serve to bind PMN via a sialyl-Lewis-X structure (Phillips et al. 1990). ELAM-1 is not present on unstimulated endothelial cells in vitro and may transiently be induced (peak at 4–6 h) by LPS, IL-1, or TNF. We have previously shown massive cytokine release in sepsis, while cytokine release was minimal in the polytraumatic situation in (baboon) models (Redl et al. 1991a). This might explain the differences in endothelial activation. Furthermore, concentrations of endotoxin circulating in the plasma are several log steps higher in sepsis than in trauma (S. Bahrami, unpubl. results). Nevertheless, small amounts of endotoxin seen after polytrauma due to bacterial translocation might account for the few positive endothelial stainings encountered in the trauma animals (Redl et al. 1991c) (Table 1).

ELAM-1 expression may serve as a marker of endothelial activation (as suggested by Pober 1990) and is one of the possible preconditions that lead to leukocyte-related

Table 1. Immunohistochemical identification of the adherence molecule ELAM-1 in baboon tissue from different organs after sepsis or polytrauma. (Redl et al. 1991c)

	Immunohistochemical localization of ELAM-1 (grades–3)			
	Septic shock		*Traumatic shock*	
	No. examined	*Average score*	*No. examined*	*Average score*
Lung	6	2.2+	13	0.03+
Liver	5	3.0+	12	0.16+
Kidney	3	2.7+	9	0.8+
Skin	2	3.0+	2	0.5+

endothelial damage. Our data certainly do not permit the clear conclusion that ELAM-1 antigen expression causes EC leakiness. However, a recent study on immune complex-induced alveolitis suggests that in vivo blocking of ELAM-1 with MoAb has favorable therapeutic effects (Ward et al. 1991).

Activated and endothelial-bound leukocytes may be responsible for vascular leakiness associated with traumatic and septic shock. In a series of animals we observed endothelial leakiness leading to edema formation in the lung.

ELAM-1 kinetics are different from ICAM-1 kinetics (studied with MoAb and other agents), as ICAM-1 is already present on unstimulated endothelial cells and is upregulated by cytokines in vitro. Preliminary evidence obtained in our baboon model (unpubl. results) suggests that the same holds true for septic shock conditions in vivo. However, the observed upregulation is protracted as compared to ELAM-1, similar to the findings of Pober after local LPS injection (Munro et al. 1989).

Based on various in vitro and in vivo studies, the following time sequence of EC involvement in adhesion phenomena can be used as a working hypothesis (Table 2, based on Pober et al. 1990).

The complementary molecules on the leukocyte surface are summarized in Fig. 4 (Springer et al. 1990) and are either constitutive (e.g., Lewis-X, LFA-1), lost during PMN activation (e.g., LAM-1), or upregulated, e.g., Mac-1 (CD11b/CD18) (Fig. 4).

Table 2. Time sequence of endothelial cell stimulation (by, e.g., thrombin) and activation (by endotoxin or cytokines)

Times post-stimulus	Minutes	1–12 h	12 h to several days
Molecules involved	GMP-140, PAF	ELAM-1, ICAM-1	ICAM-1, HLA-DR
	"EC stimulation"	"EC activation"	

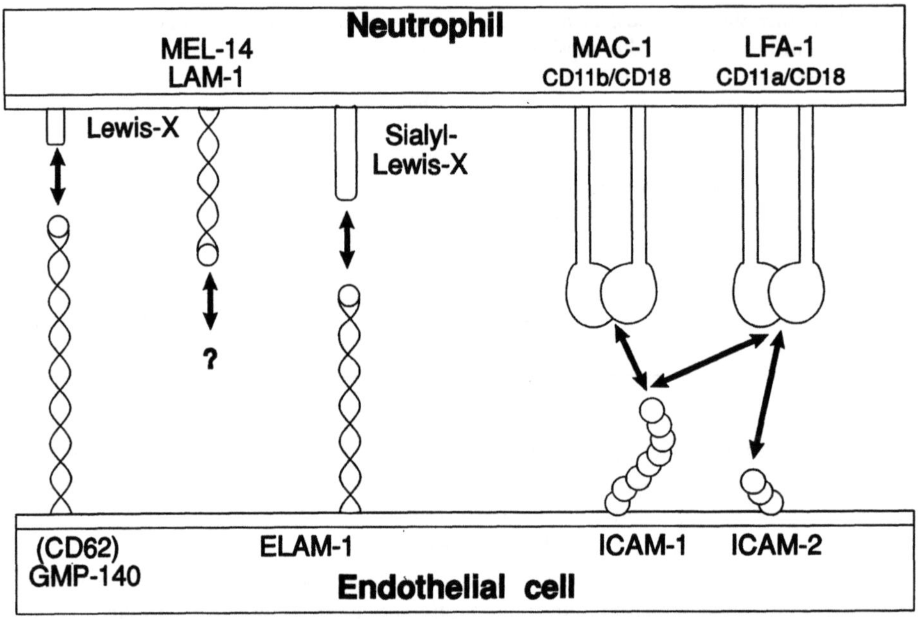

Fig. 4. Schematic representation of endothelial-neutrophil interactions. (After Springer et al. 1990)

2.6 Endothelial Damage

All these EC-leukocyte interactions seem to constitute the prerequisite for both PMN extravasation and EC damage. Since EC alteration or damage causes permeability changes and ultimately edema formation, the EC-leukocyte interactions are of central importance in shock-induced organ damage. Adhering and activated PMN might be crucial in EC damage, since a microenvironment is formed, mediators are released locally, and damage occurs without a sufficient neutralizing capacity (antiproteases, antioxidants). This type of damage is attributed to the effect of oxygen radicals, proteases, or both (Varani et al. 1989).

There is additional involvement of endothelium and monocytes in endotoxemia via the formation of disseminated intravascular coagulation (DIC) caused by the procoagulant tissue factor (see Taylor, this Vol.). Besides DIC and permeability changes, metabolic alterations occur, which lead to severe catabolic states and aggravate the course of organ failure. All the above-described inflammatory mediators are responsible for the numerous interactions of different cell types, which cause amplification effects through feedback cycles. This leads to vascular permeability changes, DIC, anoxia, and cell death, and may terminate as (multi-) organ failure.

Acknowledgments. We would like to thank all our collaborators and staff involved in the different studies, who helped to substantiate this hypothesis. We are indebted to Dr. U. Obertacke, Essen, for his valuable input and Mrs. Eva Schwarz for typing. Several studies we have quoted were supported by the Lorenz Böhler Fond.

References

Ayala A, Perrin MM, Meldrum DR, Ertel W & Chaudry IH (1990) Hemorrhage induces an increase in serum TNF which is not associated with elevated levels of endotoxin. Cytokine 2:170–174

Bengtson A, Lannsjo W & Heideman M (1987) Complement and anaphylatoxin responses to cross clamping of the aorta studies during general anesthesia with or without extradural blockade. Br J Anaesth 59:1093–1097

Bühren V, Maier B, Hower R, Holzmann A, Redl H & Marzi I (1991) PAF antagonist BN52021 reduces hepatic leukocyte adhesion following intestinal ischemia. Circ Shock 34:134–135

Chaudry IH, Ayala A, Ertel W & Stephan RN (1990) Hemorrhage and resuscitation: immunological aspects. Am J Physiol 259:R663–R678

Colletti LM, Burtch GD, Remick DG, Kunkel SL, Strieter RM & Campbell DA (1990) Role of tumor necrosis factor alpha in the pathophysiologic alterations after hepatic ischemia reperfusion injury in the rat. Transplantation 49:268–272

Deitch EA, Berg R & Specian R (1987) Endotoxin promotes the translocation of bacteria from the gut Arch Surg 122:185–190

Faist E, Ertel W, Mewes A, Strasser T, Walz A & Alkan S (1988) Mediators and the trauma induced cascade of immunologic defects Prog Clin Biol Res 308:495

Fine J, Ruteburg SH & Schweinburg FB (1959) The role of the RES in hemorrhagic shock. J Exp Med 110:547–551

Freudenberg MA, Kleine B & Galanos C (1984) The fate of lipopolysaccharide in rats: evidence for chemical alteration in the molecule. Rev Infect Dis 6:483–487

Friedl HP, Smith DJ, Till GO, Thomas PD, Louis DS & Ward PA (1990) Ischemia reperfusion in humans: appearance of xanthine oxidase activity. Am J Pathol 136:491–495

Geng JG, Bevilacqua MP, Moore KL, McIntyre TM, Prescott SM, Kim JM, Bliss GA, Zimmerman GA & McEver RP (1990) Rapid neutrophil adhesion to activated endothelium mediated by GMP 140. Nature (London) 343:757–760

Goris RJA, te Boekhorst TPA, Nuytinck JKS & Gimbrere JSF (1985) Multiple organ failure: generalized autodestructive inflammation. Arch Surg 120:1109–1115

Goris RJA, Boekholtz WKF, van Bebber IPT, Nuytinck JKS & Schillings PHM (1986) Multiple organ failure and sepsis without bacteria. Arch Surg 121:897–901

Granger DN, Rutili G & McCord JM (1981) Superoxide radicals in feline intestinal ischemia. Gastroenterology 81:22–29

Grum CM, Ragsdale RA, Ketai LH & Simon RH (1987) Plasma xanthine oxidase activity in patients with adult respiratory distress syndrome. J Crit Care 2:22–26

Harlan JM, Killen PD, Snecal F, Schwartz BR, Yee EK, Taylor RF, Beatty PG, Price T & Ochs HD (1985) The role of neutrophil membrane glycoprotein GP-150 in neutrophil adherence to endothelium in vitro. Blood 66:167–178

Heideman M, Kaijser B & Gelin LE (1978) Complement activation and hematologic hemodynamic and respiratory reactions early after soft tissue injury. J Trauma 18:696–700

Heinrich PC, Castell JV & Andus T (1990) Interleukin 6 and the acute phase response. Biochem J 265:621–636

Himmler A, Maurer-Fogy I & Adolf GR (1991) TNF binding protein the soluble form of the TNF receptor p60 chain as a natural inhibitor of TNF activity. Circ Shock 34:166

Lang H, Jochum M, Fritz H & Redl H (1989) Validity of the elastase assay in intensive care medicine. Prog Clin Biol Res 308:701–706

Lewis MS, Whatley RE, Cain P, McIntyre TM, Prescott SM & Zimmerman GA (1988) Hydrogen peroxide stimulates the synthesis of platelet activating factor by endothelium and induces endothelial cell dependent neutrophil adhesion. J Clin Invest 82:2045–2055

Marzi I, Rehkopf A, Hower R, Bühren V & Trentz O (1991) Superoxide dismutase reduces leukocyte adherence in hepatic sinusoid following hemorrhagic shock in the rat. Circ Shock 34:145

McCord J M (1985) Oxygen derived free radicals in postischemic tissue injury. N Engl J Med 312:159–163

Meakins JL & Marshall JC (1986) Multi organ failure syndrome The gastrointestinal tract: the motor of MOF. Arch Surg 121:196–208

Munro JM, Pober JS & Cotran RS (1989) Tumor necrosis factor and interferon gamma induce distinct patterns of endothelial activation and associated leukocyte accumulation in skin of papio anubis. Am J Pathol 135:1–13

Neumann S, Hunzer G, Heinrich N & Lang H (1984) PMN elastase assay: enzyme immunoassay for human polymorphonuclear elastase complexes with alpha-1 proteinase inhibitor. J Clin Chem Clin Biochem 22:693–697

Nuytinck JKS, Goris RJA, Redl H, Schlag G, van Munster PJJ (1986) Posttraumatic complications and inflammatory mediators. Arch Surg 121:886–890

Ohlsson K, Bjrk P, Bergenfeldt M, Hageman R & Thompson RC (1990) Interleukin 1 receptor antagonist reduces mortality from endotoxin shock. Nature (London) 348:550–552

Ozkan AN, Hoyt DB, Tompkins S, Ninnemann JL & Sullivan JJ (1988) Immunosuppressive effects of a trauma induced suppressor active peptide. J Trauma 28:589–592

Phillips ML, Nudelman E, Gaeta FCA, Perez M, Singhal AK, Hakomori SI & Paulson JC (1990) ELAM 1 mediates cell adhesion by recognition of a carbohydrate ligand Sialyl Le. Science 250:1130–1135

Pober JS & Cotran RS (1990) The role of endothelial cells in inflammation. Transplantation 50:537–544

Pretorius JP, Schlag G, Redl H, Botha WS, Goosen DJ, Bosman H & VanEeden AF (1987) The lung in shock as a result of hypovolemic-traumatic shock in baboons. J Trauma 27:1344–1352

Puyana JC, Rode H, Gordon J, Meakins JL, Chartrand L & Christou NV (1988) Lack of cytokine induced skin reaction correlates with acute physiology score and mortality in patients receiving intensive care. Arch Surg 123:1474–1476

Redl H & Schlag G (1990) Markers of endotoxin related leukocyte activation and injury mechanisms. Circ Shock 31:243

Redl H, Hammerschmidt DE & Schlag G (1983) Augmentation by platelets of granulocyte aggregation in response to chemotaxins: studies utilizing an improved cell preparation technique. Blood 61:125–131

Redl H, Hallström S, Lieners C, Fürst W & Schlag G (1988) In: Hörl W H & Heidland A (eds) Proteases II – potential role in health and disease pp 449–455

Redl H, Schlag G, Thurnher M, Traber LD & Traber DL (1989) Cardiovascular reaction pattern during endotoxin or peptidoglycan application in awake sheep. Circ Shock 28:101–108

Redl H, Schlag G, Bahrami S, Schade U, Ceska M & Stütz P (1991a) Plasma neutrophil activating peptide 1 interleukin 8 and neutrophil elastase in a primate bacteremia model. J Infect Dis 164:383–388

Redl H, Schlag G, Dinges HP, Buurman WA, Ceska M & Davies J (1991b) TNF dependent ELAM 1 expression and IL 8 release in baboon septicemia. Circ Shock 34:92

Redl H, Dinges HP, Buurman WA, van der Linden CJ, Pober JS, Cotran RS & Schlag G (1991c) Expression of endothelial leukocyte adhesion molecule 1 in septic but not traumatic hypovolemic shock in the baboon. Am J Pathol 139:461–466

Rush BF, Sori AJ, Murphy TF, Smith S, Flanagan JJ & Machiedo GW (1988) Endotoxemia and bacteremia during hemorrhage shock. The link between trauma and sepsis. Ann Surg 207:549–554

Saugstad OD & Ostrem T (1977) Hypoxanthine and urate levels of plasma during and after hemorrhagic hypotension in dogs. Eur Surg Res 9:48–56

Schlag G, Voigt WH, Schnells G & Glatzl A (1976) Die Ultrastruktur der menschlichen Lunge im Schock. I Anaesthesist 25:512–521

Schlag G, Redl H, Dinges HP, Davies J & Radmore K (1991a) Bacterial translocation in baboon model of hypovolemic traumatic shock. In: Schlag G & Redl H (eds.) Shock, sepsis, and organ failure, Second Wiggers Bernard Conference, Dürnstein. Springer Berlin Heidelberg New York pp 53–83

Schlag G, Redl H & Davies J (1991b) TNF antibodies CB0006 in a subchronic septic model in baboons to prevent multi organ failure. Circ Shock 34:164

Scholz W, McClurg MR, Cardenas GJ, Smith M, Noonan DJ, Hugli TE & Morgan EL (1990) C5a mediated release of interleukin 6 by human monocytes. Clin Immunol Immunopathol 57:297–307

Schraufstätter IU, Revak SD & Cochrane CG (1984) Proteases and oxidants in experimental pulmonary inflammatory injury. J Clin Invest 73:1175–1184

Springer TA (1990) Adhesion receptors of the immune system. Nature (London) 346:425–434

Strohmaier W, Redl H, Schlag G & Inthorn D (1987) D-erythro-neopterin plasma levels in intensive care patients with and without septic complications. Crit Care Med 15:757–760

Tracey KJ, Fong Y, Hesse DG, Monogue KR, Lee AT, Kuo GC, Lowry SF & Cerami A (1987) Anti cachectin TNF monoclonal antibodies prevent septic shock during lethal bacteremia. Nature (London) 330:662–664

Varani J, Ginsburg I, Schuger L, Gibbs F, Bromberg J, Johnson KJ & Ward PA (1989) Endothelial cell killing by neutrophils. Synergistic interaction of oxygen products and proteases. Am J Pathol 135:435–438

Von Asmuth EJU, Maessen JG, van der Linden CJ, Buurman WA (1990) Tumour necrosis factor alpha and interleukin 6 in a zymosan induced shock model. Scand J Immunol 32:313–319

Ward P (1991) Role of ELAM in inflammation in vivo. In: 2nd Int Congr on the Immune consequences of trauma, shock and sepsis mechanisms and therapeutic approaches. München Abstr p 22

Ward PA, Johnson KJ & Till GO (1986) Animal models of oxidant lung injury. Respiration 50 (Suppl. 1):5–12

Yokoyama Y & Parks DA (1988) Circulating xanthine oxidase: release of xanthine oxidase from isolated rat liver. Gastroenterology 94:607

Yokoyama Y, Beckman JS, Beckman TN, Wheat JK, Cash TG, Freeman BA & Parks DA (1990) Circulating xanthine oxidase: potential mediator of ischemic injury. Am J Physiol 258:G564–G570

Zilow G, Naser W, Rutz R & Burger R (1989) Quantitation of the anaphylatoxin C3a in the presence of C3 by a nove sandwich ELISA using monoclonal antibody to a C3a neoepitope. J Immunol Meth 121:261–268

Leukotrienes, Oxygen Radicals, and Cytokines in Septicemic Mice

A. WENDEL[1], M. NIEHÖRSTER[1], and G. TIEGS[1]

1 Introduction

When microorganisms or certain parts of their cell walls invade a mammalian circulation system, the risk of the pathological state of sepsis arises. This situation is characterized by fever, blood pressure changes, and a general reaction of leukocytes. An overshoot reaction toward bacterial infection may lead to the life-threatening condition of septic shock, which even in modern intensive care units is difficult to clinically manage (Goris 1987). Under these given circumstances, the development of therapeutic strategies against septic shock is a primary research goal. The fulfillment of this need, however, depends on the availability and detailed mechanistic knowledge of the reactions involved in sepsis – an area to be explored in suitable model systems. Recent advances in the identification, isolation, and characterization of lipopolysaccharides (endotoxins), as the part of the cell wall of gram-negative bacteria responsible for inducing shock (Rietschel et al. 1983), allowed the induction of septicemic states in animals by administration of purified endotoxins. When this experimental protocol is applied, a dose-dependent increase in endotoxin-induced lethality is observed. An alternative to this all-or-none reaction model is also available: rodents can be sensitized to endotoxin several 1000-fold when they have been retreated with galactosamine. This amino sugar is metabolized exclusively in the liver, leads to an indirect protein biosynthesis block (Keppler and Decker 1969), and thus renders the liver as the prime target organ of endotoxin-induced injury (Galanos et al. 1979).

The work reviewed here uses combinations of galactosamine with putative mediators of septic shock as a tool to explore the nature and order of pathogenic action of these mediators in the search for new drugs effective against shock.

2 Results and Discussion

2.1 The Role of Leukotrienes

Administration of endotoxin to mice which had been pretreated with D-galactosamine resulted in a dramatic increase in plasma enzyme activities indicative for severe liver injury 8 h after challenge (Fig. 1). This rise in enzyme activity was preceded by a

[1] University of Konstanz, Biochemical Pharmacology, POB 5560, D-7750 Konstanz, FRG

42. Colloquium Mosbach 1991
Molecular Aspects of Inflammation

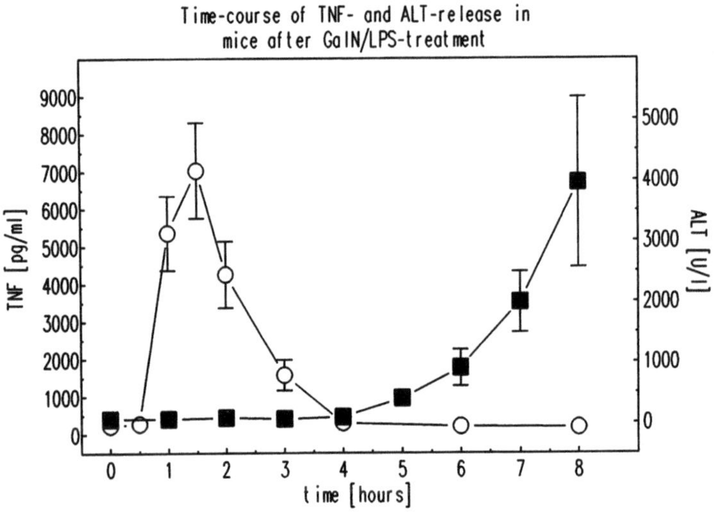

Fig. 1. Time course of the development of hepatitis (assessed by serum alanine amino transferase increase, *filled squares*) and release of tumor necrosis factor into the circulation (*open circles*) in male NMRI albino mice sensitized with 700 mg/kg D-galactosamine and challenged with 5 µg/kg *Salmonella abortus equi* endotoxin

dramatic transient increase in TNF in serum peaking 90 min after endotoxin administration. Previous work by Hagmann et al. (1984) showed that after endotoxin injection, an increase in peptidoleukotrienes in the bile occurs. These authors hypothesized that leukotriene formation and secretion may be early consequences of endotoxemia. They also showed that endotoxin induced an inhibition of LTD_4 elimination into bile. These authors proposed that the impairment of the leukotriene deactivation pathway may represent a kêy mechanism in the lethal action of endotoxin.

On the basis of these previous studies, we investigated whether replacement of endotoxin injection by various leukotrienes could bring about similar symptoms as induces by endotoxin. Therefore, different leukotrienes, i.e., LTC_4, LTD_4, LTE_4, or LTB_4, were intravenously injected into mice that had been sensitized with galactosamine. The data in Table 1 show that leukotriene D_4 had the ability to induce hepatitis, i.e., to substitute endotoxin. In a series of similarly designed experiments, intravenous injection of murine recombinant TNFα induced liver injury in galactosamine-sensitized mice. The results obtained by direct administration of alleged mediators of sepsis show that a similar pathophysiological reaction as seen after endotoxin can be produced in vivo by leukotriene or cytokine administration.

Several recent reports investigated the uptake, the production, the metabolism, and the biliary excretion of cysteinyl leukotrienes in the isolated perfused rat liver (Hagmann et al. 1989; Wettstein et al. 1989). Furthermore, it was shown that leukotrienes exhibit marked metabolic and hemodynamic changes in the isolated organ (Häussinger et al. 1988; Iwai and Jungermann 1988). Interestingly, not only endotoxin administration but also in vivo tissue trauma induced a secretion of peptidoleukotrienes into the bile (Denzlinger et al. 1985). A recent paper by Hagmann et

Table 1. Alternative experimental means of inducing fulminant hepatitis as assessed by serum aspartate amino transferase (E.C.2.6.1.1.) increase in mice sensitized by administration of 700 mg/kg D-galactosamine

Treatment	Serum ALT[a]
Untreated control	40 ± 4
Only galactosamine	50 ± 6
30 μg/kg Endotoxin i.p.	3860 ± 1440*
50 μg/kg LTD_4 i.v.	1150 ± 490*
15 μg/kg rmu TNFα i.v.	3510 ± 900*
Anti-TNFα antiserum plus TNFα	62 ± 12

* $p < 0.05$ compared to controls.
[a] Determined 8 h after challenge, data ± SEM (Tiegs and Wendel 1988; Tiegs et al. 1989 a, b).

al. (1991) considers liver mast cells in cooperation with liver macrophages, i.e., Kupffer cells, as candidates for hepatic leukotriene generation.

In general, macrophages seem to be potent cells of cysteinyl leukotriene production upon endotoxin stimulation (Lüderitz et al. 1989). It was found that lipopolysaccharides of rough mutant bacteria stimulated mouse peritoneal macrophages to release significant amounts of LTC_4. Rosenstreich et al. (1989) identified macrophages as mediator cells of endotoxin-induced host responses. Macrophages are also the predominant producers of cytokines upon LPS stimulation. Beutler et al. (1985a) first reported the synthesis of cachectin/tumor necrosis factor by a macrophage cell line after endotoxin incubation. Beutler's group also showed (1985b) that passive immunization against TNFα protected mice from lethal effects of endotoxin.

Several reports are available showing a link between leukotriene synthesis and cytokine production in vitro: leukotrienes augment the interleukin 1 production by monocytes (Rola-Pleszczynski and Lemaire 1985). These findings were extended by showing that leukotrienes activate macrophages also to release TNF (Gagnon et al. 1989).

Detailed pharmacological intervention studies allowed one to obtain further information about the sequential order of mediator action. It was shown that in vivo inhibitors of 5-lipooxygenase activity (Wendel and Tiegs 1986) or inhibitors of leukotriene D_4 synthesis (Tiegs and Wendel 1988) protected against endotoxin. This protection was due to a suppression of endotoxin-induced TNF release (Schade et al. 1990). Accordingly, these agents failed to protect against injury induced directly by TNFα administration (Tiegs et al. 1989b). On the other hand, these agents had no protective effect when the lesion was induced by LTD_4 administration; in this case, only LTD_4 antagonists were effective (Tiegs and Wendel 1988). Since compounds exhibiting vasodilatative activities such as Iloprost or Taprosten protected also against LPS (Tiegs et al. 1989a), even when given simultaneously with endotoxin, it seems that the basic principle of the leukotriene action is likely to consist in a vasoconstriction, resulting in a transient ischemia.

2.2 The Role of Reactive Oxygen

In the past decade, overwhelming experimental evidence has accumulated, suggesting that the ischemia/reflow syndrome involves oxygen radicals as a primary mechanism of organ injury (McCord 1985). The biochemical events of this ischemia/reperfusion syndrome are characterized by a degradation of intracellular ATP within ischemic areas accompanied by a conversion of xanthine dehydrogenase to xanthine oxidase activity. Following reflow caused by either degradation of vasoconstrictive activity (such as leukotrienes) or by removal of mechanical obstacles in the vessels (such as thrombotic clots), a deleterious situation arises: readmission of oxygen, the presence of formerly absent substrate, i.e., hypoxanthine, and an altered enzyme activity, i.e., an oxidase, result in the formation of O_2^- from O_2 at the expense of the conversion of hypoxanthine to xanthine. Septicemia may thus increase the production of reactive oxygen species via leukotriene-mediated ischemia/reflow episodes. The question arises whether reactive oxygen species are involved in the pathomechanism of bacterial endotoxin. This seems possible since it was shown in vivo that *hepatic* ischemia/reperfusion injury results in enhanced TNF levels with subsequent local and systemic tissue injury (Coletti et al. 1990).

We therefore investigated whether in vivo scavengers of reactive oxygen formation would affect endotoxin-induced hepatitis.

The pharmacological intervention profiles compiled in Table 2 show that pretreatment of mice with either allopurinol or superoxide dismutase protected against endotoxin-induced liver injury, whereas it failed to protect against hepatitis induced by TNFα. In the protected animals, no serum TNF was detectable, indicating that reactive oxygen scavengers interfere with the production of TNFα without affecting the toxicity of this endogenous mediator.

2.3 The Role of the Protease/Antiprotease Balance

A similar pattern of protection against LPS via inhibition of TNF production and lack of protection against directly administered TNFα was obtained following administration of the antiprotease α_1PI. This analogy raises the obvious question of how reactive oxygen production and protease activity may be linked to the production and secretion of TNFα. In order to answer it, published work from different fields of scientific interest have to be put into a joint perspective.

The primary response of the organism to an inflammatory stimulus is directed against the initiating "foreign" material. An early event in this host-defense mechanism is the release and activation of proteases of leukocytic origin hereby playing a pivotal role in inflammation (Weiss 1989) and shock (Jochum and Fritz 1989). Stimulated inflammatory cells, particularly polymorphonuclear granulocytes (PMNs) are able to release large amounts of proteolytic enzymes, accumulated in intracellular

Table 2. Pharmacological intervention profiles against galactosamine/LPS- or galactosamine/TNFα-induced hepatitis following pretreatment with antiprotease or with compounds interfering with reactive oxygen. (Wendel et al. 1987; Tiegs et al. 1989; Niehörster et al. 1990)

Pretreatment	LPS challenge (ALT after 8 h)	TNFα challenge (ALT after 8 h)	Serum TNF 1 h after LPS
Allopurinol	220 ± 50*	2230 ± 360	≤ 50
Superoxide dismutase	140 ± 60*	3560 ± 720	≤ 50
α_1PI	110 ± 20*	4290 ± 1420	≤ 50
Injury controls	2710 ± 1000	2700 ± 1150	6200 ± 1240*
Solvent controls	30 ± 10	30 ± 10	≤ 50

Doses: Human α_1PI, 42 mg/kg i.v. 1 h prior to challenge; bovine recombinant SOD, 10 mg/kg i.v. 1 h prior to challenge Allopurinol: 100 mg/kg i.p., 1 and 24 h prior to challenge.
Data ± SEM, n = 6, *$p < 0.05$ compared to controls
ALT in U/l, TNF in pg/ml serum.

granules, into the surrounding environment. Besides the gelatinase from the "specific" granules, the neutral proteases (e.g., elastase and cathepsin G) from the "azurophilic" granules stand out as being significant among the lysosomal enzymes currently recognized (Travis et al. 1980). It is under debate, however, whether protease release is a process leading to unspecific degradation of injured tissues or whether it is actively involved in specific mediation of the host-defense mechanism.

In vitro experiments showed that a membrane-bound form of TNFα exists in cells known to release TNFα (Kriegler et al. 1988). Independently, it was reported that TNFα release upon LPS stimulus in mononuclear cells was blocked by a low molecular weight serine protease inhibitor (Scuderi 1989), or by preincubation of the cells with α-globulins including α_1PI itself (Scuderi et al. 1991).

3. Synopsis

On the basis of these published facts, the entire set of our results (Fig. 2) may be explained by a model of the pathogenic sequence of LPS-induced shock which is diagrammatically presented in Fig. 3. When endotoxin is present, the lipopolysaccharide binds to a monocyte and induces the expression of membrane-bound TNFα. In addition to this slow process, an immediate synthesis and release of leukotrienes from macrophages into the bloodstream is triggered by endotoxin. These eicosanoids are very potent vasoconstrictors (Lefer 1986) and lead to a local and transient restriction of blood flow. In fact, when peptidoleukotrienes were perfused into the isolated liver, a reduction of flow or an increase in pressure at constant flow was observed (Häussinger et al. 1988). After elimination of the short-lived leukotriene, reflow and reoxygenation occur, giving rise to extracellular superoxide production and oxidative inactivation of antiprotease activity, thereby allowing the corresponding protease acti-

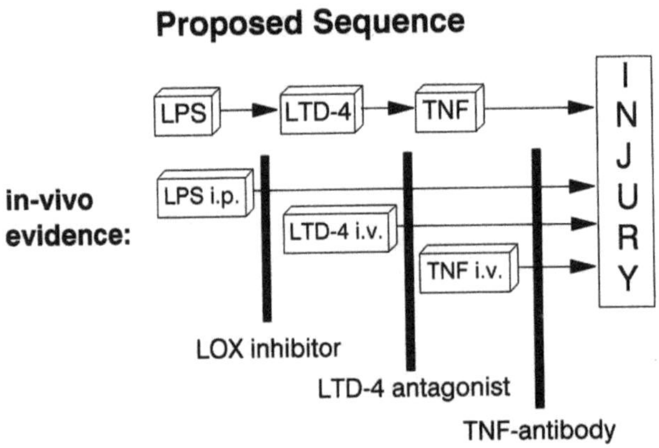

Fig. 2. Working hypothesis of the alignment of mediators of sepsis and shock (*upper part*) and the available in vivo evidence in the galactosamine-sensitized mouse. *LPS* lipopolysaccharide, i.e., endotoxin; *LTD_4* leukotriene D_4; *TNF* tumor necrosis factor; *LOX* lipoxygenase; *i.v.* intravenously administered; *i.p.* intraperitoneally administered

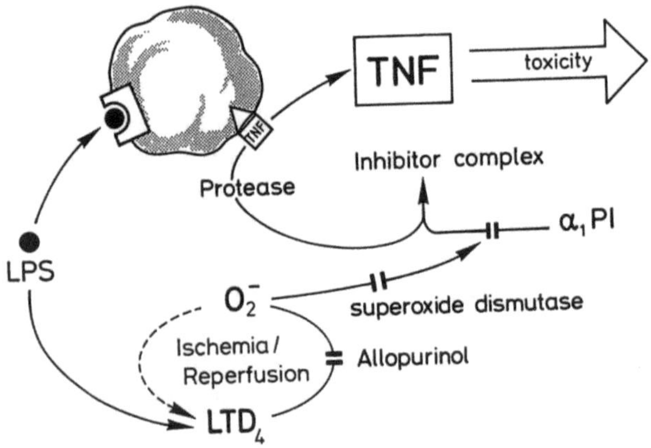

Fig. 3. Proposed interactive network of mediators of shock and sepsis in the galactosamine-sensitized mouse model showing pathway blocked by agent indicated and feedback amplification; *α_1PI* α_1 proteinase inhibitor, i.e., α_1 antitrypsin; Δ membrane anchor of surface bound TNFα precursor

vity to become active. In this situation, the cleavage of previously expressed membrane-bound TNF to its soluble form becomes possible and the terminal events of TNFα-induced shock will take place. Indeed, this latter step has also been very recently demonstrated in an in vitro system where excess α_1PI prevented the endotoxin-induced release of TNFα from monocytes (Scuderi et al. 1991).

This concept of key events in inflammation and shock reconciles several independently recognized pieces of evidence, i.e., the involvement of eicosanoids, the government of the protease/antiprotease balance by oxygen radicals, and the processing of a pivotal cytokine such as TNFα, and bridges some gaps between a pathogenic event which had been previously separated from each other.

Acknowledgment. This work was supported by grant We 686/13–1 from the SFB 156 of the Deutsche Forschungsgemeinschaft.

References

Beutler B, Mahoney J, Le Trang N, Pekala P & Cerami A (1985a) Purification of cachectin, a lipoprotein lipase-suppressing hormone secreted by endotoxin-induced raw 264.7 cells. J Exp Med 161:984–995
Beutler B, Milsark IW & Cerami AC (1985b) Passive immunization against cachectin/tumor necrosis factor protects mice from lethal effect of endotoxin. Science 229:869–871
Colletti LM, Remick DG, Burtch GD, Kunkel SL, Strieter RM & Campbell Jr DA (1990) Role of tumor necrosis factor-α in the pathophysiologic alterations after hepatic ischemia/reperfusion injury in the rat. J Clin Invest 85:1936–1943
Denzlinger C, Rapp S, Hagmann W & Keppler D (1985) Leukotrienes as mediators in tissue trauma. Science 230:330–332
Gagnon L, Filion LG, Dubois C & Rola-Pleszczynski M (1989) Leukotrienes and macrophage activation: augmented cytotoxic activity and enhanced interleukin 1, tumor necrosis factor and hydrogen peroxide production. Agents Actions 26:141–147
Galanos C, Freudenberg MA & Reutter W (1979) Galactosamine-induced sensitization to the lethal effects of endotoxin. Proc Natl Acad Sci USA 76:5939–5943
Goris RJA (1987) The adult respiratory distress syndrome and multiple-organ failure. Intense Care News 1:1–7
Hagmann W, Denzlinger C & Keppler D (1984) Role of peptide leukotrienes and their hepatobiliary elimination in endotoxin action. Circ Shock 14:223–235
Hagmann W, Parthé S & Kaiser I (1989) Uptake, production and metabolism of cysteinyl leukotrienes in the isolated perfused rat liver. Biochem J 261:611–616
Hagmann W, Kaiser I & Jakschik BA (1991) The sensitized liver represents a rich source of endogenous leukotrienes. Hepatology 13:482–488
Häussinger D, Stehle T & Gerok W (1988) Effects of leukotrienes and the thromboxane A_2 analogue U-46619 in isolated perfused rat liver: metabolic, hemodynamic and ion-flux responses. Biol Chem Hoppe-Seyler 369:97–107
Iwai M & Jungermann K (1988) Leukotrienes increase glucose and lactate output and decrease flow in perfused rat liver. Biochem Biophys Res Commun 151:283–290
Jochum M & Fritz H 1989) Pathobiochemical mechanisms in inflammation. In: Faist E, Ninnemann J L & Green D R (eds.) Immune consequences of trauma, shock and sepsis. Springer Berlin Heidelberg New York pp. 165–172
Keppler D & Decker K (1969) Studies on the mechanism of glactosamine hepatitis: accumulation of galactosamine-1-phosphate and its inhibition of UDP-glucose pyrophosphorylase. Eur J Biochem 10:219–225
Kriegler M, Perez C, DeFay K, Albert I & Lu SD (1988) A novel form of TNF/cachectin is a cell surface cytotoxic transmembrane protein: ramifications for the complex physiology of TNF. Cell 53:4553
Lefer AM (1986) Leukotrienes as mediators of ischemia and shock. Biochem Pharmacol 35:123–127
Lüderitz T, Brandenburg K, Seydel U, Roth A, Galanos C & Rietschel ET (1989) Structural and physicochemical requirements of endotoxins for the activation or arachidonic acid metabolism in mouse peritoneal macrophages in vitro Eur J Biochem 179:11–16
McCord JM (1985) Oxygen-derived free radicals in postischemic tissue injury. New Engl J Med 312:159
Niehörster M, Tiegs G, Schade UF & Wendel A (1990) In vivo evidence for protease-catalysed mechanism providing bioactive tumor necrosis factor α. Biochem Pharmacol 40:1601–1603
Rietschel ET, Zähringer U, Wollenweber HW, Miragliotta G, Musehold J, Lüderitz T & Schade U (1983) Bacterial endotoxins: chemical structure and biologic activity. Am J Emergency Med 2:60–69

Rola-Pleszczynski M & Lemaire (1985) Leukotrienes augment interleukin 1 production by human monocytes. J Immunol 135:3958–3961

Rosenstreich DL & Vogel SN (1980) Central role of macrophages in the host response to endotoxin. In: Schlessinger D (ed) Microbiology – 1980. Am Soc Microbiol Washington DC pp 11–15

Schade UF, Burmeister I, Engel R, Reinke M & Wolter DT (1989) Lipoxygenase inhibitors suppress formation of tumor necrosis factor in vitro and in vivo. Lymphokine Res 8:245–249

Scuderi P (1989) Suppression of human leukocyte tumor necrosis factor secretion by the serine protease inhibitor p-Toluene-sulfonyl-L-Arginine Methyl Ester (TAME). J Immunol 143:168–173

Scuderi P, Dorr RT, Lidill JD, Finley PR, Meltzer P; Raitano AB & Rybski J (1991) Alpha-globulins suppress human leukocyte tumor necrosis factor secretion. Eur J Immunol 19:939–942

Tiegs G & Wendel A (1988) Leukotriene-mediated liver injury. Biochem Pharmacol 37:2569–2573

Tiegs G, Werner C & Wendel A (1989a) Involvement of reactive oxygen species in endotoxin-induced hepatitis in galactosamine-sensitized mice. Medical biochemical and chemical aspects of free radicals. In: Hayaishi O, Niki E, Kondo M & Yoshikawa T (eds) Proc 4th Biennial Meet of the Society for Free Radical Research, Kyoto. Elsevier Amsterdam pp 1379–1383

Tiegs G, Wolter M & Wendel A (1989b) Tumor necrosis factor is a terminal mediator in galactosamine/endotoxin-induced hepatitis in mice Biochem Pharmacol 38:627–631

Travis J, Giles PJ, Porcelli L, Reily CF, Baugh R & Powers J (1980) Human leucocyte elastase and cathepsin G: structural and functional characteristics. In: Protein degradation in health and disease. Ciba Found Symp 75 Exerpta Medica Amsterdam pp 51–68

Weiss SJ (1989) Tissue destruction by neutrophils, mechanisms of disease. N Engl J Med 320:365–376

Wendel A & Tiegs G (1986) A novel biologically active seleno-organic compound-VI. Protection by ebselen (PZ51) against galactosamine/endotoxin-induced hepatitis in mice. Biochem Pharmacol 35:2115–2118

Wendel A, Tiegs G & Werner C (1987) Evidence for the involvement of a reperfusion injury in galactosamine/endotoxin-induced hepatitis in mice. Biochem Pharmacol 36:2637–2639

Wettstein M; Gerok W & Häussinger D (1989) Metabolism of cysteinyl leukotrienes in nonrecirculating rat liver perfusion: hepatocyte heterogeneity in uptake and biliary excretion. Eur J Biochem 181:115–124

Baboon Model of *E. Coli* Sepsis: Summary of Staging, Mechanism, and Diagnostic Markers

F. B. TAYLOR Jr.[1]

1 Introduction

Septic shock can present a cruel paradox, for in spite of watchful care it occurs most often in the hospital. It occurs in a wide variety of patients ranging from premature neonates with necrotizing enterocolitis to the aged with burns limited to only 20 to 30% of the body surface. It affects 30,000 out of 1 million admissions per year in the VA system. Approximately half of these are due to gram-positive and the other half to gram-negative organisms. Approximately 15,000 of this group of 30,000 patients die [1]. This is in spite of diagnostic monitoring and treatment with appropriate antibiotics.

Sepsis due to gram-positive *Staphylococcus aureus* requires that the organisms be alive. It requires that they breach the vascular wall and multiply in the tissues, bringing the neutrophils after them. If uncontrolled, the live bacteria and neutrophils digest tissues, form abscesses, and in some unknown manner produce shock, multiple organ failure, and death.

Sepsis due to gram-negative *E. coli* organisms, on the other hand, does not require that the organisms be alive, nor that they gain access to the tissues. Instead, from their position in the cardiovascular reticuloendothelial system, they kill the host by turning the host's inflammatory and coagulant systems against the host.

In numerous *in vitro* and *in vivo* studies, investigators have focused on the effects of the endotoxin from these gram-negative organisms on the plasma and cellular components of the blood and blood vessels and on the inflammatory mediators released from these cells [2]. So much new information has been uncovered that it has become necessary to bring it into perspective by reexamining the natural history of *E. coli* sepsis in primates.

This review therefore will cover the following topics. First, the four stages of experimental *E. coli* shock as observed in the baboon will be described. Second, the hypothesis that gram-negative shock is primarily an acute inflammatory disease of the vascular endothelium will be offered together with an examination of key aspects of its mechanism. Third, variations of the four-stage response to *E. coli*, ranging from capillary leak to microvascular thrombosis, will be described.

[1] Oklahoma Medical Research Foundation Oklahoma City, Oklahoma 73104, USA

42. Colloquium Mosbach 1991
Molecular Aspects of Inflammation

2 Description of the Four Stages of the Baboon Model of *E. coli* Sepsis

1. Stage I (0–2 h). Figure 1 shows that as the concentration of *E. coli* organisms of plasma rises during the 2-h infusion of LD_{100}*E. coli*, the neutrophil concentration falls sharply. Coincident with this fall, at least some of the neutrophils adhere to the venous endothelium. This includes the veins and sinusoids of the major target organs, including the adrenals, liver, and kidney. This corresponds to that seen in humans.

2. Stage II (2 to 6 h). Figure 1 shows that this neutrophil response is followed by a sharp fall in the concentration of *E. coli* organisms and fibrinogen. Coincidentally, and after these responses, blebs appear on the aortic endothelium of smaller vessels. The fibrin deposition is not extensive. Fibrin degradation products rise sharply and the platelet concentration falls gradually (not shown), while the white cell concentration remains low during this interval.

3. Stage III (6 to 10 h). Figure 1 shows that the fall in fibrinogen is followed by a rise in serum glutamine pyruvate transferase (SGPT) concentration. Coincidentally, the

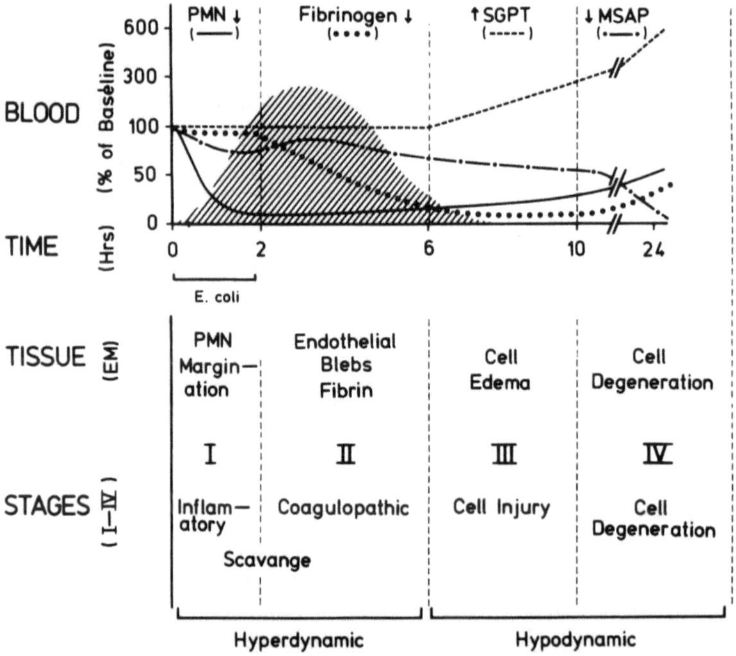

Fig. 1. Four-stage (I–IV) response of the baboon to lethal *E. coli*. The *E. coli* are infused over a 2-h period (4 x 10^{10} organisms/kg). The *hatched area* represents *E. coli* organisms in the blood which are cleared after 8 to 10 h. The polymorphonuclear leukocyte *(PMN)* count of the blood falls during stage I, followed by a fall in the fibrinogen level of plasma during stage II. The plasma levels of intracellular enzymes, such as serum glutamine pyruvate transferase *(SGPT)* rise during stages III and IV, reflecting cell injury and degeneration. This latter response is accompanied by a steady fall in mean systemic arterial blood pressure *(MSAP)*. Death usually follows in 12 to 30 h

endothelial fenestra of the adrenal and hepatic sinusoids are closed. Fluid accumulates in and between the parenchymal cells of the kidney and liver. the white cell concentrations rise slightly. The fibrinogen concentration remains low. The vascular endothelium remains intact except in the adrenals, where there is focal extravasation of red blood corpuscles. During this interval, the platelet concentration and the mean blood pressure decline slowly and steadily to approximately 50% of their baseline values.

4. *Stage IV (10 + h).* During this interval, death usually occurs at any time between 12 and 30 h. The white cell and fibrinogen concentrations rise to 50 % of baseline values. The SGPT and other markers of cell injury continue to rise dramatically. At this point, there is irreversible degeneration of the organelles of hepatocytes and particularly of the epithelial cells of the proximal kidney tubules, findings similar to those reported in rats and monkeys. Finally, during this stage, the mean blood pressure and platelet concentration continue to fall until the time of death.

The four divisions of the baboon response to *E. coli* infusion: inflammatory stage I, coagulopathic stage II, cell injury stage III, and cell degeneration stage IV are also shown in Fig. 1; also, the more familiar cardiovascular hyperdynamic and hypodynamic stages are shown in relation to these four stages.

These four stages in relation to events in the blood vessels and tissues are illustrated in Fig. 2.

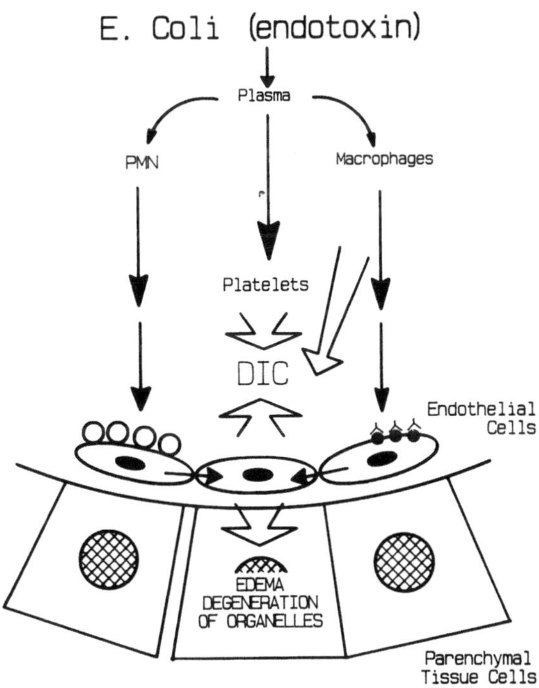

Fig. 2. Stages I to IV in relation to the hypothetical vascular endothelium and underlying parenchymal tissue. The inflammatory stage I involves neutrophils and macrophages/monocytes which release mediators. These mediators perturb the vascular endothelium and initiate the coagulopathic stage II. The sum of platelet, macrophage, and endothelial procoagulant activity leads to a massive systemic disseminated intravascular coagulation *(DIC)* response. Finally, in response to these explosive events, occurring within the vasculature, the adjacent parenchymal tissue accumulates fluid (stage III) and ultimately undergoes degeneration (stage IV)

3 General Observations on the Baboon Model of *E. coli* Sepsis

First, *E. coli* organisms induce a massive intravascular inflammatory coagulant response, which then affects the tissues (hypoxia, multiple organ failure). Unlike *Staphylococcus aureus* organisms, which must remain alive and colonize tissue, the *E. coli* organisms need not be alive nor colonize tissue in order to produce septic shock. Thus, the principal field of battle in the case of *E. coli* is within the vasculature. Once within the baboon vasculature, there is a narrow range between the sublethal and lethal concentrations of *E. coli* (10^6 vs 10^7 organisms/ml plasma at 2 h). At some point within this narrow range, endotoxin appears in the plasma and the normally protective neutrophil and macrophage responses go out of control.

Second, the sequence of events leading up to the coagulopathic response suggests that it is caused by the preceding inflammatory response.

Third, during cell injury stage III and cell degeneration stage IV responses, the vascular endothelium remains largely intact and the white cell and fibrinogen concentrations return to their normal baseline values. This occurs in the face of cardiovascular collapse, a falling platelet count, and extensive injury and degeneration of the parenchymal tissue cells which underlie the intact endothelium.

4 Intervention and Mediation Studies of the Baboon Model of *E. coli* Sepsis

1. Stage I (inflammatory stage). The most noteworthy general observation during this stage is the fall in white cell count and the cell's adherence to the endothelium of selected vascular beds (Fig. 1). Tumor necrosis factor (TNF) also appears and reaches a peak in the plasma during this 2-h inflammatory stage. This is a mark of macrophage activation: Markers of neutrophil (elastase) and endothelial cell (tissue plasminogen activator) activation also appear and reach their peaks during this stage. Table 1 summarizes our observations on the role of TNF during this stage. We believe that TNF is a necessary but not sufficient mediator of the inflammatory response in baboons. AntiTNF protects against the lethal effects of LD_{100}*E. coli* (top of Table 1). Infusion of TNF alone (100 to 200 µg/kg), however, does not reproduce these lethal effects. Infusion of sublethal *E. coli* followed by TNF, on the other hand, does produce a lethal response (bottom of table 1). The TNF must be given 60 to 90 min after the sublethal *E. coli* in order to produce this effect. These in vivo observations coincide with the in vitro observations of Nathan in which neutrophils were activated by TNF only if they were adherent to a surface [3]. In these in vivo studies we believe that *E. coli* primes the baboons' response to TNF by inducing the neutrophils to adhere to the vessel wall. This is supported by the fact that the optimal time for infusion of TNF (60 to 90 min) corresponds to the time at which the fall in peripheral white cell count has reached its nadir.

Other inflammatory components besides TNF therefore must come into play during this first stage. We propose that as was the case with infusion of antiTNF, coinfusion of monoclonal antibodies against these other inflammatory components with *E. coli* will yield more information as to which of them are most important. The list

Table 1. Role of TNF in host response to *E. coli*

Intervention	Stage I	Stage II	Stage III–IV	Survivors
LD_{100} *E. coli*	+	+	+	16–26 h (0/5)
Anti-TNF (15 mg/kg) + LD_{100} *E. coli*	+ PMN fall (+) TNF app. (–)	+	–	Permanent (5/5)
Mediation				
Sublethal *E. coli* alone	+ PMN fall (+) TNF app. (–)	–	–	Permanent (5/5)
TNF alone (100 μg/kg)	+	–	–	Permanent (5/5)
Sublethal *E. coli* + TNF at T + 60')	+	±	+	< 12 h (0/3)

ranges from CD18 and ELAM-1 receptors to components of the complement and kinin systems (Table 2).

2. *Stage II (coagulopathic stage).* The most noteworthy general observations during this stage are the fall in fibrinogen concentration and deposition of fibrin on an intact viable vascular endothelium in the absence of platelets (Fig. 1). This strongly suggests that a functional vascular endothelium, in which there is upregulation of tissue factor and downregulation of thrombomodulin expression [4], may play a role in producing the consumptive coagulopathy that characterizes this stage. The observations summarized in Table 3 support this. The coagulant response, as reflected by the consumption of fibrinogen, was attenuated by coinfusion of antitissue factor or activated protein C with LD_{100}*E. coli* (top of Table 3), whereas coinfusion of antibodies to protein C or protein S with sublethal *E. coli* (bottom of Table 3), converted the response into a coagulant, lethal response.

The fact that coinfusion of activated protein C or of antitissue factor with LD_{100}*E. coli* (top of Table 3) not only attenuated the coagulant response but also protected the primate from its lethal effects raises the possibility that the coagulant response plays a key role in mediating the lethal effects of *E. coli.* Table 3, however, shows that coinfusion of active site-inhibited factor Xa (DEGR-Xa) with LD_{100}*E. coli*, while completely inhibiting the consumption of fibrinogen, had no effect on lethality. The fact that antiTNF had no effect on the coagulant response while fully protecting the animal also supports the view that in this primate model the coagulant response per se does not contribute to lethality.

We are still left with the fact, however, that the endogenous protein C anticoagulant system plays an important protective role. A clue may be seen in Table 3 (bottom) in which coinfusion of inhibitors of endogenous protein C or protein S with

Table 2. A short list of examples of factors which may participate in host response to *E. coli*

1. Stage I	
Lipoprotein polysaccharide	(plasma)
Lipoprotein binding factor	(plasma)
Complement/kinin	(plasma)
CD_{14} receptor	(neutrophils)
CD_{18} receptor	(neutrophils)
GMP-140 receptor	(platelets & endothelium)
Elastase/cathepsin G	(neutrophils)
TNF/Il-1	(monocytes and macrophages)
t-PA	(endothelium)
2. Stage II	
Tissue factor	(endothelium and macrophages)
ELAM	(endothelium)
ICAM	(endothelium)
Products of coagulation	
Thrombomodulin	(endothelium)
Protein C	(plasma)
Protein S	(plasma)
C4bBP/immune complex	(hepatocytes, plasma)
3. Stage I–III	
Free radical and other low molecular weight cell products (PAF, etc.)	(neutrophils, endothelium, platelets, etc.)

sublethal *E. coli* is followed not only by a full coagulant response but also by the appearance of TNF in plasma and a lethal outcome. Thus, activation of early coagulant factors (FVa; tissue factor) may be linked to production or release of inflammatory mediators. It is important to note, however, that infusion of coagulant factors such as XaPCPS in sufficient amounts as to induce a full consumption of fibrinogen is not accompanied by the appearance of TNF or any other inflammatory mediators. Thus, if activation of the early coagulant factors plays a role in determining lethality, it is through amplification rather than initiation of the inflammatory response. If this link exists, coinfusion of activated protein C with LD_{100}*E. coli* should also attenuate the appearance of TNF in plasma. The range of TNF concentrations varies so widely, however, that one cannot establish wether or not this is the case.

3. Stages III and IV (cell injury, cell degeneration). The most noteworthy general observations during this stage are the rise of markers of cell injury (SGPT) and edema,

Table 3. Role of coagulant/anti-coagulant factors in host response to *E. coli*

Intervention	Stage I	Stage II	Stage III–IV	Survival
Activated protein C (20–80 mg) + LD_{100} *E. coli*	+	–	–	5/5
Anti-tissue factor (500 mg/kg) + LD_{100} *E. coli*	+	±	–	5/5
Anti-TNF (15 mg/kg) + LD_{100} *E. coli*	+ PMN fall (+) TNF app. (–)	+	–	5/5
DEGR-Xa + LD_{100} *E. coli*	+	–	+	0/3
Mediation				
Sublethal *E. coli* alone	+ PMN fall (+) TNF app. (–)	–	–	5/5
Sublethal *E. coli* + anti-protein C, Anti-protein S, or C4bBP	+ PMN fall (+) TNF app. (+)	+	+	0/4 in each group

and then the degeneration of parenchymal tissue in the presence of an intact endothelium and in the absence of adherent neutrophils on fibrin clots (Fig. 1). Blebs appear at the endothelial cell junctions. There is, however, no evidence of detachment from the subendothelium or of degeneration of organelles which in contrast is so prominent in the underlying parenchyma of hepatic or proximal tubular epithelial cells.

This observation raises the possibility that the hypoxia characteristics of septic shock may not affect the endothelium to the extent it affects the parenchyma, particularly of the tissues with a high metabolic rate such as proximal tubular epithelium. As with the clot found adherent to intact endothelium, this observation raises the question of a dysfunctional endothelium with respect to structures on its antiluminal side as well as with respect to the blood on its luminal side.

5 Variants of the Baboon Model of *E. coli* Sepsis

Depending on the status of the host ("primed," "partially protected"), a certain percentage of animals exhibits an immediate or delayed response to LD_{100}*E. coli* that differs from the "normal" response.

The *immediate* response is characterized by death in 12 h or less, in which the inflammatory response completely dominates, as reflected by a full-blown picture of capillary leak such as might be seen in necrotizing enterocolitis. This response might be viewed as a hyper-response of the inflammatory components, chief of which would by products of neutrophils (elastase) and macrophages (TNF/IL-1). This is supported by the fact that this immediate lethal response with capillary leak can be reproduced by a coinfusion of TNF with sublethal *E. coli* (Table 1, bottom).

The delayed response is characterized by death in 30 h or more (i.e., 100 h) in which the inflammatory and coagulopathic responses are withstood, only to be followed by a massive microvascular thrombosis such as might be seen in hemolytic uremic syndrome. This response might be viewed as a delayed hyper-response of the coagulant system which, however, paradoxically occurs after the initial massive disseminated coagulopathy has resolved. This response might be viewed, therefore, as a second-generation coagulant response, which although more limited in intensity than the initial consumptive coagulopathy, is unaccompanied by a fibrinolytic response which otherwise would keep the microvasculature clear. The chief components of this delayed response include the acute-phase proteins, including C4bBP, which neutral-

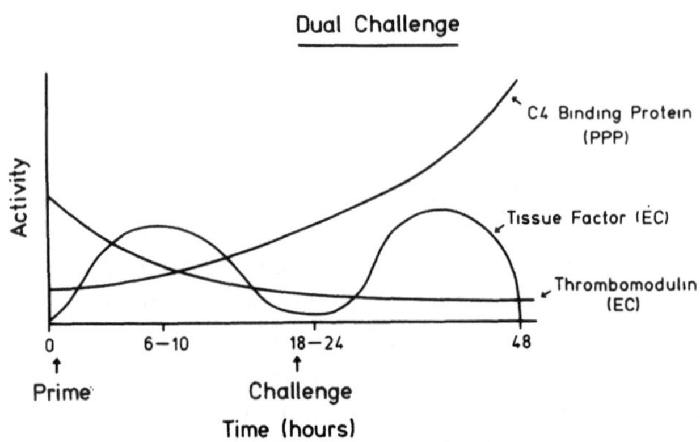

Fig. 3. The primary stimulus (TNF, LPS) upregulates the expression of the procoagulant receptors, tissue factor, and downregulates the expression of the anticoagulant receptor, thrombomodulin, by the endothelium. The increase in tissue factor expression is temporary and peaks at 6 to 10 h, while the decrease in thrombomodulin expression is sustained. In addition, this primary stimulus induces a rise in the level of C4bBP and other acute phase proteins which continue to rise over the 48-h period and beyond. C4bBP inhibits the protein S cofactor necessary for expression of anticoagulant activity by protein C (protein C is activated by thrombin-thrombomodulin). Thus, at the time of the challenge stimulus conditions are set for an extensive microvascular thrombosis. That is, tissue factor expression is elevated for the second time under conditions in which thrombomodulin expression is still depressed and in which protein S function is inhibited by the high levels of C4bBP. This challenge stimulus could range from translocated *E. coli* organisms to virus particles

izes protein S (cofactor for the anticoagulant, activated protein C). This is supported by the fact that this delayed lethal response with microvascular thrombosis can be reproduced by coinfusion of C4bBP with sublethal *E. coli* (Table 3, bottom).

Figure 3 illustrates the principle which may be operative in the delayed (two-stage) response. There is a priming event which sets up conditions which favor development of microvascular thrombosis following a challenge or second stimulus. The conditions as hypothesized for this two-step model require that tissue factor expression by endothelium be upregulated and that thrombomodulin be downregulated while the acute phase proteins in plasma, including C4bBP levels, are rising. The fact that this Schwartzman-like syndrome can be reproduced in one step by coinfusion of sublethal *E. coli* with C4bBP suggests that the elevated levels of the protein are the key factor in the triad (i.e., depressed thrombomodulin, elevated tissue factor, elevated acute phase C4bBP) which produces microvascular thrombosis.

6 Diagnostic Studies Which Differentiate Between Lethal and Sublethal Concentrations of *E. coli*

The four stages of the response to lethal *E. coli* have been described in detail. Now the responses to lethal *E. coli* can be compared with those to sublethal *E. coli*. Certain responses are unique to animals challenged with sublethal concentrations of *E. coli*. The responses of those animals which are compensated can be compared with the responses of animals challenged with lethal *E. coli* which are decompensated. These results in turn can be discussed in terms of markers which might be of early diagnostic or prognostic value. Table 4 summarizes the differences between the responses to lethal and sublethal *E. coli*.

Neither C5b-9, elastase, α1AT complexes, nor interleukin 6 (Il-6) appear in plasma following sublethal *E. coli*, whereas they all appear following a lethal *E. coli* infusion. Thrombin-anti-thrombin complex (TAT) appears in plasma after sublethal

Table 4. Comparison of the responses of components of the inflammatory, coagulant, and fibrinolytic systems to sublethal vs lethal *E. coli* challenge. From studies done in collaboration with DeBoer et al. (5).

Components Assayed	Sublethal	Lethal
C5b-9/s	–	↑
elastase/α/AT,	–	↑
IL-6	–	↑
TAT	↑	↑↑
Fibrinogen	–	↓
t-PA/PAI	↑	↑↑
PAP	↑	↑↑
FDP	–	↑

E. coli as well as after lethal *E. coli*, whereas fibrinogen consumption occurs only after an infusion of lethal *E. coli*. The case is the same with tissue plasminogen activator/plasminogen activator inhibitor complex (t-PA/PAI) and for the plasmin anti-plasmin complex (PAP) vs fibrin degradation products (FDP). The t-PA/PAI and PAP complexes appear in plasma after sublethal as well as lethal *E. coli*, whereas FDP appears only after an infusion of lethal *E. coli*. Assays of combinations of components might be used to establish an early diagnosis or as a prognostic tool from which to distinguish a severe uncompensated response from a less severe compensated response.

For instance, the appearance of TAT, t-PA/PA, or PAP complexes in the absence of fibrinogen consumption and production of FDP might be of value in early diagnosis and as a marker of compensated response to an inflammatory stimulus, whereas the appearance of these markers together with C5b-9/S, elastase/α1AT complexes, or Il-6 might be a marker of a more severe decompensated response.

Thrombin-antithrombin (TAT) complexes appear early (30 min) following infusion of both lethal and sublethal *E. coli* [5] and yet this is followed by fibrinogen consumption only after the lethal *E. coli* infusion (see Table 4). Since both TAT complexes and activated protein C-inhibitor complexes appear following sublethal *E. coli* [6], we conclude that thrombin is generated (see Table 4). The amounts are sufficient to generate activated protein C via the complex it forms with thrombomodulin. The excess activated protein C which is not engaged in the inactivation of factor Va and VIIIa is complexed by serine protease inhibitors (i.e., $\alpha_1$antitrypsin) to form the activated protein C inhibitor complex. The thrombin which is complexed to thrombomodulin, in turn, reacts with antithrombin III to produce the TAT complexes shown in Table 4. In this highly regulated response to sublethal *E. coli*, there is insufficient thrombin produced to override the protein C and antithrombin III systems, or to induce disseminated intravascular coagulation. Infusion of lethal *E. coli*, on the other hand, overrides these regulatory systems, resulting in the consumption of fibrinogen and disseminated intravascular coagulation as well as generation of TAT complexes (see Table 4).

Similarly, tissue plasminogen activator/plasminogen activator inhibitor (t-PA/PAI) and plasmin anti-plasmin (PAP) complexes also appear early [5] following infusion of both lethal and sublethal concentrations of *E. coli* and yet this is followed by the appearance of fibrin degradation products (FDP) only after the lethal *E. coli* infusion (see Table 4). These results raise two questions: (1) might the appearance of TAT or t-PA/PAI, PAP complexes mark *early* stages of perturbation of the host by foreign material (i.e., *E. coli*)?; and (2) might the consumption of fibrinogen and the appearance of FDP mark the *late* stages or a more severe level of perturbation of the host?

Similarly, since C5b-9/S, elastase/α1AT complexes, and Il-6 appear in large amounts only after infusion of lethal *E. coli* ([5]; see Table 4), might their appearance also mark a later or more severe level of perturbation of the host? Numerous papers have been written regarding the prognostic value of single markers of the host response including C3 [7], C3a and C4a [8], TNF [9], endotoxins [10], and elastase [11]. None of these studies, however, have addressed the possible value of a panel of assays by which both the stage (early, late) and severity of the host response might be evaluated to achieve both an early diagnosis and a prognosis (see Table 4).

The relevance of the above observations to clinical diagnosis and prognostication remains to be examined. These observations, however, do illustrate certain principles and problems which should be kept in mind in the development of improved methods of early diagnosis and prognostication. *First*, the diagnostic or prognostic value of any given marker may differ with the stage of the disease (I–IV) and the form of expression of this disease (see Sect. 5). *Second*, the T 1/2 of any given marker should be such that it persist long enough for detection. From these studies it appears that the half-lives of enzyme inhibitor complexes (i.e., TAT, t-PA/PAI, PAP, and C5b-9/S and elastase/α1 AT) are relatively long (hours) compared to those of C5a, TNF, and Il-1 (minutes). *Third*, the markers should reflect proximal events such as the formation of regulatory complexes, or the expression of receptors rather than more distal transitory events such as release of activation peptides or secretion of small molecules. Thus, assay of monocyte expression of tissue factor or leukocyte expression of Mac-(12) by flow cytometry with fluorescent antibodies might be preferable to assays of the release products from these cells. Thus, assays of TAT, C5b-9/S complex formation might be more preferable to assays of TNF, Il-1, LTD_4, PAF, etc., even though the latter may be among the principal mediators or amplifiers of sepsis. *Fourth*, it should be economically and temporally feasible to screen patients with assays for any given diagnostic or prognostic markers. Flow cytometry would be suitable as an assay because it is rapid and require only small amounts of blood.

Acknowledgments. The baboon model of *E. coli* sepsis was developed by Dr. Lerner B. Hinshaw, and the two-step hypothesis for microvascular thrombosis (Fig. 3)was introduced by Dr. Charles T. Esmon. Much of what is presented here has grown from the practical and intellectual input of these two individuals. I also wish to acknowledge the sustained commitment and excellence of Mr. A.C.K. Chang, Mr. Glenn Peer, and Ms. M. Lockhart in the laboratory, and Ms. B. Irish in the office. It is of historical interest to me to note that the role of the protein C system in regulating the host response to *E. coli* was discovered through the observations of Dr. Ray Fletcher. Extracorporeal perfusion without the aid of heparin protected animals from endotoxin shock. This was later found to be due to activation of the protein C system.

References

1. Hinshaw LB, Peduzzi P, Wilson M et al. (1987) The Veterans Administration Systemic Sepsis Cooperative Study Group: effect of high dose glucocorticoid therapy on mortality in patients with clinical signs of systemic sepsis. N Engl J Med 817:659
2. Taylor FB (1989) Baboon model of *E. coli* septic shock: staging and observation on the role of the vascular endothelium. In: Fuhrman BP, Shoemaker WC (eds) Critical care: stage of the ARt, vol. 10. Fullerton, CA, Society of Critical Care Medicine, pp 251–284
3. Nathan CF (1987) Neutrophil activation on biologic surfaces. J Clin Invest 80:1550–1560
4. Nawroth PP, Stern DM (1986) Modulation of endothelial cell hemostatic properties by tumor necrosis factor. J Exp Med 163:740–745
5. De Boer JP, Creasey AA, Roem D, Eerenberg AJM, Taylor FB et al. (1991) Mediators in a baboon model of *E. coli* sepsis. Thrombosis and haemostasis Congress, Amsterdam, July, 1991 (Abstr.)
6. Heeb MJ, Gruber A, Griffin JH (1990) Metal ion-dependent inhibition of activated protein C (APC) by α_2-Macroglobulin (α_2M) and α_2-Antiplasmin (α_2AP) in human blood. Circulation 82:III, 305

7. McCabe WR (1973) Serum complement levels in bacteremia due to gram-negative organisms. N Egl J Med 288:21
8. Hack CE, Nuijens JH, Felt-Bersma RJ, Schreuder WO, Eerenberg-Belmer AJ et al. (1981) Elevated plasma levels of the anaphylatoxins C3a and C4a are associated with a fatal outcome in sepsis. Am J Med 86:20
9. Debets JM, Kampimeijer R, van der Linden MP, Buurman WA, van der Linden CJ (1989) Plasma tumor necrosis factor and mortality in critically ill septic patients. Crit Care Med 17:489
10. Marra MN, Wilde CG, Griffith JE, Snable JL, Scott RW (1990) Bactericidal/permeability increasing protein has endotoxin neutralizing activity. J Immuno 144:662
11. Oswald KH, Jochum M, Schramm W, Fritz H (1985) Released granulocytic elastase: an indicator of pathobiochemical alterations in septicemia after abdominal surgery. Surgery 98:982
12. Moore FD Jr, Moss NA, Revhaug A, Wilmore D, Mannick JA et al. (1987) single dose of endotoxin activates neutrophils without activating complement. Surgery 102:201

G. Hauska, R. K. Thauer (Eds.)

The Molecular Basis of Bacterial Metabolism

41. Colloquium, 5.–7. April 1990

1990. IX, 183 pp. 84 figs. 15 tabs. (41. Colloquium der Gesellschaft für Biologische Chemie in Mosbach/Baden) Hardcover
ISBN 3-540-52996-9

This is not a comprehensive account of the **Molecular Basis of Bacterial Metabolism,** but rather a presentation of new, exciting progress on sometimes exotic reactions of specifically bacterial, mainly anaerobic metabolism.
The topics of the 17 contributions reach from the biochemistry of methanogenesis, via anaerobic radical reactions, metal biochemistry in hydrogen and nitrogen metabolism, conversions of light- and redox energy, to the regulation of metabolic adaptation, and the attempts to bioengineer novel pathways for the degradation of xeno-biotica.
These proceedings represent a highly progressive field of overlapping disciplines, comprising microbiology and molecular genetics, chemistry of biomimetic interest, and biophysics.
The reader will gain insight into the impact modern technologies have on microbiological research today.

U. Gehring, E. J. M. Helmreich, G. Schultz (Eds.)

Molecular Mechanisms of Hormone Action

40. Colloquium, 6.–8. April 1989

1989. XII, 204 pp. 77 figs. (40. Colloquium der Gesellschaft für Biologische Chemie in Mosbach/Baden) Hardcover
ISBN 3-540-51607-7

Recent years have seen tremendous progress in the field of hormone action and consequent signal transduction.
The 40th Colloquium Mosbach was devoted to the discussion of results concerning the molecular progress of hormone action, especially the processes following hormone binding to the corresponding receptors.
Structural and functional aspects of steroid hormone receptors as well as ion-channel-coupled and enzyme-linked receptors were treated in detail. Particular interest focussed on the latest results concerning transcriptional control, protein phosphorylation, the role of G-Proteins, oncogene proteins, involvement of phospholipases and the regulation of ion channels.

MIX
Papier aus verantwortungsvollen Quellen
Paper from responsible sources
FSC® C105338

If you have any concerns about our products,
you can contact us on
ProductSafety@springernature.com

In case Publisher is established outside the EU,
the EU authorized representative is:
Springer Nature Customer Service Center GmbH
Europaplatz 3, 69115 Heidelberg, Germany

Printed by Libri Plureos GmbH
in Hamburg, Germany